CAMBRIDGE TRACTS IN MATHEMATICS

General Editors

207 Auxiliary Polynomials in Number Theory

CAMBRIDGE TRACTS IN MATHEMATICS

A complete list of books in the series can be found at www.cambridge.org/mathematics.
Recent titles include the following:

171. Orbifolds and Stringy Topology. By A. Adem, J. Leida, and Y. Ruan
172. Rigid Cohomology. By B. Le Stum
173. Enumeration of Finite Groups. By S. R. Blackburn, P. M. Neumann, and G. Venkataraman
174. Forcing Idealized. By J. Zapletal
175. The Large Sieve and its Applications. By E. Kowalski
176. The Monster Group and Majorana Involutions. By A. A. Ivanov
177. A Higher-Dimensional Sieve Method. By H. G. Diamond, H. Halberstam, and W. F. Galway
178. Analysis in Positive Characteristic. By A. N. Kochubei
179. Dynamics of Linear Operators. By F. Bayart and É. Matheron
180. Synthetic Geometry of Manifolds. By A. Kock
181. Totally Positive Matrices. By A. Pinkus
182. Nonlinear Markov Processes and Kinetic Equations. By V. N. Kolokoltsov
183. Period Domains over Finite and p-adic Fields. By J.-F. Dat, S. Orlik, and M. Rapoport
184. Algebraic Theories. By J. Adámek, J. Rosický, and E. M. Vitale
185. Rigidity in Higher Rank Abelian Group Actions I: Introduction and Cocycle Problem. By A. Katok and V. Niţică
186. Dimensions, Embeddings, and Attractors. By J. C. Robinson
187. Convexity: An Analytic Viewpoint. By B. Simon
188. Modern Approaches to the Invariant Subspace Problem. By I. Chalendar and J. R. Partington
189. Nonlinear Perron-Frobenius Theory. By B. Lemmens and R. Nussbaum
190. Jordan Structures in Geometry and Analysis. By C.-H. Chu
191. Malliavin Calculus for Lévy Processes and Infinite-Dimensional Brownian Motion. By H. Osswald
192. Normal Approximations with Malliavin Calculus. By I. Nourdin and G. Peccati
193. Distribution Modulo One and Diophantine Approximation. By Y. Bugeaud
194. Mathematics of Two-Dimensional Turbulence. By S. Kuksin and A. Shirikyan
195. A Universal Construction for Groups Acting Freely on Real Trees. By I. Chiswell and T. Müller
196. The Theory of Hardy's Z-Function. By A. Ivić
197. Induced Representations of Locally Compact Groups. By E. Kaniuth and K. F. Taylor
198. Topics in Critical Point Theory. By K. Perera and M. Schechter
199. Combinatorics of Minuscule Representations. By R. M. Green
200. Singularities of the Minimal Model Program. By J. Kollár
201. Coherence in Three-Dimensional Category Theory. By N. Gurski
202. Canonical Ramsey Theory on Polish Spaces. By V. Kanovei, M. Sabok, and J. Zapletal
203. A Primer on the Dirichlet Space. By O. El-Fallah, K. Kellay, J. Mashreghi, and T. Ransford
204. Group Cohomology and Algebraic Cycles. By B. Totaro
205. Ridge Functions. By A. Pinkus
206. Probability on Real Lie Algebras. By U. Franz and N. Privault
207. Auxiliary Polynomials in Number Theory. By D. Masser

Auxiliary Polynomials in Number Theory

DAVID MASSER

University of Basle, Switzerland

CAMBRIDGE
UNIVERSITY PRESS

CAMBRIDGE
UNIVERSITY PRESS

University Printing House, Cambridge CB2 8BS, United Kingdom

Cambridge University Press is part of the University of Cambridge.

It furthers the University's mission by disseminating knowledge in the pursuit of education, learning, and research at the highest international levels of excellence.

www.cambridge.org
Information on this title: www.cambridge.org/9781107061576

First published 2016

A catalogue record for this publication is available from the British Library.

Library of Congress Cataloguing in Publication Data
Names: Masser, David William, 1948–
Title: Auxiliary polynomials in number theory / David Masser, Universitat Basel, Switzerland.
Description: Cambridge : Cambridge University Press, 2016. | Series: Cambridge tracts in mathematics ; 207 | Includes bibliographical references and index.
Identifiers: LCCN 2015050947 | ISBN 9781107061576 (Hardback : alk. paper)
Subjects: LCSH: Number theory. | Polynomials.
Classification: LCC QA241 .M395 2016 | DDC 512.7/4–dc23 LC record available at http://lccn.loc.gov/2015050947

ISBN 978-1-107-06157-6 Hardback

Contents

Introduction

Ever since it was invented, arguably by Runge, the method of auxiliary polynomials has been vital to (and of unreasonable effectiveness in) the modern development of key aspects of number theory. The aim of this book is to give an account of the method in many of its forms, focusing almost exclusively on those polynomials which cannot be written down explicitly.

I well remember (standing in Heffers bookshop Cambridge around 1970) reading about this method in the foreword to Lang's book on transcendental numbers, and experiencing disbelief that anything so far-fetched could work at all. So I will not attempt any explanation at this point.

Instead, I (or from now on, the authorial we) treat the method as the union of its examples, and there is no shortage of these.

Here is the plan of this book (Mike Tyson said that everyone has a plan until you punch them in the face – then they don't have a plan). The general strategy is to present in each chapter an application of the method to a different sort of problem, often the simplest in its area. Then at the end of each chapter we give a brief account of subsequent developments in the area.

We start with a short Prologue (Chapter 1) where we show that the basic idea can be used in rather simple situations which have nothing to do with number theory.

Then in Chapter 2 we commence our diophantine considerations with a discussion of irrationality. We quickly dispose of the number e by the standard truncation argument and we show also that e is not a quadratic irrational. Here we meet a small problem, which can be rather quickly solved; however, it is typical of the problems that arise in later applications and in some examples its solution can be distinctly non-trivial. Thus Roth, in showing that irrational algebraic numbers cannot be approximated to within an order of $q^{-2-\epsilon}$ by rationals p/q, had to solve such a problem. The solution is called

Roth's Lemma, and it was certainly one of the achievements that gained him a Fields Medal. We do not prove Roth's Theorem here but we do treat Thue's Method in Chapter 12.

We postpone to Chapter 13 a proof that

$$e^{\alpha} = \sum_{k=0}^{\infty} \frac{\alpha^k}{k!} = 1 + \alpha + \frac{1}{2}\alpha^2 + \frac{1}{6}\alpha^3 + \frac{1}{24}\alpha^4 + \cdots$$

is irrational and even transcendental for all rational and even algebraic $\alpha \neq 0$, as this requires elements of algebraic number theory. Truncation gives only irrationality and that essentially only for $\alpha = 1, 2, 4$ (and slightly surprisingly $\alpha = \sqrt{2}$).

In the same Chapter 2 we construct our first auxiliary polynomial with a diophantine purpose: namely, to show that for any rational $\alpha \neq 0$ the classical series

$$\sum_{k=0}^{\infty} \frac{\alpha^k}{2^{k(k-1)/2}} = 1 + \alpha + \frac{1}{2}\alpha^2 + \frac{1}{8}\alpha^3 + \frac{1}{64}\alpha^4 + \cdots$$

is irrational. This is somewhat related to theta functions. Although it converges quite rapidly, the speed is also insufficient for mere truncation. The result itself is not so fundamental, but it provides a good introduction to the use of auxiliary polynomials; that used here is probably the simplest of its kind, and we calculate a few examples. One needs also some elementary complex analysis, which will be much developed later on.

In Chapter 3 we then progress to the similar but more elaborate Mahler's Method, still sticking just to irrationality; the results here are historically important and they led to the solution of the Mahler–Manin Conjecture and then to Nesterenko's Theorem on the algebraic independence of π and e^{π}. The irrationality here will be generalized to full-blooded transcendence in a later chapter. Here we treat just

$$\sum_{k=0}^{\infty} \alpha^{2^k} = \alpha + \alpha^2 + \alpha^4 + \alpha^8 + \alpha^{16} + \cdots$$

for every rational α with $0 < |\alpha| < 1$. The proof is quite similar to that in Chapter 2 but a little more analysis is needed, and further the auxiliary polynomial is more complicated, in fact already of a fairly typical sort; still we calculate some more examples. Mahler's Method has been greatly developed and some recent applications refer to the famous Mandelbrot set. It also played a transient role in proving that the decimal digits of numbers like $\sqrt{2}$ cannot be generated by a finite automaton.

In Chapter 4 we prove that certain diophantine equations in two variables have at most finitely many solutions, using the auxiliary polynomial pioneered by Runge. The method enables all solutions to be found in principle. A typical example is that there are at most finitely many integers x, y with

$$x(x^3 - 2y^3) = y.$$

Or, coming from Cassels's well-known result on the Catalan Equation recently solved completely by Mihăilescu, there are at most finitely many integers x, y with $x^5 - y^7 = 1$ provided y is not divisible by 5 (we do not prove Mihăilescu's Theorem here). Of course equations like

$$x^3 - 2y^3 = m$$

for fixed m are more natural, and these will be considered in Chapter 12. For the proofs here, we need to know that the large complex solutions are given by Puiseux (or better Laurent) series. It seems that this is not so easy to find in the literature, especially regarding the crucial convergence properties, so we provide quite a few details.

Then in Chapter 5 we prove some results similar to the classical Hilbert Irreducibility Theorem, usually abbreviated to HIT, by using the machinery of the preceding chapter. They are not so general as HIT, but when they do work, they deliver more information. The results were first found by Sprindzhuk also using auxiliary polynomials, but in a more elaborate way. Nowadays this sort of thing can be done with heights machinery, but that is not so elementary. A typical example, related to that of the previous chapter, is that there are at most finitely many integers y such that the polynomial

$$X(X^3 - 2y^3) - y$$

in $\mathbb{Q}[X]$ is reducible over the rationals, and in principle these can all be found. A literal application of HIT would show only that there are infinitely many rational y such that the polynomial is not reducible. So sometimes we get a Strong Hilbert Irreducibility Theorem; but we refrain from abbreviating this. Here we need resultants; these can be found almost anywhere, but because we use them frequently in this book we provide a self-contained account.

In Chapter 6 we jump to a different topic. We prove that the number N of points modulo a prime p on an affine elliptic curve satisfies

$$|N - p| \leq 8\sqrt{p};$$

this is slightly weaker than the classical result of Hasse involving $2\sqrt{p}$. The proof uses the simplest non-trivial example of the auxiliary polynomial introduced in a surprising way by Stepanov in 1969; here we attempt to motivate the proof with the help of some easier intermediate results. Not even the definition of elliptic curve is needed here, let alone any properties. Thus all we do is count the integer solutions (x, y) modulo p of an equation $y^2 = x^3 + ax^2 + bx + c$ modulo p. There are many generalizations (and Schmidt wrote an entire book about them) but none has quite the same appeal. With rather little extra effort one can treat $y^2 = x^5 + \cdots$ and worse; in the geometric context this is far from simple because it would involve curves of genus 2 and worse.

In Chapter 7 we make another jump which seems even bigger, to the topic of exponential sums. The best known is Gauss's

$$\sum_{k=1}^{p} \exp\left(\frac{2\pi i k^2}{p}\right),$$

also for prime p, whose absolute value $p^{1/2}$ is much smaller than the number of its terms. One of these sums, due to Heilbronn, resisted for some time all efforts to prove its smallness until Heath-Brown in 1996 achieved this. His beautiful proof imitated Stepanov's auxiliary polynomial in a kind of analytic context involving a logarithm-like function. Some arguments had been anticipated by Mitkin in 1992. We give the details. Specifically

$$\left|\sum_{k=1}^{p} \exp\left(\frac{2\pi i k^p}{p^2}\right)\right| \leq 4p^{11/12}.$$

As far as I know, these arguments have not been developed very far since then, despite some interesting features involving differential equations.

In Chapter 3 we proved the irrationality of the values μ of Mahler's series at non-zero rationals. Thus the quantity $|\mu - p/q|$ is positive for all integers p and $q \geq 1$. A natural question is: "How small can this quantity get?" Indeed with an algebraic irrational in place of μ this question is fundamental in the theory of diophantine equations, as we will see in Chapter 12. Our answer in Chapter 8 requires refining the arguments of Chapter 3. There are two key steps. One is a "zero estimate" asserting that not too many things can vanish; such estimates play a major role in more recent developments. The other, more classical, is an estimate for the coefficients of the auxiliary polynomial; this involves the famous Siegel Lemma, which will be used over and over again in the sequel. We also make a simple application of the maximum modulus principle for analytic functions. This too will be used frequently later, under the popular name of the Schwarz Lemma. In this way we will prove that there exist $c = c(\mu) > 0$ and $\kappa = \kappa(\mu)$ such that

$$\left|\mu - \frac{p}{q}\right| \geq \frac{c}{q^\kappa}.$$

For example with $\mu = \sum_{k=0}^{\infty}(2/3)^{2^k}$ we can take $\kappa = 77$.

There is a famous result of Pólya on entire functions mapping the natural numbers to the rational integers; this may have influenced Gelfond in his pioneering work on the transcendence of α^β (see Chapter 19). Pólya's original proof used interpolation formulae and gave the best possible constant. Much later Waldschmidt gave a version by auxiliary polynomials, which sadly gives a worse constant. The proof is nevertheless illuminating; it needs binomial coefficients to avoid factorials, one of the key ideas in Thue's famous proof (see Chapter 12). More precisely, we show in Chapter 9 that an entire function f with

$$f(0), f(1), f(2), \ldots, f(n), \ldots$$

all in $\mathbb{Z}$ must be a polynomial if $|f(z)|$ grows of order at most $C^{|z|}$ for a certain $C > 1$. Pólya could take any $C < 2$; and the standard example 2^z shows that nothing better is possible. Or reformulated: if a non-polynomial entire function f has this growth, then at least one of $f(0), f(1), f(2), \ldots$ must be non-integral. Gelfond's step from non-integrality to transcendence needed many more ideas, all of which will be developed in this book.

The rather natural generalization to the Gaussian integers $\mathbb{G} = \mathbb{Z} + \mathbb{Z}i$ with f mapping $\mathbb{G}$ into itself also played a similar historical role; for example it probably directly inspired Gelfond's proof of the transcendence of e^π. But the best possible constant did not appear until a relatively recent paper of Gramain; paradoxically enough, his proof involves an auxiliary polynomial (or better an auxiliary function). More precisely, f itself must be a polynomial if $|f(z)|$ now grows of order at most $C^{|z|^2}$ for a certain $C > 1$. Gelfond considered this problem too, and obtained the notorious value

$$C = \exp\left(\frac{\pi}{2(1+\exp(164/\pi))^2}\right) < 1 + 10^{-45}$$

(modestly not mentioned in his book). In the late 1970s, I obtained a constant, extremely difficult to compute, which later turned out to be about 1.181; and I conjectured that the best possible constant was $\exp(\frac{\pi}{2e})$ about 1.782. This Gramain proved, and so do we in Chapter 10.

In Chapter 11 we present our first transcendence result. We extend Mahler's Method in Chapter 3 to prove the transcendence of his $\sum_{k=0}^{\infty} \alpha^{2^k}$ for all algebraic α with $0 < |\alpha| < 1$. That is apparently how he tested his recovery while convalescing at home from an illness. No more ideas are needed, but to go beyond irrationality requires some rudimentary notion of "size" of an algebraic number, with some sort of "Liouville estimate". This sort of technicality is fundamental to all transcendence proofs. The concept will be developed later into the more sophisticated "height", which will then be studied for its own sake, for example with reference to Lehmer's Question of 1933 in connexion with factorization problems.

At last in Chapter 12 we prove the famous Thue improvement of Liouville's classical result. The proofs here start getting more elaborate, and another key element is dealing with the dangerously heavy factorials that threaten to sink the method; however this problem has been solved in Chapter 9. Yet another feature is a simple form of zero estimate. These have proved crucial in later developments involving Roth, Schmidt, Schlickewei, and others. More precisely, given any algebraic number α of degree $d \geq 3$ and any $\kappa > \frac{d}{2} + 1$, we show that there is a positive constant $c = c(\alpha, \kappa)$ such that

$$\left|\alpha - \frac{p}{q}\right| \geq \frac{c}{q^{\kappa}}$$

for all integers p and $q \geq 1$. The Liouville result was for $\kappa = d$, and the later Roth estimate was for any $\kappa > 2$. Here we try to break the proof into molecules, and we also speculate on how Thue may have arrived at his proof; there are interesting connexions with Newton's Method in numerical analysis and later improvements by Halley and others. We also give the applications to diophantine equations. Here we encounter the uncomfortable phenomenon of ineffectivity for the first time.

Then in Chapter 13, using the machinery of the previous chapter, we prove the Hermite–Lindemann result on the transcendence of the values of the exponential function at algebraic numbers; thus e^{α} is transcendental for every algebraic $\alpha \neq 0$. Our proof is a kind of *ad hoc* development of the auxiliary polynomial techniques introduced so far; we have by now illustrated so many of these techniques that several proofs are available. We choose the one most suited for generalization to the Schneider–Lang Theorem later on in Chapter 19.

Chapter 14 is where we develop the size in Chapter 11 to the absolute height $H(\alpha) \geq 1$ or the logarithmic version $h(\alpha) = \log H(\alpha) \geq 0$. This is rather easy to define, but to establish properties like $H(\alpha^2) = H(\alpha)^2$, we need quite a bit of algebraic number theory, and we will sketch the details. The motivation is two-fold: first, the results of the next two chapters are about heights *per se*, and second, the proof of the later Schneider–Lang result then becomes fairly streamlined. We also give a version of the Siegel Lemma in the heights language. This requires essentially defining the height of a vector $(\alpha_1, \ldots, \alpha_n)$ of algebraic numbers. To break the monotony, we prove on the way some easy results on lower and upper bounds for heights that have led to some lively modern developments.

Then in Chapter 15 we prove Bilu's Theorem on the distribution of the conjugates of an algebraic number, using an auxiliary polynomial due to Mignotte as well as the Siegel Lemma from the previous chapter. As a matter of fact, our version is completely explicit numerically. But there is a problem: this explicitness is based on the Erdős–Turán Theorem, and there seems to be no easy proof of that. So at this point the book is definitely not self-contained; however we find this didactically permissible, as the present chapter serves as a natural springboard for the next one, and Bilu's Theorem is not further used in the book. More precisely, if α is an algebraic number of degree d and absolute logarithmic height h, we show that the number n of its conjugates in any sector of angle θ based at the origin satisfies

$$\left|n - \frac{\theta}{2\pi}d\right| \leq 24(d^{2/3}(\log 2d)^{1/3} + dh^{1/3}).$$

That n is asymptotically $\frac{\theta}{2\pi}d$ as $h \to 0$ is the main content of Bilu's result (which is expressed more felicitously in terms of weak approximation).

Then in Chapter 16 using the machinery developed in the previous chapter, we prove up to logarithms the famous Dobrowolski Theorem, which is to this day the best approach to the classical Lehmer Question, using essentially the original auxiliary polynomial. The result is exceptionally useful and, as far as I know, none of the applications actually need the logarithms. Providing the best known logarithms is an exercise on the Prime Number Theorem, which is carried out in several books. Thus we prove here that for any $\kappa > 1$ there is a positive constant $c = c(\kappa)$ such that every non-zero algebraic $\alpha \neq 0$ of degree d which is not a root of unity satisfies

$$h(\alpha) \geq \frac{c}{d^{\kappa}}.$$

Admittedly there are quicker proofs without auxiliary polynomials, but these don't generalize to the higher dimensional results such as the Amoroso–David Theorem that are very important today in diophantine geometry.

In Chapter 17 we restore some symmetry by giving a non-trivial height upper bound. This concerns the algebraic numbers α with $\alpha^n + (1-\alpha)^n = 1$ for some integer $n \geq 2$. In a relatively recent investigation connected with irreducibility, Beukers showed that $H(\alpha) \leq 216$. His proof used hypergeometric functions. Using instead the powerful method of auxiliary polynomials, we get $H(\alpha) \leq 10^{120}$ (in the style of Stephen Leacock "ten years ago the deficit on my farm was about a hundred dollars; but by well-designed capital expenditure, by drainage and greater attention to detail, I have got it into the thousands"). However this method generalizes considerably, as current work of Amoroso, Zannier and the author shows.

In Chapter 18 we use some of the ideas developed so far to give a generalization to algebraic points of the 1989 Bombieri–Pila Theorem on counting rational points on analytic curves. The original proof, although not fundamentally different from ours, is based on identities related to the confluent Lagrange Interpolation Formulae and not on an auxiliary polynomial. Such counting results (usually in higher dimensions) are nowadays being applied to prove a variety of special cases of the general Zilber–Pink Conjectures about unlikely intersections. We will prove something implying the following. Let f be a transcendental function analytic on an open set containing the real interval $[0, 1]$. Then for any $\epsilon > 0$ there exists $c = c(f, \epsilon)$ such that, for every positive integer n, at most cn^ϵ of the values

$$f(0), f\left(\frac{1}{n}\right), f\left(\frac{2}{n}\right), \ldots, f(1)$$

are in $\mathbb{Z}/n$. This vaguely resembles the reformulation of Pólya's Theorem.

Then in Chapter 19 we prove the famous Schneider–Lang Theorem, which includes Hermite–Lindemann in Chapter 13 as well as several other things involving elliptic and abelian functions. Thanks to the preceding chapters the proof is now reasonably smooth. It is a natural climax to the book; however the next chapter follows on quite naturally, and so does the one after that. Thus we prove the Gelfond–Schneider Theorem on the transcendence of $\alpha^\beta = \exp(\beta \log \alpha)$ whenever $\alpha \neq 0$ and irrational β are algebraic, which includes the transcendence of $2^{\sqrt{2}}$ as specified by Hilbert in his Seventh Problem. A key technical trick is the use of "large radius" in the Schwarz Lemma.

In Chapter 20 we systematically consider the elliptic analogues, motivated partly by the need to prove the transcendence of integrals like

$$\int_0^1 \frac{\mathrm{d}X}{\sqrt{X - X^3}}, \quad \int_4^5 \frac{(X - 8)\mathrm{d}X}{\sqrt{X^3 - 7X + 6}}.$$

The results involve a Weierstrass function $\wp(z)$ with invariants g_2, g_3 that are themselves algebraic; the analogue of Hermite–Lindemann then asserts the transcendence of $\wp(\alpha)$ for any algebraic $\alpha \neq 0$. Already the elliptic analogue of Gelfond–Schneider has consequences for the modular function $j(\tau)$ defined on the upper half-plane: namely that $j(\alpha)$ is transcendental whenever α is algebraic but not quadratic; this we postpone to the next chapter. But as Schneider discovered, there are several other interesting consequences; and even he overlooked one of them. This chapter is the longest in the book, due to our supplying the main details for the proofs of most of these consequences. It might get shorter if we could use facts about commutative group varieties, but that would introduce too much algebraic geometry not in the elementary spirit of the book.

In 1969 Mahler conjectured that the alternative modular function

$$J(q) = q^{-1} + 744 + 196884q + 21493760q^2 + 864299970q^3 \\ +20245856256q^4 + \cdots$$

(the pattern of coefficients is quite self-evident), defined for all q in the unit disc, takes transcendental values at non-zero algebraic q. In 1996 Barré, Diaz, Gramain and Philippon proved this using an auxiliary polynomial directly on J itself. This was not only the first auxiliary polynomial of its kind (as Schneider had wanted many years ago in his Second Problem), but it led soon afterwards to Nesterenko's unexpected breakthrough implying the algebraic independence of π and e^{π}. This is a most attractive area where aspects of elliptic, modular and exponential functions blend into each other. We give a proof of Mahler's Conjecture in Chapter 21, after deducing the analogous result for $j(\tau) = J(e^{2\pi i\tau})$ from the results of the previous chapter.

Up to now we never discussed problems of algebraic independence. Maybe the reader's curiosity for this topic has been awakened at the end of the previous chapter, and now she gets a classical example. After the famous Lindemann–Weierstrass result (not covered in this book), which was generalized to E-functions by Siegel and Shidlovsky (likewise not here), the most spectacular was the algebraic independence of α^{β} and α^{β^2}, for algebraic $\alpha \neq 0$ and cubic β, due to Gelfond in 1949. But as a lot of the machinery is already available, our proof in Chapter 22 will not be too long. Here too one needs "large radius".

Finally in an Appendix we prove exotic height results like

$$h\left(\frac{3\xi - 4\sqrt{\xi^3 + 3\xi + 4} + 8}{\xi^2}\right) \leq h(\xi) + 10000(\sqrt{h(\xi)} + 1)$$

where a crude estimate would give at least $2h(\xi)$ on the right-hand side. Indeed if we replace 8 in the numerator by 7 this is unavoidable. The square root here, traditionally associated with the quadratic nature of Néron–Tate heights on abelian varieties, is actually needed.

Let us mention here yet another use for auxiliary polynomials: to show that certain algebraic numbers arising from commutative group varieties have "large degree". It is well-known that the root of unity $e^{2\pi i/n}$ has degree $\phi(n)$ the Euler ϕ-function, and also that for any $\theta < 1$ there is a positive constant c, of course effectively computable, with $\phi(n) \geq cn^{\theta}$. By a famous result of Serre the elliptic analogue has any $\theta < 2$, but only recently has this been made effective, in an elaborate proof involving, among other things, isogeny estimates. Using an auxiliary polynomial directly, in the functions $\wp(z), \wp(Nz)$ which are "almost algebraically independent", one can quickly obtain an effective lower bound for any $\theta < 1$. Furthermore this method works also for abelian varieties, where the analogue of Serre's Theorem is still not yet fully known. The resulting estimates have recently been very useful in problems of unlikely intersections. We omitted any detailed account, first for lack of space and second because one needs more theory, such as Néron–Tate heights. See Masser (1977) and also Appendix D of Zannier (2012). However in Exercise 14.92 we sketch how the lower bound $cn/\log n$ can be obtained in the cyclotomic case.

The reader will observe that the auxiliary polynomial usually operates in a proof by contradiction. So this book is mostly about things that don't exist! With Woody Allen we may hate reality but it's still the best place to get a decent steak. Or we may think of the Cape Town telephone company error message "the number that you have called does not exist".

The pleasant task of collecting together all these applications of auxiliary polynomials has resulted in some features that may not be familiar to all experts.

Thus I am not sure if Theorem 5.1 in Chapter 5 appears explicitly in the literature. In Chapter 6 the warm-up before the proof of Theorem 6.1 may not have appeared before in this form. In Chapter 7 the proof of Lemma 7.3 is new, although it proceeds on well-known general principles. The (rather easy) estimate (8.7) of Chapter 8 is probably new. In Chapter 10 the Proposition 10.4 might possibly be useful in other contexts. In Chapter 12 it is indeed I who must accept full responsibility for the attempt to explain the proof of Theorem 12.1 in terms of numerical analysis; also the Proposition 12.2, although known to some experts, may not have appeared explicitly before. Our explicit estimate in Theorem 15.2 of Chapter 15 could be new, although its shape is fairly well-known. Some of the preliminary discussion in Chapter 16 may not be familiar. The method introduced in Chapter 17 is new, due to Amoroso, Zannier and myself. In Chapter 18 the main result Theorem 18.2 for $\mathbb{Q}(i)$ is not in the published work that I have seen, although here too its shape for $\mathbb{Q}$ is fairly well-known. Lemma 20.7 from Chapter 20 might look familiar, but it is not; also some of the details towards the end of this chapter have never appeared in print, although this may well be due to the alternative approach, more conceptual to some, through group varieties. In particular the proof of Theorem 20.11 might well be a "desperately-needed gap in the literature". Here also the (again rather easy) remark about the gamma function is new. And in Chapter 21 the Lemma 21.8 enables us to avoid an appeal to certain estimates for coefficients of modular transformation polynomials, whose (non-classical) proofs are somewhat elaborate. In Chapter 22 the proof of Proposition 22.5 is a small variation of a proof that I have seen. Finally in the Appendix the Theorem A.1, although presented only for a particular example, is also new, arising from the above work of Amoroso, Zannier and myself.

What are the prerequisites for a happy reading of this book? The first thirteen chapters could be understood by third-year university students or good second-year students (and indeed in 2013/2014 they were – and I thank this class, especially Gabriel Dill, who examined with a fine-toothed comb the first ten, although I may well have invented new mistakes during revision). The proofs are elementary (but that does not always imply that they are easy). Here there are elements of algebra such as the concept of transcendence, the fact that $\mathbb{Z}[X]$ is a unique factorization domain, or the integral closure of a ring R in a larger ring S (which I like to denote by R_S); elements of analysis such as order of vanishing, Cauchy's Theorem or the Maximum Modulus Principle; and elements of algebraic number theory such as field embeddings, conjugates or rings of integers. There is a jump at Chapter 14, where we need slightly more advanced algebraic number theory, which we explain without full proofs, freely using concepts like prime ideals and valuations. This enables us to get all the way to Chapter 20, where we then need some theory of elliptic functions, which again we explain without full proofs. Similarly in Chapter 21 we need some theory of modular functions. Finally in Chapter 22 we need a bit about transcendence degree. By contrast in the Appendix, although it has a considerable whiff of diophantine geometry, we develop from scratch the rudiments of algebraic curve theory that we need. And oh yes, it will be good to bear in mind that our

$$\mathbb{N} = \{1, 2, 3, \ldots\}$$

does not contain 0 as it might in some other cultures. But $\mathbb{Z}, \mathbb{Q}, \mathbb{R}, \mathbb{C}$ and the fields $\mathbb{F}_p = \{0, 1, \ldots, p-1\}$ are unambiguous.

And what about the exercises? These are at the end of each chapter, where they are divided into two parts by a starred line. Those above the line need only the prerequisites above and what's in the book so far, and they are essentially what were fed to students as homework reinforcing the lectures. Those below the line go further, and sometimes need extra knowledge; they are of varying levels of difficulty, sometimes hinted at. Concerning the lectures themselves, there are many possibilities; for example I covered Chapters 1, 2, 3, 4, 5 together with some algebraic number theory in a single semester, then followed up with Chapters 6, 8, 9, 11, 12, 13 and more algebraic number theory, and so was able to start a third semester with Chapter 14 in detail, then Chapters 15, 16, 18, 19 and bits of Chapter 20.

There is also a bibliography, but this has no pretence of being comprehensive. Instead I have tried to restrict it to books, especially those that give a good overview of the subsequent development of some of the topics treated here; but I have also included some key original papers.

I conjecture, but have no time to prove, that every mathematics book with at least 100 pages contains at least one misprint (possibly apart from those that have gone through several editions – however in a 2008 seminar we did find a mistake in Landau's "Elementary Number Theory" (Chelsea 1958), despite the author, according to Littlewood, reading proof sheets seven times, once for each sort of error – curiously we could not find it again later, this "Lost Mistake"). Boas has a conjecture that is shockingly stronger, and (continued p. 94).

The book you are now reading is certainly no countexerample, and I apologize in advance for my misprints, howlers and blunders (and my King Charles's Head of continued fractions). In fact I was once thanked in print by a non-English author for "teaching him mistakes". I hope to be able to pass on these skills to my readers.

And also to convey to them the joys of "doing transcendence" rather than merely "doing mathematics".

I gladly express my great gratitude to David Tranah of Cambridge University Press, for his warm initial encouragement to write the book, for his gentle reminders about actually writing it, and, once I gave in and started in earnest, for his regular enquiries about its progress and his rapid and detailed answers to my many questions.

1

Prologue

In this chapter we give a couple of examples where the method of auxiliary polynomials is used for problems that have no diophantine character. Thus we are not following Sam Goldwyn's advice to start with an earthquake and work up to a climax.

Here is maybe one of the simplest examples.

There is an old chestnut which often turns up in problem-solving sessions: given a polynomial F in a variable X, can one always multiply it by a non-zero polynomial to get a product involving only powers X^p for p prime?

For example with $F = X^{100} + 1$ we have $X^3F = X^{103} + X^3$. But what about $F = X^{100} + X^3$? Here multiplying by some $P = aX^d$ will not do. However

$$(X^{111} - X^{14})F = X^{11}(X^{100} - X^3)F = X^{11}(X^{200} - X^6) = X^{211} - X^{17}.$$

At first sight it appears to be a difficult problem about primes, possibly in arithmetic progressions. So what about $F = X^{1000} + X^{100} + X^3$?

Let us consider multiplying F by some

$$P = \sum_{i=0}^{L} p_i X^i \tag{1.1}$$

for unknown L and undetermined coefficients p_i (not necessarily primes, but they might be). Then PF has degree at most $L + 1000$, and its coefficients are linear forms in the p_i. We would like to eliminate the terms X^n for n not prime with $0 \leq n \leq L + 1000$. There are

$$L + 1001 - \pi(L + 1000)$$

of these, where $\pi(x) = \sum_{p \leq x} 1$ is the standard prime-counting function. If we equate the corresponding coefficients to zero, then we get a system of

$L+1001-\pi(L+1000)$ homogeneous linear equations in the $L+1$ unknowns p_i. By linear algebra this system is solvable non-trivially, provided

$$L+1 \; > \; L+1001-\pi(L+1000);$$

that is, $\pi(L+1000) > 1000$. Every schoolgirl knows that there are infinitely many prime numbers, so $\pi(x)$ tends to infinity with x and there exists such an L; for example $L = 6927$ (with Maple). So the answer is yes for this F; the trouble of course is that we have to solve 6927 equations in 6928 unknowns to get P explicitly.

The reader may now see first that this works for any F, and second that the primes are irrelevant, in the sense that we may demand only powers X^m in PF with m in any prescribed infinite set; for example the elements of the sequence $4, 27, 3125, 823543, \ldots$ of all $m = p^p$.

This is perhaps the simplest application of the method of auxiliary polynomials.

Here is a second example.

Consider the expressions

$$x = t^2 + t, \;\; y = t^2 + 1. \tag{1.2}$$

How can we eliminate t? Common sense, or a general consideration of transcendence degree, shows that there must be an algebraic relation between x, y not involving t. And indeed a moment's thought gives

$$x^2 - 2xy + y^2 + 2x - 3y + 2 = 0. \tag{1.3}$$

But what about

$$x = t^3 + t, \;\; y = t^4 + t?$$

We could solve the first equation by radicals for t, and then substitute into the second equation, and finally somehow clear the radicals. We get

$$x^4 + 3x^3 - 4x^2y - y^3 + 3x^2 - 5xy + 2y^2 + 2x - 2y = 0 \tag{1.4}$$

but I confess that here I just used Maple to calculate the resultant (see Chapter 5) of

$$t^3 + t - x, \;\; t^4 + t - y$$

with respect to t.

What about

$$x = t^{1948} + t^{666} + 1, \;\; y = t^{1291} + t^{163} + t? \tag{1.5}$$

Here the degrees are my year of birth and the (traditional) year of birth of the earliest part of the Swiss Federation, where I first gave these lectures (in Basle, after earlier attempts in Ann Arbor, Constance, Hong Kong, Heraklion and Vienna).

Maple doesn't respond for 47 seconds; and then gives an incomprehensible error message (the resultant that you are seeking does not exist). How can we find this relation $P(x, y) = 0$?

Let us write

$$P(X, Y) = \sum_{i=0}^{L} \sum_{j=0}^{L} p_{ij} X^i Y^j \tag{1.6}$$

for unknown L and undetermined coefficients p_{ij} which are presumably integers as in (1.3) and (1.4). If we substitute (1.5) into $P(x, y)$, then we obtain a polynomial in t of degree at most $3239L$. Its coefficients are linear forms in the p_{ij}. If we equate these coefficients to zero, then we get a system of $3239L + 1$ homogeneous linear equations in the $(L + 1)^2$ unknowns p_{ij}. By linear algebra this system is solvable non-trivially, provided

$$(L + 1)^2 > 3239L + 1$$

for example if $L = 3238$.

This proves something: namely that there is a non-trivial relation of degree at most 3238 in each variable. There may well be $(L + 1)^2 = 10491121$ terms in the relation, which accounts for Maple's chickening out. And the trouble for anyone, of course, is that we now have to solve 10487883 equations in 10491121 unknowns.

The reader may see first that this works for any two polynomials F, G in t instead of (1.5), and second that it generalizes to more variables; for example between any three polynomials in two variables there is a non-trivial algebraic relation (as would follow more simply by consideration of transcendence degree).

This example is perhaps more typical of those to follow in these pages. After the substitution (1.5) we may regard the function $P(x, y)$ as having a large order of vanishing at $t = 0$; so large, indeed, that it must vanish identically.

In both examples the goal is practically the auxiliary polynomial itself; so it is hardly "auxiliary".

We will see many more and subtler applications in this book (sadly not Siegel's Theorem (Siegel, 1955) about functions on compact manifolds, which is a very sophisticated generalization of the second example – see however Lemma 20.4 and Exercise 20.67). But before we start, let us ask: since there appear to be so many terms in the relation connecting (1.5), what are the coefficients like? If we normalize them to be integers, how big are they? We note that the Cramér formulae for solving linear equations involve determinants whose size is the number n of unknowns. Such determinants already have $n!$ terms, so their values are likely to be somewhat larger. Thus it would be surprising if the integers in our relation were substantially less than the factorial 10491121!. And it can be seen that some of the entries of the determinants are almost as large as 3^{6476} (as in Exercise 1.17). This means that we could expect some coefficients in P to have thirty thousand million (American thirty billion, Swiss dreissig Milliarden) decimal digits (see however Exercise 8.13). So it is doubtful if P could ever be expressed explicitly.

Exercises

1.1 Show that there is $P \neq 0$ in $\mathbb{C}[X]$ such that $P(X)(X^{1000} + X^{100} + X^3)$ has the form $\sum_{n=0}^{N} a_n X^{n^2}$.

1.2 Let F be in $\mathbb{C}[X]$ with degree at most D. Show that there is $P \neq 0$ in $\mathbb{C}[X]$ with degree at most $D^2 - D$ such that PF has the form $\sum_{n=0}^{N} a_n X^{n^2}$.

1.3 Let F be in $\mathbb{C}[X]$. Find $P \neq 0$ in $\mathbb{C}[X]$ such that PF has the form $\sum_{n=0}^{N} a_n X^{2n}$.

1.4 Let F be in $\mathbb{C}[t]$ with degree at most $D \geq 1$ and let G be in $\mathbb{C}[t]$ with degree at most $E \geq 1$. Show that there is $P \neq 0$ in $\mathbb{C}[X,Y]$ with degree at most $D + E - 1$ in each variable such that $P(F,G) = 0$.

1.5 Let F be in $\mathbb{C}[t]$ with degree at most $D \geq 1$ and let G be in $\mathbb{C}[t]$ with degree at most $E \geq 1$. Show that there is $P \neq 0$ in $\mathbb{C}[X,Y]$ with degree at most E in X and degree at most $DE - E + 1$ in Y such that $P(F,G) = 0$.

1.6 Let t, u be independent variables, and let F, G, H be in $\mathbb{C}[t,u]$. Show that there is $P \neq 0$ in $\mathbb{C}[X,Y,Z]$ such that $P(F,G,H) = 0$.

* * * * * * * * * * * * * * * * * * *

1.7 Show that there is an absolute constant c (that is, not depending on any parameters) with the following property. Let F be in $\mathbb{C}[X]$ with degree at most $D \geq 2$. Then there is $P \neq 0$ in $\mathbb{C}[X]$ with degree at most $cD \log D$ such that PF has the form $\sum_p a_p X^p$, where p runs over the set of primes.

1.8 Let F be in $\mathbb{C}[X]$. Find $P \neq 0$ in $\mathbb{C}[X]$ such that PF has the form $\sum_{n=0}^{N} a_n X^{3n}$.

1.9 Let F be in $\mathbb{F}_p[X]$. Show that there is $P \neq 0$ in $\mathbb{F}_p[X]$ such that $G = PF$ satisfies $G(X_1 + X_2) = G(X_1) + G(X_2)$ in $\mathbb{F}_p[X_1, X_2]$.

1.10 Let F be in $\mathbb{C}[t]$ with degree at most $D \geq 1$ and let G be in $\mathbb{C}[t]$ with degree at most $E \geq 1$. Show that there is $P \neq 0$ in $\mathbb{C}[X,Y]$ with degree at most E in X and degree at most D in Y such that $P(F,G) = 0$ [*Hint*: resultants].

1.11 Let $P \neq 0$ in $\mathbb{C}[X,Y]$ be such that $P(t^{1948} + t^{666} + 1, t^{1291} + t^{163} + t) = 0$. Show that P has degree at least 1291 in X and degree at least 1948 in Y (compare Exercise 5.7).

1.12 Can one essentially improve the $D^2 - D$ in Exercise 1.2? I don't know.

1.13 Let F, G be in $\mathbb{C}(t)$ (rational functions). Show that there is $P \neq 0$ in $\mathbb{C}[X,Y]$ such that $P(F,G) = 0$.

1.14 Let

$$F = 256\frac{(t^2-t+1)^3}{t^2(t-1)^2}, \quad G = 256\frac{(t^2+t+1)^3}{t^2(t+1)^2}.$$

Show that $P(F,G) = 0$ for

$$P = X^3Y - 2X^2Y^2 + XY^3 - 1728(X^3+Y^3) + 1216(X^2Y + XY^2)$$

$$+3538944(X^2+Y^2) - 2752512XY - 2415919104(X+Y) + 549755813888.$$

(This is related to the simultaneous complex multiplication of two different elliptic curves and also to the André–Oort Conjecture – see Exercise 21.19. Of course the present exercise and the following are illustrations of Littlewood's Principle that "All identities are trivial (once they have been written down by someone else)" as quoted by Cassels.)

1.15 Let

$$F = tu(t^{10} + 11t^5u^5 - u^{10}), \quad G = -t^{20} - u^{20} + 228(t^{15}u^5 - t^5u^{15}) - 494t^{10}u^{10},$$

$$H = t^{30} + u^{30} + 522(t^{25}u^5 - t^5u^{25}) - 10005(t^{20}u^{10} + t^{10}u^{20}).$$

Show that $G^3 + H^2 = 1728F^5$. (This is related to the icosahedron – see Klein, 1956.)

1.16 Let

$$F = 1728\frac{u^3}{u^3-v^2}, \quad G = -1728\frac{u^2v}{u^3-v^2}, \quad H = -288\frac{u(tuv - 3u^3 - 4v^2)}{u^3-v^2},$$

$$K = -24\frac{3t^2u^2v - 18tu^4 - 24tuv^2 + 95u^3v + 16v^3}{u^3-v^2}.$$

Show that

$$2F^2(F-1728)^2GK - 3F^2(F-1728)^2H^2 + (F^2 - 1968F + 2654208)G^4 = 0.$$

(This is related to the differential equation for the modular function – see Exercise 21.15.)

1.17 With $L = 3238$ show that $(t^{1948} + t^{666} + 1)^L(t^{1291} + t^{163} + t)^L$ has a coefficient at least

$$\frac{3^{2L}}{L^2 + L + 1} > 10^{3082}.$$

1.18 Show that there is $P \neq 0$ in $\mathbb{Q}[X]$ for Exercise 1.1.

1.19 Show that there is $P \neq 0$ in $\mathbb{Z}[X]$ for Exercise 1.1.

1.20 Show that there is $P \neq 0$ in $\mathbb{Z}[X, Y]$ for Exercise 1.10.

1.21 Find $P \neq 0$ in $\mathbb{Z}[X, Y]$ with $P(t + i, t - i) = 0$.

1.22 If x, y are in $\mathbb{C}$ with (1.3), must there exist t in $\mathbb{C}$ with (1.2)?

1.23 What about Exercise 1.22 with $\mathbb{C}$ replaced by $\mathbb{R}, \mathbb{Q}, \mathbb{Z}, \mathbb{F}_p$?

2

Irrationality I

The main application of the method of auxiliary polynomials is in diophantine approximation and transcendence. But before these topics comes irrationality: one seeks to prove that a given number is not in $\mathbb{Q}$. One of the earliest examples is of course

$$e = \sum_{k=0}^{\infty} \frac{1}{k!} = \frac{1}{0!} + \frac{1}{1!} + \frac{1}{2!} + \frac{1}{3!} + \cdots = 1 + 1 + \frac{1}{2} + \frac{1}{6} + \cdots .$$

It is proved in any elementary text on number theory that e is irrational; the proof is based on the rapid convergence of the series together with the reasonable behaviour of the denominators. We give a proof nevertheless.

Consider the truncation

$$f_n = e - \sum_{k=0}^{n} \frac{1}{k!} = \sum_{k=n+1}^{\infty} \frac{1}{k!}$$

for $n = 0, 1, 2, \ldots$. The first term on the extreme right-hand side dominates and indeed

$$\sum_{k=n+1}^{\infty} \frac{1}{k!} = \frac{1}{(n+1)!}\left(1 + \frac{1}{n+2} + \cdots\right) < \frac{2}{(n+1)!}.$$

Thus $0 < f_n < 2/(n+1)!$ and

$$0 < |n! f_n| < \frac{2}{n+1}. \tag{2.1}$$

Now if s is a denominator for the rational e, then multiplying by s and making n tend to infinity gives a contradiction to the so-called Fundamental Theorem of Transcendence that every non-zero integer has absolute value at least 1.

The proof is slightly easier for the alternating series

$$e^{-1} = \sum_{k=0}^{\infty} \frac{(-1)^k}{k!} = 1 - 1 + \frac{1}{2} - \frac{1}{6} + \cdots,$$

because we no longer need the dominance.

The proofs extend to give the linear independence over $\mathbb{Q}$ of $1, e, e^{-1}$, which amounts to the fact that e cannot be quadratic over $\mathbb{Q}$. In particular e^2 is irrational. But there is a minor snag. We assume that $r + se + te^{-1} = 0$ for integers r, s, t not all zero, and then for

$$f_n = r + s\sum_{k=0}^{n} \frac{1}{k!} + t\sum_{k=0}^{n} \frac{(-1)^k}{k!} = -s\sum_{k=n+1}^{\infty} \frac{1}{k!} - t\sum_{k=n+1}^{\infty} \frac{(-1)^k}{k!},$$

we get

$$|n!f_n| \leq |s|\frac{2}{n+1} + |t|\frac{2}{n+1}.$$

Hence the $n!f_n$ are integers tending to zero as n tends to infinity. But we no longer know that these integers are non-zero as in (2.1).

In fact it is not too hard to show that

$$f_n = f_{n+1} = f_{n+2} = 0 \tag{2.2}$$

is impossible for any n. Namely,

$$(n+1)!(f_{n+1} - f_n) = s + (-1)^{n+1}t$$

and so

$$(n+2)!(f_{n+2} - f_{n+1}) = s - (-1)^{n+1}t.$$

Thus (2.2) would imply $s = t = 0$ so $r = 0$ too, a contradiction.

Here the problem makes its first appearance but is relatively harmless; however in later chapters we will see it getting more and more dangerous.

But as soon as we consider $e^2 = \sum_{k=0}^{\infty} 2^k/k!$ directly, some other difficulties arise. The convergence is practically just as fast, but after truncating at $k = n$ and multiplying by a denominator $n!$, we get a term $2^{n+1}/(n+1)$ which no longer tends to zero.

The proof can be fixed by calculating the power of 2 dividing $n!$ to yield a smaller denominator; this involves restricting n to a special form like 2^m. Such a trick can be extended to give the linear independence over $\mathbb{Q}$ of $1, e^2, e^{-2}$ (see Exercise 2.13); in particular e^4 is irrational. But it is amusing (I learnt it from the wonderful book of Conway and Guy (1996), p. 253) that these ideas also give the irrationality of $\lambda = e^{\sqrt{2}}$ via

$$\lambda + \frac{1}{\lambda} = 2\sum_{l=0}^{\infty} \frac{2^l}{(2l)!}$$

(see Exercise 2.4).

Actually I know of no such simple proof that e^3 is irrational, although this is not difficult to establish by considering $\int_0^1 e^{3t} t^n (1-t)^n \mathrm{d}t$ (see Exercise 2.14). It can also be done with auxiliary polynomials of the type mentioned in Chapter 1, most efficiently by introducing derivatives as in the differential equation $(e^z)' = e^z$. However some extra arithmetic and analytic machinery is needed. This will be introduced step by step in the following chapters. By these means we will show in Chapter 13 that

$$e^{\alpha} = \sum_{k=0}^{\infty} \frac{\alpha^k}{k!}$$

is irrational and even transcendental for any rational and even algebraic $\alpha \neq 0$.

In the present chapter we will consider the series

$$\sum_{k=0}^{\infty} \frac{z^k}{2^{k(k-1)/2}} = 1 + z + \frac{1}{2}z^2 + \frac{1}{8}z^3 + \frac{1}{64}z^4 + \cdots, \tag{2.3}$$

which similarly converges for all real and even complex z to a function $f(z)$. It converges faster than the series of e^z, so let us see what truncation of $f(\alpha)$ gives.

Let us start with $\alpha = a/b$, for simplicity taking $a \geq 1, b \geq 1$ in $\mathbb{Z}$. The truncations are

$$f_n = f(\alpha) - \sum_{k=0}^{n} \frac{\alpha^k}{2^{k(k-1)/2}} = \sum_{k=n+1}^{\infty} \frac{\alpha^k}{2^{k(k-1)/2}}.$$

Again there is domination on the extreme right, and if n is large enough we get

$$|f_n| \leq 2\frac{\alpha^{n+1}}{2^{n(n+1)/2}} = \frac{2a^{n+1}}{2^{n(n+1)/2}b^{n+1}}. \tag{2.4}$$

Taking into account a common denominator $2^{n(n-1)/2}b^n$, we deduce

$$|2^{n(n-1)/2}b^n f_n| \leq \frac{2a^{n+1}}{2^n b} = \frac{2a}{b}\left(\frac{a}{2}\right)^n.$$

The proof works if the estimate tends to zero as n tends to infinity (assuming we can rule out the snag $f_n = 0$ as in (2.2) above). Unfortunately this is the case only for $a = 1$.

So all we get by these means is the irrationality of the $f(1/b)$.

In fact we can scrape by with $a = 2$ by estimating the denominators more carefully as with e^2, thanks to the numerators. But we get stuck at $a = 3$, for example, with the irrationality of

$$f(3) = 13.401244574556308427693105053675595778707177552368\ldots$$

For the value $\zeta(3)$ of the Riemann zeta function one had help from Apéry. Instead of his original proof one can consult Beukers's version (Beukers, 1979) – see also Exercise 2.22 – and also the account in van der Poorten (1979). Here an extra ingredient enables us to prove the irrationality of $f(\alpha)$ for all rational $\alpha \neq 0$ using an auxiliary polynomial; and that with only a modicum of extra machinery. Since

$$\frac{(2z)^k}{2^{k(k-1)/2}} = 2z\frac{z^{k-1}}{2^{(k-1)(k-2)/2}},$$

we have the functional equation

$$f(2z) = 1 + 2zf(z) \tag{2.5}$$

and this provides a substitute for the differential equation of e^z.

Namely after replacing z by $z/2$ we get $f(z) = 1 + zf(z/2)$, and we can iterate this to give

$$f(z) = 1 + zf\left(\frac{z}{2}\right) = 1 + z\left(1 + \frac{z}{2}f\left(\frac{z}{4}\right)\right) = \cdots$$

until

$$f(z) = 1 + z + \frac{1}{2}z^2 + \cdots + \frac{1}{2^{(n-1)(n-2)/2}}z^{n-1} + \frac{z^n}{2^{n(n-1)/2}}f\left(\frac{z}{2^n}\right). \tag{2.6}$$

Thus the truncation $f(z) - \sum_{k=0}^{n} z^k/2^{k(k-1)/2}$ is none other than

$$\frac{z^n}{2^{n(n-1)/2}}\left(f\left(\frac{z}{2^n}\right) - 1\right). \tag{2.7}$$

This involves the function $\phi_0(z) = f(z) - 1 = z + \cdots$ (if we take the liberty to change z for a moment). This visibly has a zero of order 1 at $z = 0$, and so for all sufficiently small z we have say

$$|f(z) - 1| \leq 2|z|. \tag{2.8}$$

If we now estimate the truncation f_n at $z = a/b$ using (2.7) and (2.8) (with z changed back again – sorry), then we get exactly the same estimate (2.4) as before. So what if anything has been achieved?

Well, suppose we could find a function $\phi(z)$ instead of $f(z) - 1$ above which has an order of vanishing greater than 1 at $z = 0$. Then we could think about

$\phi(z/2^n)$ instead of $f(z/2^n) - 1$ in (2.7) and hope that is small enough with respect to its denominator.

Finding ϕ looks easy: for example already

$$\phi(z) = f(z) - 1 - z = \frac{1}{2}z^2 + \cdots$$

has a zero of order 2 (the exponent in the first non-zero term in the power series). But on the other hand this is $z\phi_0(z/2)$, and so $\phi(z/2^n)$ is essentially $\phi_0(z/2^{n+1})$, so we have slipped backwards into the previous proof and cannot expect any advance on $a = 1$ (or $a = 2$). We must be a bit cleverer.

Accordingly out of a top hat I pluck with a flourish

$$\phi(z) = (2 - z)f(z) - (2 + z) \tag{2.9}$$

(whose existence is indisputable). On expanding, we find

$$\phi(z) = -\frac{1}{4}z^3 - \frac{3}{32}z^4 + \cdots \tag{2.10}$$

with a zero of order 3. Thus for sufficiently small z we have say

$$|\phi(z)| \leq |z|^3. \tag{2.11}$$

Substitute $z = \alpha = a/b$ as before, to get

$$\left|\phi\left(\frac{\alpha}{2^n}\right)\right| \leq \frac{a^3}{2^{3n}b^3} \tag{2.12}$$

where the exponent $3n$ will be crucial, reflecting the zero of order 3.

But what about a denominator?

Here

$$\phi\left(\frac{\alpha}{2^n}\right) = \left(2 - \frac{\alpha}{2^n}\right)f\left(\frac{\alpha}{2^n}\right) - 2 - \frac{\alpha}{2^n}$$

in which we certainly see a contribution $2^n b$. And by (2.6)

$$f\left(\frac{\alpha}{2^n}\right) = \frac{2^{n(n-1)/2}}{\alpha^n}\left(f(\alpha) - 1 - \alpha - \cdots - \frac{1}{2^{(n-1)(n-2)/2}}\alpha^{n-1}\right). \tag{2.13}$$

Assuming $f(\alpha) = r/s$ for integers r and $s \geq 1$ and recalling $\alpha = a/b$, we see that this has a denominator $a^n s$.

So $N = 2^n a^n bs\phi(\alpha/2^n)$ is a rational integer. By (2.12) it satisfies

$$|N| \leq \frac{sa^{n+3}}{2^{2n}b^2} = \frac{sa^3}{b^2}\left(\frac{a}{4}\right)^n.$$

This tends to zero if $a = 1, 2, 3$. So it would appear that we have established the irrationality of say $f(3)$!

As usual there is a snag. How do we know that $N \neq 0$? Or equivalently, why is it impossible that

$$\phi\left(\frac{\alpha}{2^n}\right) = 0 \tag{2.14}$$

for all sufficiently large n?

This is most easily seen by (2.10); as well as the upper bound (2.11) we have the lower bound say

$$|\phi(z)| \geq \frac{1}{5}|z|^3 \tag{2.15}$$

for all sufficiently small z.

And how was the conjuring trick done? I just took

$$\phi(z) = (p_0 + p_1 z)f(z) + (q_0 + q_1 z) \tag{2.16}$$

for undetermined p_0, p_1, q_0, q_1 and tried to find values such that the order of vanishing is as large as possible. We have four unknowns and we can expect from our linear algebra considerations to make the coefficients of $1, z, z^2$ vanish. And indeed these lead to the three equations

$$p_0 + q_0 = p_0 + p_1 + q_1 = \frac{1}{2}p_0 + p_1 = 0.$$

Here

$$p_0 = 2, \ \ p_1 = -1, \ \ q_0 = -2, \ \ q_1 = -1$$

is a simple solution, giving (2.9).

How do we go further, for example with $f(4)$? Well, again this can be done by truncation and careful denominator estimates (or by linking it to $f(2)$ using the functional equation), so we had better aim for $f(5)$. Incidentally, the irrationality of $\zeta(5)$ has still not been proved.

At least the approach should be clear: we try

$$\phi(z) = (p_0 + p_1 z + p_2 z^2)f(z) + (q_0 + q_1 z + q_2 z^2), \tag{2.17}$$

which can accommodate a zero of order at least 5. Then, in place of (2.12), we get

$$\left|\phi\left(\frac{\alpha}{2^n}\right)\right| \leq c\frac{a^5}{2^{5n}b^5}$$

with some c independent of n. As for denominators, the former contribution $2^n b$ increases to $(2^n b)^2$. But the other part $a^n s$ remains the same. We get for the rational integer $2^{2n}a^n b^2 s\phi(\alpha/2^n)$ the upper bound

$$c\frac{a^{n+5}s}{2^{3n}b^3} = c'\left(\frac{a}{8}\right)^n$$

with $c' = csa^5/b^3$ also independent of n. So in one swell foop with Reverend Spooner we get $a = 1, 2, 3, 4, 5, 6, 7$.

However the snag has not quietly gone away. Presumably there is still some lower bound like (2.15), most probably of the form $|\phi(z)| \geq c''|z|^5$ for some $c'' > 0$. Here the c'' would come from the coefficient of z^5 in $\phi(z)$.

But how do we know in advance that this coefficient is non-zero? We could just calculate it, of course; but we would rather prefer not to have to solve all these linear equations. If by bad luck the coefficient is zero (and we will see in the next chapter that such luck can occur), then the coefficient of z^6 will do just as well to disprove (2.14).

But what if this too vanishes? And the next one, and the next? Do we really know this to be impossible?

It would mean that $\phi(z)$ is identically zero, and of course with such a function one can prove nothing at all. But in that case $f(z)$ would be the rational function

$$-\frac{q_0 + q_1 z + q_2 z^2}{p_0 + p_1 z + p_2 z^2}$$

in $\mathbb{Q}(z)$. In fact $f(z)$ is not even in $\mathbb{C}(z)$; but we have to prove that.

We are now ready to give a full proof of our main result.

Theorem 2.1 *If $\alpha \neq 0$ is rational, then $f(\alpha) = \sum_{k=0}^{\infty} \frac{\alpha^k}{2^{k(k-1)/2}}$ is irrational.*

We need three preliminary lemmas, of which the first has been foreseen.

Lemma 2.2 *The function $f(z)$ does not lie in $\mathbb{C}(z)$.*

Proof We give two proofs, one algebraic and the other analytic; in this game the two viewpoints are sometimes hard to keep apart.

First, we deduce a contradiction from the existence of polynomials P, Q in $\mathbb{C}[z]$, not both zero, with

$$P(z)f(z) + Q(z) = 0. \tag{2.18}$$

We may assume that $P(z), Q(z)$ are coprime.

Changing z into $2z$ and using the functional equation, we deduce also

$$P(2z)(1 + 2zf(z)) + Q(2z) = 0. \tag{2.19}$$

We can eliminate f to get

$$P(z)(P(2z) + Q(2z)) = 2zP(2z)Q(z). \tag{2.20}$$

This implies that $P(2z)$ divides $P(z)Q(2z)$. But clearly $P(2z), Q(2z)$ stay coprime; and it follows that $P(2z)$ divides $P(z)$. Looking at degrees we see that

$P(z) = cP(2z)$ for some c in $\mathbb{C}$. It is easy to see that this implies $P(z) = pz^d$ for some p in $\mathbb{C}$. If $p = 0$, then $P = 0$ and so by (2.18) also $Q = 0$, a (local) contradiction.

Now (2.20) reduces to

$$Q(2z) - 2^{d+1}zQ(z) = -2^d pz^d.$$

The second term on the left-hand side dominates the degree and it follows that Q has degree exactly $d - 1$. Thus the term in $Q(z)$ with smallest degree has the form cz^e with $e \le d - 1$. Therefore the term on the left-hand side with smallest degree is $c(2z)^e$; however this cannot be the term of smallest degree on the right-hand side. This finishes the first purely algebraic proof.

For the second (apparently shorter) proof we continue to use the language of complex function theory, and we deduce a contradiction if $f(z)$ lies in $\mathbb{C}(z)$. Because the series (2.3) converges everywhere, $f(z)$ is also entire. Therefore it must be a polynomial. But (2.3) is an infinite sum. This finishes the analytic proof. □

We remark also that if this lemma were false, then the rationality of the Taylor coefficients of $f(z)$ would force it to be in $\mathbb{Q}(z)$. But then Theorem 2.1 could not be true! So Lemma 2.2 is somehow unavoidable.

Now here come two auxiliary polynomials, or rather a single auxiliary polynomial in z and f. When $\phi(z)$ is analytic on a neighbourhood of w and not identically zero there, we denote its (finite) order of vanishing at w by $\mathrm{ord}_{z=w}\,\phi(z)$. As hinted above, this means that the power series about w has the form $\phi(z) = c(z-w)^e + \cdots$ for $e = \mathrm{ord}_{z=w}\,\phi(z)$ and $c \neq 0$.

Lemma 2.3 *For any $L \ge 0$ in $\mathbb{Z}$ there are P, Q in $\mathbb{Z}[X]$ of degree at most L such that*

$$\phi(z) = P(z)f(z) + Q(z) \tag{2.21}$$

is not identically zero and

$$\mathrm{ord}_{z=0}\,\phi(z) \ge 2L + 1. \tag{2.22}$$

Proof We write

$$P(X) = \sum_{i=0}^{L} p_i X^i, \quad Q(X) = \sum_{i=0}^{L} q_i X^i$$

as in (1.1). The condition (2.22) tells us that the coefficients of $1, z, z^2, \ldots, z^{2L}$ in the Taylor expansion of ϕ are zero. These coefficients are linear forms in the p_i, q_i with rational coefficients. Thus we have a system of $2L+1$ homogeneous linear equations in $2L+2$ unknowns. By linear algebra this system is solvable non-trivially over the rationals, which gives us P and Q not both zero in $\mathbb{Q}[X]$. We inveigle them into $\mathbb{Z}[X]$ by multiplying up by a common denominator. Now the non-vanishing of (2.21) follows from Lemma 2.2. □

For each L in Lemma 2.3 we do not know that the ϕ there is unique, and indeed we can multiply it by any positive integer without changing its properties. Also the linear algebra supplies a space of solutions of positive dimension, and this dimension may be greater than 1. But we will suppose for simplicity that for each L some ϕ has been fixed.

Lemma 2.4 *For any real $Z \geq 0$ and any $L \geq 0$ in $\mathbb{Z}$ there is $C = C(Z, L)$ such that*

$$|\phi(z)| \leq C|z|^{2L+1}$$

for all complex z with $|z| \leq Z$.

Proof Because of (2.22), we can write $\phi(z) = z^{2L+1}\psi(z)$ with a power series $\psi(z)$ also everywhere convergent. So ψ is continuous (or even analytic) on the set of complex z with $|z| \leq Z$. Therefore it is bounded on this set, say by a constant C which depends only on Z and of course ϕ, so only on Z and L. Now we conclude

$$|\phi(z)| = |z|^{2L+1}|\psi(z)| \leq C|z|^{2L+1}$$

as desired. □

We remark here that this lemma shows that a function with a high order of vanishing at $z = 0$ must be small in a neighbourhood of $z = 0$. Such a principle will be developed and used frequently in this book. It has been foreshadowed in (2.8) and (2.11).

We can now prove Theorem 2.1. The proof is not long, but without our preamble it may have seemed somewhat mysterious.

Accordingly take $\alpha = a/b$ for $a \neq 0, b \neq 0$ in $\mathbb{Z}$, for convenience with $a > 0$. It will suffice to deduce a contradiction from the existence of r, s in $\mathbb{Z}$ with $s > 0$ and $f(\alpha) = r/s$.

We choose $L = L(a, b, r, s)$ in Lemma 2.3 and we calculate this later. Now for any $n \geq 0$ in $\mathbb{Z}$ we can write by (2.21)

$$\phi\left(\frac{\alpha}{2^n}\right) = p(n)u(n) + q(n) \tag{2.23}$$

with

$$p(n) = P\left(\frac{\alpha}{2^n}\right), \quad q(n) = Q\left(\frac{\alpha}{2^n}\right), \quad u(n) = f\left(\frac{\alpha}{2^n}\right).$$

It is clear that $2^{nL}b^Lp(n), 2^{nL}b^Lq(n)$ are in $\mathbb{Z}$. And from (2.13) we see exactly as before that $a^n su(n)$ is in $\mathbb{Z}$.

It follows from (2.23) that $2^{nL}b^La^n s\phi(2^{-n}\alpha)$ is in $\mathbb{Z}$. But from Lemma 2.4 with $Z = |\alpha|$, we can estimate the absolute value from above by

$$2^{nL}|b|^La^n sC\left(\frac{|\alpha|}{2^n}\right)^{2L+1} = C'\left(\frac{a}{2^{L+1}}\right)^n$$

where C depends only on α and L, and so $C' = |b|^{-L-1}sCa^{2L+1}$ depends only on a, b, r, s and L but not n.

We now choose L so that

$$2^{L+1} > a, \tag{2.24}$$

and make n tend to infinity. We find something in $\mathbb{Z}$ with absolute value tending to zero. So by the Fundamental Theorem of Transcendence it must be zero for all large enough n. This implies that

$$\phi\left(\frac{\alpha}{2^n}\right) = 0 \tag{2.25}$$

for all large enough n.

This is exactly the snag (2.14) encountered earlier, and we already saw how to deal with it, even though we have no complete control over ϕ, which depends in a complicated way on L. Still, there are several methods. For example we know from Lemma 2.3 that $\phi(z)$ is not identically zero, so that $D = \mathrm{ord}_{z=0}\,\phi(z)$ is finite (and by the way at least $2L+1$), and this means $\phi(z) = z^D\tilde{\psi}(z)$ for a power series $\tilde{\psi}(z)$ with $\tilde{\psi}(0) \neq 0$. Now

$$\lim_{0\neq z\to 0} \frac{\phi(z)}{z^D} = \tilde{\psi}(0) \neq 0; \tag{2.26}$$

but (2.25) would imply that this limit is 0.

This completes the proof of Theorem 2.1.

We illustrate with some examples.

First take $a = 1$. Then the crucial (2.24) already holds for $L = 0$. This means $P(X) = p_0, Q(X) = q_0$ are constants, and clearly $p_0 = 1, q_0 = -1$ will do in Lemma 2.3, because then $\phi_0(z) = f(z) - 1$ has order at least 1 at $z = 0$. So we recover the argument of (2.7), as we saw here being equivalent to truncation.

Next take $a = 2$. Now (2.24) holds for $L = 1$, which means $P(X) = p_0 + p_1X,\ Q(X) = q_0 + q_1X$, so we recover the argument with (2.16). And already (2.24) for $L = 1$ covers $a = 3$.

And then (2.24) for $L = 2$ covers $a = 4, 5, 6, 7$, as we also found.

Next jump to $a = 1948$, so that $L = 10$ suffices. Thus to prove the irrationality of

$$f(1948) = 136473792159447560036.28081649031298655017721 3\dots$$

we have to solve a system of 21 equations in 22 unknowns. It is fortunate that we don't have to do this explicitly!

But if we want to prove that $f(1948)$ is not too near a rational number, then we must know that the system has a solution which is not too large. This fact will be established in Chapter 8, and it will be used frequently after that, notably in Chapter 12, when we show that $\sqrt[3]{2}$ is not too near a rational number as a special case of Thue's Theorem.

By contrast with e^{α}, it is not known if $f(\alpha)$ is transcendental for all algebraic or even rational $\alpha \neq 0$. Essentially one cannot even prove that it is not quadratic (but see Exercise 2.10 below). But there are some results of linear independence over $\mathbb{Q}$.

The function $f(z)$ usually occurs in the literature with a parameter q satisfying $|q| > 1$, as $f(z,q) = \sum_{k=0}^{\infty} z^k/q^{k(k-1)/2}$, now with $f(qz,q) = 1 + qzf(z,q)$. Then it can be regarded as a one-sided theta function. Perhaps the most fundamental theta function is

$$\theta(w,\tau) = \sum_{k\in\mathbb{Z}} e^{\pi i k^2 \tau}\cos(2\pi k w) = \sum_{k\in\mathbb{Z}} e^{\pi i k^2\tau} e^{2\pi i k w} \tag{2.27}$$

defined for τ in the upper half plane. If we put

$$q = e^{-2\pi i\tau}, \quad z = e^{2\pi i w + \pi i \tau}$$

(the first in violent contradiction to the more usual notation, for example in Exercise 20.99 or Chapter 21), then (2.27) becomes

$$F(z,q) = \sum_{k\in\mathbb{Z}} \frac{z^k}{q^{k(k-1)/2}},$$

which is a double-sided version of our $f(z,q)$. Now we have

$$F(qz,q) = qzF(z,q), \tag{2.28}$$

which is equivalent to the periodicity

$$\theta(w-\tau,\tau) = e^{2\pi i w - \pi i \tau}\theta(w,\tau).$$

The functions $f(z,q)$ have a respectable history going back to Bernstein and Szász in 1915 and are also associated with the name of Tschakaloff. It was Duverney who introduced auxiliary polynomials into the game; see Duverney

(1997) for more references. But the functions have not attracted as much attention as those that Mahler introduced, and which are the subject of the next chapter.

Exercises

Here $f(z) = \sum_{k=0}^{\infty} z^k/2^{k(k-1)/2}$.

2.1 Let $t_0 \geq 2,\ t_1 \geq 2, \ldots$ be unbounded in $\mathbb{N}$. Show that $\tau = \sum_{k=0}^{\infty} 1/(t_0 \cdots t_k)$ is irrational.

2.2 Let $t_0 \geq 2,\ t_1 \geq 2, \ldots$ be unbounded in $\mathbb{N}$, and let $c_0, c_1, \ldots$ be bounded in $\mathbb{N} = \{1, 2, \ldots\}$. Show that $\tau = \sum_{k=0}^{\infty} c_k/(t_0 \cdots t_k)$ is irrational.

2.3 Let $t_0 \geq 2,\ t_1 \geq 2, \ldots$ be unbounded in $\mathbb{N}$, and let $c_0, c_1, \ldots$ be in $\{1, -1\}$. Show that $\tau = \sum_{k=0}^{\infty} c_k/(t_0 \cdots t_k)$ is irrational.

2.4 (a) Show that $\operatorname{ord}_2 (2n)!/(2l)! \geq n - l \ \ (0 \leq l \leq n)$.
(b) Show that $\lambda = e^{\sqrt{2}}$ is irrational [*Hint*: $\frac{1}{2}(\lambda + 1/\lambda) = \mu = \sum_{l=0}^{\infty} 2^l/(2l)!$].

2.5 Let $|e - r/s| = \delta$ for r, s in $\mathbb{Z}$ with $s \geq 1$.
(a) For $\tilde{f}_n = r/s - \sum_{k=0}^{n} 1/k!$ show that $|sn!\tilde{f}_n| \leq 2s/(n+1) + sn!\delta$.
(b) Show that either $\tilde{f}_{4s-1} \neq 0$ or $\tilde{f}_{4s} \neq 0$.
(c) Conclude that $|e - \frac{r}{s}| \geq \frac{1}{2s(4s)!}$ (irrationality measure).

2.6 Let R be a subring of $\mathbb{C}$. Let $P(X, Y)$ in $R[X, Y]$ be of degree at most $L \geq 0$ in X and degree at most $M \geq 0$ in Y. Let $\phi(z) = P(z, f(z))$. Let a, b, s in $\mathbb{N}$ and r in R be such that $f(\alpha) = r/s$ for $\alpha = a/b$. Find D in $\mathbb{N}$ with $D\phi(\alpha/2^n)$ in R.

2.7 Find explicit ϕ in Exercise 2.6 which suffices to prove that $f(1/7)$ is irrational.

2.8 What sort of ϕ in Exercise 2.6 suffices to prove that $f(1/9)$ is irrational?

2.9 (a) Show that every α in $\mathbb{Q}(i)$ has the form ω'/ω for $\omega', \omega \neq 0$ in $\mathbb{Z}[i]$.
(b) Let $\omega \neq 0$ be in $\mathbb{Z}[i]$. Show that $|\omega| \geq 1$.

2.10 Let $\alpha \neq 0$ be in $\mathbb{Q}(i)$. Show that $f(\alpha)$ is not in $\mathbb{Q}(i)$.

* * * * * * * * * * * * * * * * * * * *

2.11 Let $t_0 \geq 2,\ t_1 \geq 2, \ldots$ be unbounded in $\mathbb{N}$ and let $c_0 \neq 0, c_1 \neq 0, \ldots$ be bounded in $\mathbb{Z}$. Must $\sum_{k=0}^{\infty} c_k/(t_0 \cdots t_k)$ be irrational?

2.12 Let $m \geq 0$ and let $f_n = \sum_{i=0}^{m} r_i \sum_{k=0}^{n} i^k/k!$. If there is $n \geq 0$ with $f_n = f_{n+1} = \cdots = f_{n+m} = 0$, show that $r_0 = r_1 = \cdots = r_m = 0$.

2.13 (a) Show that $\operatorname{ord}_2 k! = k - \beta(k)$, where $\beta(k)$ is the number of ones in the binary expansion of k.

(b) Deduce that e^2 is neither rational nor quadratic irrational.

2.14 (a) For a in $\mathbb{N}$ show that

$$\frac{a^{2n}}{n!}\int_0^1 e^{at}t^n(1-t)^n dt \quad (n = 0, 1, 2, \ldots)$$

are in $\mathbb{Z} + \mathbb{Z}e^a$ [*Hint*: binomial coefficients – see also Chapter 9].

(b) Deduce that e^a is irrational.

(c) Conclude that e^α is irrational for every rational $\alpha \neq 0$.

2.15 Find $p_0, \ldots, p_n$ in $\mathbb{Z}$, not all 0, with $\operatorname{ord}_{z=0}(p_0 e^{nz} + \cdots + p_n) \geq n$.

2.16 Let p_0, p_1, q_0, q_1 be in $\mathbb{C}$ not all 0. Show that $\operatorname{ord}_{z=0}\phi(z) \leq 3$ with $\phi(z) = (p_0 + p_1 z)f(z) + (q_0 + q_1 z)$.

2.17 Show that there are c, κ such that $|f(z)| \leq cZ^{\kappa \log Z}$ for all z with $|z| \leq Z$ $(Z \geq 2)$.

2.18 Let $F(z) = \sum_{k=-\infty}^{+\infty} z^k/2^{k(k-1)/2}$ $(z \neq 0)$. Show that $F(2z) = 2zF(z)$. What is $\operatorname{ord}_{z=0} F(z)$?

2.19 Let τ be in $\mathbb{R}$ with $1/3 < \tau \leq 1$. Show that there are $t_0, t_1, \ldots$ in $\{2, 3, 4\}$ with $\sum_{k=0}^{\infty} 1/(t_0 \cdots t_k) = \tau$ (Hančl and Tijdeman, 2004).

2.20 Show that there is a function ω from $\mathbb{N}^2$ to $\mathbb{R}$ such that $\operatorname{ord}_{z=0}\phi(z) \leq \omega(L, M)$ for all polynomials $P(X, Y)$ in $\mathbb{C}[X, Y]$, with degree at most $L \geq 1$ in X and degree at most $M \geq 1$ in Y, and $\phi(z) = P(z, f(z))$ not identically 0.

2.21 Show that z and $f(z)$ are algebraically independent over $\mathbb{C}$.

2.22 Show that

$$\int_0^1\int_0^1\int_0^1 \frac{(t(1-t)u(1-u)v(1-v))^n}{(1-(1-tu)v)^{n+1}} dt du dv \quad (n = 0, 1, 2, \ldots)$$

are in $\mathbb{Q} + \mathbb{Q}\zeta(3)$ (Beukers 1979).

3
Irrationality II – Mahler's Method

Here we prove an analogue of Theorem 2.1 for the function

$$f(z) = \sum_{k=0}^{\infty} z^{2^k} = z + z^2 + z^4 + z^8 + z^{16} + \cdots$$

which was considered by Mahler. Now this series converges in the complex plane only on the unit disc $\mathbb{D}$ defined by $|z| < 1$; and indeed it has a so-called natural boundary on $|z| = 1$ which precludes any further analytic continuation. Near $z = 0$ it converges rather more rapidly than $\sum_{k=0}^{\infty} z^k/2^{k(k-1)/2}$ and so we might expect things to be easier here. So let us again see what truncation of $f(\alpha)$ gives.

Let us start with $\alpha = a/b$, for simplicity taking $a \geq 1, b \geq 1$; but now $a < b$. The truncations are

$$f_n = f(\alpha) - \sum_{k=0}^{n} \alpha^{2^k} = \sum_{k=n+1}^{\infty} \alpha^{2^k}.$$

Once more there is domination on the extreme right and if n is large enough we get

$$|f_n| \leq 2\alpha^{2^{n+1}} = \frac{2a^{2^{n+1}}}{b^{2^{n+1}}}. \tag{3.1}$$

Taking into account a common denominator b^{2^n}, we deduce

$$|b^{2^n} f_n| \leq \frac{2a^{2^{n+1}}}{b^{2^n}} = 2\left(\frac{a^2}{b}\right)^{2^n}.$$

The proof works if this tends to zero as n tends to infinity (assuming we can rule out the usual snag $f_n = 0$). That is the case only for

$$a < b^{1/2}, \tag{3.2}$$

which is at least better than just $a = 1$ in the previous chapter.

As before, we go further by exploiting a functional equation, which in this situation is clearly

$$f(z^2) = f(z) - z. \tag{3.3}$$

We can iterate this to give

$$f(z) = z + f(z^2) = z + z^2 + f(z^4) = z + z^2 + z^4 + f(z^8) = \cdots$$

until

$$f(z) = z + z^2 + \cdots + z^{2^n} + f(z^{2^{n+1}}). \tag{3.4}$$

Thus the truncation $f(z) - \sum_{k=0}^{n} z^{2^k}$ is none other than simply

$$f(z^{2^{n+1}}). \tag{3.5}$$

This involves the function $f(z) = z + \cdots$ (if we take the liberty, now with precedent, to change z for a moment). This visibly has a zero of order 1 at $z = 0$, and as before the obvious estimation gives back the truncation argument.

So again we must find a function $\phi(z)$, instead of $f(z)$ above, which has an order of vanishing greater than 1 at $z = 0$. But easy examples like $\phi(z) = f(z) - z = z^2 + \cdots = f(z^2)$ will not do, of course. The reader can supply her own top hat to find

$$\phi(z) = (1 - z)f(z) - z = -z^3 + \cdots \tag{3.6}$$

in the style of (2.9). Using $f(\alpha^{2^{n+1}}) = f(\alpha) - \sum_{k=0}^{n} \alpha^{2^k}$ for $f(\alpha) = r/s$, we find for the absolute value of the rational integer $b^{2^{n+1}} b^{2^n} s\phi(\alpha^{2^{n+1}})$ the upper bound say

$$2b^{2^{n+1}} b^{2^n} s \left(\frac{a^{2^{n+1}}}{b^{2^{n+1}}} \right)^3 = 2s \left(\frac{a^6}{b^3} \right)^{2^n}.$$

But now a little shock: this tends to zero only for $a < b^{1/2}$, no improvement at all on (3.2)!

Maybe only a blip; but disturbingly the same thing happens with the scheme (2.17). And quite generally in the situation of Lemma 2.3 with P, Q of degree at most L and a zero of order at least $2L + 1$. Now we get an upper bound

$$cb^{2^{n+1}L} b^{2^n} s \left(\frac{a^{2^{n+1}}}{b^{2^{n+1}}} \right)^{2L+1} = cs \left(\frac{a^{4L+2}}{b^{2L+1}} \right)^{2^n}$$

with c independent of n; so the same condition (3.2).

We note that in the above discussion, the f contributes a denominator of order only b^{2^n}, while the implicit z^L contributes a much larger $b^{2^{n+1}L}$. So we could allow larger powers of $f(z)$. The simplest example, obtained by allowing $f(z)^2$ in (3.6), is

$$\phi(z) = (3-2z)f(z)^2 + (1-4z)f(z) - z = 2z^6 + \cdots \tag{3.7}$$

with order 6.

Incidentally this is an example of "bad luck" mentioned in the previous chapter; for the general shape

$$(p_0 + p_1 z)f(z)^2 + (q_0 + q_1 z)f(z) + (r_0 + r_1 z)$$

we have 6 coefficients, and so we can expect only order at least 5.

For us it is in fact good luck, because we get something smaller than expected. The rational integer now is bounded above by

$$cb^{2^{n+1}}(b^{2^n}s)^2\left(\frac{a^{2^{n+1}}}{b^{2^{n+1}}}\right)^6 = cs^2\left(\frac{a^{12}}{b^8}\right)^{2^n}$$

leading to the improvement

$$a < b^{2/3} \tag{3.8}$$

of (3.2). If we had used only the expected order 5, then we would have ended up with the slightly worse $a < b^{3/5}$, but still beating (3.2).

Thus we must allow an auxiliary polynomial $P(X, Y)$ of large degree in each variable, with the corresponding auxiliary function $\phi(z) = P(z, f(z))$. We are nearly ready for the formal proof, but there is an extra matter. In the previous chapter, we could rule out $\phi(z)$ being identically zero by proving that the previous $f(z)$ is not rational; that is, not in $\mathbb{C}(z)$. But we must make sure that the present $f(z)$ is not algebraic; that is, not algebraic over $\mathbb{C}(z)$. Intuitively this is excluded by the natural boundary. It is very slightly weaker than (well really the same as) the algebraic independence of $z, f(z)$ over $\mathbb{C}$; that is, if F is a polynomial in $\mathbb{C}[X, Y]$ such that $F(z, f(z))$ is identically zero, then $F = 0$.

Here is the main result of this chapter.

Theorem 3.1 *If α is rational with $0 < |\alpha| < 1$, then $f(\alpha)$ is irrational.*

We need three preliminary lemmas, of which the first has been foreseen, and the other two are of a familiar type.

Lemma 3.2 *The functions $z, f(z)$ are algebraically independent over $\mathbb{C}$.*

Proof If the lemma is false, then there is such an $F = F(X, Y) \neq 0$, and we can choose one where the degree M with respect to Y is minimal. Clearly $M \geq 1$. We can write $F(z, f) = 0$, where $f = f(z)$, as

$$f^M + R_1 f^{M-1} + \cdots + R_M = 0 \tag{3.9}$$

with $R_i = R_i(z)$ in $\mathbb{C}(z)$.

Now replace z by z^2 in 3.9, and use the functional equation (3.3). We obtain

$$(f-z)^M + S_1(f-z)^{M-1} + \cdots + S_M = 0 \tag{3.10}$$

with $S_i = S_i(z) = R_i(z^2)$ also in $\mathbb{C}(z)$.

Subtract (3.9) from (3.10) to get

$$(-Mz + S_1 - R_1)f^{M-1} + \cdots = 0.$$

If the leading coefficient is not zero, then we contradict the minimality of M. So it must vanish, which is just

$$R(z^2) = R(z) + Mz \tag{3.11}$$

for $R = R_1$. This is a functional equation very like (3.3). Only now R is rational.

As with Lemma 2.2, we have once again reached the fork in the road between algebra and analysis. So, on the advice of Yogi Berra, we take it.

Write $R = P/Q$ for coprime P, Q. Then (3.11) gives

$$P(z^2)Q(z) = P(z)Q(z^2) + MzQ(z)Q(z^2).$$

So $Q(z^2)$ divides $P(z^2)Q(z)$. We see from consideration of common zeroes that $P(z^2), Q(z^2)$ stay coprime, and it follows that $Q(z^2)$ divides $Q(z)$. Looking at degrees we deduce that Q must be constant. So in fact R is a polynomial.

We can see this analytically as follows. Suppose first that $R(z)$ has a pole at some $z = z_0 \neq 0$. Then by (3.11) it would also have a pole at two different points $\pm z_0^{1/2}$. Call these z_1, z_2 both non-zero. So also poles at $\pm z_1^{1/2}, \pm z_2^{1/2}$, which are four different points. Call these z_3, z_4, z_5, z_6. So also at $\pm z_3^{1/2}, \pm z_4^{1/2}, \pm z_5^{1/2}, \pm z_6^{1/2}$, which are eight different points. And so on for ever. This is impossible, as no rational function can have infinitely many poles.

What if $z_0 = 0$? If there is a pole at $z = 0$, then it has some order $m \geq 1$. Now the right-hand side of (3.11) has a pole of order m, but the left-hand side has a pole of order $2m$. This is a contradiction.

Thus $R(z)$ has no poles at all. This means it must be a polynomial, as already proved by algebraic means.

But now if it has degree $m \geq 0$, then the right-hand side of (3.11) has degree at most m and the left-hand side degree $2m$, another contradiction if $m \neq 0$. And finally as $M \neq 0$, the case $m = 0$ is also impossible. □

Contrary to the situation in Lemma 2.2, the present lemma does not seem unavoidable, for if it happened for example that $f(z)$ was $\sqrt{z^2+2z^3} = z + z^2 + \cdots$ not in $\mathbb{C}(z)$, then Theorem 3.1 would not be trivially ruled out. But here in fact $\sqrt{z^2+2z^3} = 60/343$ for $z = 15/98$; and Lemma 3.2 will be even more unavoidable when we come to prove the transcendence of $f(\alpha)$ in Chapter 11.

Now here comes the main auxiliary polynomial; we allow it to have different degrees in the two variables, even though the proof of Theorem 3.1 goes through with equal degrees. As before, it gives rise to a function with a high order of zero.

Lemma 3.3 *For any $L \geq 0, M \geq 0$ in $\mathbb{Z}$ there is P in $\mathbb{Z}[X,Y]$ of degree at most L in X and of degree at most M in Y such that*

$$\phi(z) = P(z, f(z)) \quad . \tag{3.12}$$

is not identically zero and

$$\mathrm{ord}_{z=0}\,\phi(z) \geq LM + L + M. \tag{3.13}$$

Proof We write

$$P(X,Y) = \sum_{i=0}^{L}\sum_{j=0}^{M} p_{ij}X^iY^j$$

as in (1.6). The condition (3.13) tells us that the coefficients of

$$1, z, z^2, \ldots, z^{LM+L+M-1}$$

in the Taylor expansion of ϕ are zero. These coefficients are linear forms in the p_{ij} with rational coefficients. Thus we have a system of $LM + L + M$ homogeneous linear equations in $(L+1)(M+1)$ unknowns. By the usual linear algebra this system is solvable non-trivially over the rationals, which gives us P not zero in $\mathbb{Q}[X,Y]$. We multiply by a common denominator, and now the non-vanishing of (3.12) follows from Lemma 3.2. Note that for $M = 1$ the order reduces to $2L + 1$ as in Lemma 2.3. □

Lemma 3.4 *For any real Z with $0 \leq Z < 1$ and any $L \geq 0, M \geq 0$ in $\mathbb{Z}$ there is $C = C(Z,L,M)$ such that*

$$|\phi(z)| \leq C|z|^{LM+L+M}$$

for all complex z with $|z| \leq Z$.

Proof This is essentially the same as that of Lemma 2.4, but locally. Because of (3.13), we can write $\phi(z) = z^{LM+L+M}\psi(z)$ with a power series $\psi(z)$ also convergent on $\mathbb{D}$. So ψ is continuous (or even analytic) on the set of complex z with $|z| \le Z$. Thus it is bounded on this set, say by a constant C which depends only on Z and of course ϕ, so only on Z, L, M. Then we conclude

$$|\phi(z)| = |z|^{LM+L+M}|\psi(z)| \le C|z|^{LM+L+M}$$

as desired. □

We can now prove Theorem 3.1.

Accordingly take $\alpha = a/b$ for a, b in $\mathbb{Z}$ with $0 < |a| < b$. It will suffice to deduce a contradiction from the existence of r, s in $\mathbb{Z}$ with $s > 0$ and $f(\alpha) = r/s$.

We choose $L = L(a,b,r,s), M = M(a,b,r,s)$ in Lemma 3.3 and we calculate these later. Now for any $n \ge 0$ in $\mathbb{Z}$ we can write by (3.12)

$$\phi(\alpha^{2^{n+1}}) = P(\alpha^{2^{n+1}}, f(\alpha^{2^{n+1}})). \tag{3.14}$$

Here $b^{2^{n+1}}\alpha^{2^{n+1}}$ is in $\mathbb{Z}$. And from (3.4) we see that

$$sb^{2^n} f(\alpha^{2^{n+1}}) = sb^{2^n}(f(\alpha) - \alpha - \alpha^2 - \cdots - \alpha^{2^n})$$

is also in $\mathbb{Z}$.

It follows from (3.14) that $(b^{2^{n+1}})^L(sb^{2^n})^M\phi(\alpha^{2^{n+1}})$ is in $\mathbb{Z}$. But from Lemma 3.4 with $Z = |\alpha|$, we can estimate the absolute value from above by

$$(b^{2^{n+1}})^L(sb^{2^n})^M C|\alpha|^{2^{n+1}(LM+L+M)} = C'\left(\frac{|a|^{2(LM+L+M)}}{b^{2(LM+L+M)-2L-M}}\right)^{2^n}$$

where C depends only on α, L, M and so $C' = s^M C$ depends only on s, α, L, M and so only on a, b, r, s; but not n.

All this is a bit messy; but now on making n tend to infinity, we see that for given a, b the proof will work provided

$$|a| < b^{1-\delta} \tag{3.15}$$

with

$$\delta = \delta(L, M) = \frac{2L+M}{2(LM+L+M)},$$

provided we exclude the case $L = M = 0$. Note that we overcome the snag $\phi(\alpha^{2^{n+1}}) = 0$ just as in the previous chapter by considering the Taylor expansion of $\phi \ne 0$ at $z = 0$ as in (2.26).

Now our a, b satisfy $|a| < b$, so we can certainly find a $\delta_0 > 0$ satisfying $|a| < b^{1-\delta_0}$; and then we merely have to choose L, M with $\delta \le \delta_0$. This is possible because δ tends to zero as L, M both tend to infinity. We can even choose $L = M$, say. This completes the proof.

We can check the above $\delta(L, M)$ by going back to the original truncation proof, which we saw amounts to $\phi(z) = f(z)$ and so $L = 0, M = 1$. Now $\delta = 1/2$, consistent with (3.2). Or the example (3.6) with $L = 1, M = 1$ giving also (3.2), as well as the example $M = 1$ with arbitrary L which convinced us that we must take f to higher degree. This gave us the example (3.7) with $L = 1, M = 2$ and $\delta = 2/5$ consistent with $a < b^{3/5}$ just after (3.8); in fact we could get the latter because we had some luck.

For example to prove the irrationality of

$$f\left(1 - \frac{1}{1948}\right) = 10.595173311537198286921958443601941 66\ldots$$

(which Maple calculated with some difficulty) we have $a = 1947, b = 1948$. Maple gives (this time with ease) $\delta_0 = 1/d$ for $d = 14752$ (the smallest such integer). The condition $\delta(L, M) \leq \delta_0$ can be rewritten as

$$2LM \geq 2L(d-2) + M(d-1).$$

From this we see easily that $2L \geq d - 1, M \geq d$, and if $L = M$ then $L = M \geq 22126$, much worse than for the $f(1948)$ of the previous chapter. But when we come to prove the present $f(1947/1948)$ transcendental in Chapter 11, we will see that both L, M have to go to infinity.

There has been a great deal of work on extending Mahler's Method to other functions, notably by Mahler himself, as well as much later by Loxton and van der Poorten, who revived interest in the method. The examples $\sum_{k=0}^{\infty} z^{d^k}$ $(d = 2, 3, \ldots)$ are relatively straightforward, with obvious functional equations. Also

$$h^-(z) = \prod_{k=0}^{\infty}(1 - z^{2^k}), \quad h^+(z) = \prod_{k=0}^{\infty}(1 + z^{2^k})$$

have simple functional equations, even though some denominators are involved. There is no essential difficulty in proving the irrationality of $h^-(\alpha)$ for rational α. And Mahler (1975) claimed also the irrationality and even transcendence of $h^+(\alpha)$. But he overlooked the fact that h^+ happens to be a simple rational function (see Exercise 3.3), so that the analogue of our Lemma 3.2 is false.

However one cannot handle some slightly more complicated functional equations (21.13) associated with modular functions.

See Chapter 11 for an interesting application to the Mandelbrot set.

One can even treat several variables. A particularly interesting example is

$$f(z, w) = \sum_{k=1}^{\infty} \sum_{1 \leq h < \eta k} z^k w^h$$

for real η, whose value at $w = 1$ is $\sum_{k=1}^{\infty}[k\eta]z^k$, a series considered by Hardy and Littlewood. Here $[x]$ is the integer part of x, defined by $x - 1 < [x] \leq x$. If η is quadratic irrational there are functional equations for $f(z, w)$ (see Exercise 3.17 – but none for $f(z, 1)$ oddly enough).

Unfortunately one still cannot handle the very simple functional equation (2.28) associated with theta-functions, even though this looks somewhat similar to that of Exercise 3.17.

One can even handle sequences of functional relations, without having a functional equation for any fixed function.

We will see in Chapter 11 that everything in the present chapter can be extended from irrationality to transcendence. The book Nishioka (1996) is a good modern reference for all this.

Exercises

Here $f(z) = \sum_{k=0}^{\infty} z^{2^k}$.

3.1 (a) Show that the Taylor coefficients of $(f(z))^2$ are in $\{0, 1, 2\}$.
(b) Show that $(f(\frac{1}{10}))^2$ is irrational.

3.2 (a) Show that z and $g(z) = \sum_{k=0}^{\infty} z^{3^k}$ are algebraically independent over $\mathbb{C}$.

(b) Show that $g(\alpha)$ is irrational for every rational α with $0 < |\alpha| < 1$.

3.3 Show that

$$h(z) = z\prod_{k=0}^{\infty}(1 + z^{2^k}) = \sum_{k=0}^{\infty} \frac{z^{2^k}}{1 - z^{2^{k+1}}} = \frac{z}{1-z}.$$

* * * * * * * * * * * * * * * * * * * *

3.4 Show that $\prod_{k=0}^{\infty}(1 - \alpha^{2^k})$ is irrational for every rational α with $0<|\alpha|<1$.

3.5 Show that $(f(z))^j = \sum_{n=0}^{\infty} a_{jn}z^n$ with $A_j = \sup_{n\geq 0}|a_{jn}| < \infty$ $(j = 0, 1, 2, \ldots)$.

3.6 Let $P(X) = a_0X^d + \cdots + a_{d-1}X + a_d$ be in $\mathbb{Z}[X]$ with $a_0 \geq 0, \ldots, a_{d-1} \geq 0$ not all zero. Show that $P(f(\frac{1}{10})) \neq 0$ (so $f(\frac{1}{10})$ is "practically transcendental").

3.7 Let $U_1 = 1, U_2 = 1, U_3 = 2, U_4 = 3, \ldots$ be the Fibonacci numbers $(U_n = U_{n-1} + U_{n-2})$ and $V_1 = 1, V_2 = 3, V_3 = 4, V_4 = 7, \ldots$ the not-quite-Fibonacci numbers (also $V_n = V_{n-1} + V_{n-2}$). Show that

$$\sum_{m=0}^{\infty} \frac{1}{U_{2^m}} = 2 - h\left(\frac{1-\sqrt{5}}{2}\right) = \frac{7-\sqrt{5}}{2}$$

with h as in Exercise 3.3 and

$$\prod_{m=0}^{\infty} \frac{V_{2^m}}{(\frac{1}{2}(1+\sqrt{5}))^{2^m}} = \frac{5+\sqrt{5}}{10}.$$

3.8 Let $f_1(z), f_2(z)$ be functions, analytic on a neighbourhood of $z = 0$, with functional equations $f_1(z^2) = f_1(z) + b_1(z)$, $f_2(z^2) = f_2(z) + b_2(z)$ for $b_1(z)$, $b_2(z)$ in $\mathbb{C}(z)$. If $f_1(z), f_2(z)$ are algebraically dependent over $\mathbb{C}(z)$, show that there are λ_1, λ_2 in $\mathbb{C}$, not both zero, with $\lambda_1 f_1(z) + \lambda_2 f_2(z)$ in $\mathbb{C}(z)$.

3.9 Show that there is no open set strictly containing $\mathbb{D}$ on which $f(z)$ has an analytic continuation.

3.10 Show that $\lim_{0<x\to 1} f(x)/\log(1-x) = -1/\log 2$.

3.11 For $0 < x < 1$ let

$$C(x) = f(x) + \sum_{l=1}^{\infty}(x^{1/2^l} - 1) + \frac{\log(-\log x)}{\log 2}.$$

Show that $C(x^2) = C(x)$ and $|C(x) - c| < .000002$ with

$$c = \int_0^1 C(e^{-2^t})dt = \frac{1}{\log 2}\int_{e^{-1}}^{e^{-2}} \frac{C(x)}{x\log x}dx = -.332746\ldots.$$

3.12 Show that the decimal expansions of $f(\frac{3}{10})$, $f(\frac{4}{10})$, and $f(\frac{5}{10})$ are not periodic.

3.13 Show that $A_j \leq (j!)^2$ in Exercise 3.5.

3.14 Show that the partial quotients $a_1, a_2, \ldots$ of the continued fraction of $f(\frac{1}{2})$ are given by $a_1 = 1, a_2 = 4, a_3 = 2, a_4 = 4$ and $a_{n+1} = 4, a_{n+2} = 6, a_{n+k} = a_{n-k+3}$ $(n = 2^m, m = 2, 3, 4, \ldots; k = 3, 4, 5, \ldots, n)$; thus

1, 4, 2, 4, 4, 6, 4, 2, 4, 6, 2, 4, 6, 4, 4, 2, 4, 6, 2, 4, 4, 6, 4, 2, 6, 4, 2, 4, 6, 4, 4, 2, 4, 6, 2,

4, 4, 6, 4, 2, 4, 6, 2, 4, 6, 4, 4, 2, 6, 4, 2, 4, 4, 6, 4, 2, 6, 4, 2, 4, 6, 4, 4, 2, 4, 6, 2, 4, 4, ...

with a non-periodic "reflection pattern" ("paper-folding" – Kmošek, 1979, Shallit, 1979).

3.15 Does $f(z)$ have infinitely many zeroes? I don't know, but Mahler found sixteen.

3.16 Let $\eta > 0$ be in $\mathbb{R} \setminus \mathbb{Q}$ and let $f(\eta; z, w) = \sum_{k=1}^{\infty} \sum_{1 \le h < \eta k} z^k w^h$. Show that

$$f(1/\eta; z, w) = -f(\eta; w, z) + \frac{zw}{(1-z)(1-w)}$$

and

$$f(\eta + a; z, w) = f(\eta; zw^a, w) + \frac{zw(1-w^a)}{(1-z)(1-w)(1-zw^a)} \quad (a = 0, 1, 2, \ldots).$$

3.17 In Exercise 3.16 let $\eta = \frac{\sqrt{5}-1}{2}$, with $1/\eta = \eta + 1$, and let $f(z, w) = f(\eta; z, w)$. Show that

$$f(zw, w) = -f(w, z) + \frac{zw^2}{(1-w)(1-zw)}.$$

4

Diophantine equations – Runge's Method

We make an apparent jump to solve some diophantine equations. One that often turns up in problem-solving sessions is: find all rational integers x, y with

$$y^2 = x^4 + x + 7. \tag{4.1}$$

Someone with too much knowledge might shy away, thinking that it is a question of elliptic curves and singular models into the bargain. Someone more ignorant might notice that

$$(x^2)^2 \; < \; x^4 + x + 7 \; < \; x^4 + 2x^2 + 1 \; = \; (x^2 + 1)^2$$

at least if $x \geq 0$ is sufficiently large. So y^2 would have to lie strictly between two consecutive squares – an impossibility. A similar observation

$$(x^2 - 1)^2 \; = \; x^4 - 2x^2 + 1 \; < \; x^4 + x + 7 \; < \; (x^2)^2$$

rules out $x \leq 0$ sufficiently large. On examining small x one finds just

$$(x, y) = (-7, \pm 49), \quad (1, \pm 3), \quad (2, \pm 5).$$

Someone else, idly playing about, may write

$$y \; = \; \sqrt{x^4 + x + 7} \; = \; \pm x^2 \sqrt{1 + \frac{1}{x^3} + \frac{7}{x^4}}$$

and develop the square root for large $|x|$ getting

$$y = \pm x^2 \left(1 + \frac{1}{2x^3} + \frac{7}{2x^4} - \frac{1}{8x^6} - \frac{7}{4x^7} + \cdots \right). \tag{4.2}$$

Thus

$$y \mp x^2 \; = \; \pm \frac{1}{2x} + \cdots \tag{4.3}$$

is small as $|x|$ gets large, and by the Fundamental Theorem of Transcendence this is going to force $-y \pm x^2 = 0$. On the other hand the series on the right is not identically zero, so cannot vanish if $|x|$ is sufficiently large.

Thus we are not so far from some of the things discussed in the previous two chapters.

Here is another example in connexion with Catalan's Equation. Find all rational integers x, y with

$$x^5 - y^7 = 1. \tag{4.4}$$

The person with too much knowledge gives up in disgust on seeing a curve of genus 12 with a nasty singularity. And there seems to be no possibility of arguing with consecutive fifth or seventh powers. Furthermore $y = \sqrt[7]{x^5 - 1} = x^{5/7}(1 - 1/7x^5 + \cdots)$ doesn't even have an expansion with integral powers of x. We do find that $y - x^{5/7}$ is small, but it is unlikely to be in $\mathbb{Z}$ or even $\mathbb{Q}$.

A second idler may write

$$y^7 = x^5 - 1 = (x-1)(x^4 + x^3 + x^2 + x + 1)$$

and try to prove that both factors are seventh powers. Indeed a common factor $m \geq 1$ would have $x \equiv 1 \pmod{m}$ as well as $0 \equiv x^4 + x^3 + x^2 + x + 1 \equiv 5 \pmod{m}$. Thus m divides 5 and $m = 1$ or $m = 5$.

The case $m = 1$ indeed leads to seventh powers, so we can write

$$x - 1 = x_1^7, \quad x^4 + x^3 + x^2 + x + 1 = y_1^7$$

for new integers x_1, y_1. Naturally these also satisfy a diophantine equation, namely

$$y_1^7 \;=\; (x_1^7+1)^4 + (x_1^7+1)^3 + \cdots + 1 \;=\; x_1^{28} + 5x_1^{21} + \cdots .$$

Now the first idler takes over and writes

$$y_1 \;=\; \sqrt[7]{x_1^{28} + 5x_1^{21} + \cdots} \;=\; x_1^4 + \frac{5}{7x_1^3} + \cdots \tag{4.5}$$

with only integral powers. Thus

$$y_1 - x_1^4 = \frac{5}{7x_1^3} + \cdots \tag{4.6}$$

is small but not zero as $|x_1|$ gets large.

But what if $m = 5$? We then get for example

$$x - 1 = 5^a x_1^7, \quad x^4 + x^3 + x^2 + x + 1 = 5^b y_1^7 \tag{4.7}$$

with positive integers a, b; and for example $a = 6, b = 1$ leads to

$$y_1 = \sqrt[7]{5^{23}x_1^{28} + 5^{18}x_1^{21} + \cdots} = \alpha x_1^4 + \frac{\alpha}{21875x_1^3} + \cdots \tag{4.8}$$

for $\alpha = 5^{23/7}$. But we cannot use the Fundamental Theorem on $y_1 - \alpha x_1^4$, as α is irrational!

We will see later that the mere irrationality is not the sticking point, but the fact that α has the full degree 7 over the rationals; with degree 6 there is a good chance of success. The expert muttering "Liouville" may see this more clearly.

But in the present problem the methods used so far cannot be made to work, and one needs the deep considerations from algebraic number theory used by Mihăilescu in his solution of the Catalan Equation. See for example Bilu et al. (2014).

At any rate, in the situation (4.7) we see that 5 divides $x^5 - 1 = y^7$ and so y; consequently we have proved that (4.4) has at most finitely many integer solutions with y not divisible by 5.

A similar argument with

$$y + 1 = x_1^5, \ \ y^6 - y^5 + y^4 - y^3 + y^2 - y + 1 = y_1^5$$

and the slightly more complicated

$$5y_1 - 5x_1^6 + 7x_1 = \frac{7}{5x_1^4} + \cdots \tag{4.9}$$

shows that (4.4) has at most finitely many integer solutions with x not divisible by 7.

It is clear that in each case all the solutions can be found in an effective way; that is, we could (with some hard work) make a complete list of all the solutions x, y. Cassels went a lot further and considered $x^p - y^q = 1$ for any distinct odd primes p, q, and showed that with either p not dividing y or q not dividing x there are actually no solutions at all, consistent with the original conjecture of Catalan; for any fixed p, q this is a matter of finite computation, but for arbitrary p, q one must take a great deal more trouble as already evinced by the slight complication in (4.9) over (4.6).

Let us consider another diophantine equation

$$y(y^2 - x^2) = x. \tag{4.10}$$

Are there series like (4.3), (4.5), (4.8) and (4.9)?

Now it is not so easy to solve for y in terms of x; we get a cubic equation, whose solution

$$\frac{1}{6}\sqrt[3]{\left(108x+12\sqrt{-12x^6+81x^2}\right)}+\frac{2x^2}{\sqrt[3]{\left(108x+12\sqrt{-12x^6+81x^2}\right)}}$$

by Cardano is not very helpful. But we can argue empirically as follows. Writing it as

$$y(y-x)(y+x)=x \tag{4.11}$$

and comparing with the hyperbola $(y-x)(y+x) = 1$, we may guess at least the existence of asymptotes $y = \pm x$. First taking y about x we see that $y - x = x/(y(y+x))$ is about $1/(2x)$. Then taking y about $x + 1/(2x)$ we get another term; and so on until

$$y=x+\frac{1}{2x}-\frac{3}{8x^3}+\cdots. \tag{4.12}$$

Similarly from y about $-x$ we get

$$y=-x+\frac{1}{2x}+\frac{3}{8x^3}+\cdots. \tag{4.13}$$

As there is a cubic to solve, we can expect one more expansion and this turns out to be

$$y=-\frac{1}{x}-\frac{1}{x^5}-\frac{3}{x^9}+\cdots \tag{4.14}$$

corresponding to the asymptote $y = 0$.

In fact all three series converge for sufficiently large $|x|$; moreover in that situation they converge to the three solutions of (4.10). We will prove such things later on.

Thus with the corresponding functions

$$y-x,\ \ y+x,\ \ y \tag{4.15}$$

we can show that (4.10) has at most finitely many solutions in integers x, y.

At this point we could prove without too much trouble the following result. Let F in $\mathbb{Z}[X,Y]$ be a polynomial of the shape

$$F(X,Y)\ =\ (Y-a_1X)\cdots(Y-a_dX)+f(X,Y) \tag{4.16}$$

for different rational integers $a_1,\ldots,a_d$ and a polynomial f of total degree strictly less than d. Then there are at most finitely many (x,y) in $\mathbb{Z}^2$ with $F(x,y) = 0$. For we just have to consider suitable expansions of $y - ax$ or possibly $y - ax - b$ as in (4.15). We would have to make sure that $y - ax - b$ doesn't vanish identically, for example by assuming $d \geq 2$ and the irreducibility of F over $\mathbb{Q}$.

But we prefer a challenge, not to mention auxiliary polynomials.

What about

$$y(y^3 - x^3) = x \tag{4.17}$$

in place of (4.10)? The solution by radicals would occupy the good part of a page! (And wouldn't be possible for $y(y^4 - x^4) = x$.)

The left-hand side would have to be factored as

$$y(y - x)(y - \omega_1 x)(y - \omega_2 x) = x \tag{4.18}$$

for the two primitive cube roots of unity ω_1, ω_2. So we have to deal with small values of $y - \omega_1 x$, for example. However as ω_1 is not real, it is clear that as soon as this is too small for rational integers x, y then $x = y = 0$. And indeed the graph shows only two real asymptotes, corresponding to the other factors $y, y - x$. These can be handled as above, and we will get the finiteness of the set of solutions of the diophantine equation (4.17).

And finally in this parade, what about

$$y(y^2 - 2x^2) = x \tag{4.19}$$

corresponding to

$$F(X, Y) = Y(Y^2 - 2X^2) - X?$$

We write

$$y(y - \alpha x)(y + \alpha x) = x \tag{4.20}$$

for $\alpha = \sqrt{2}$, so again three real asymptotes. We have to deal with small $y - \alpha x$, but unfortunately α is real and so we cannot argue as above. It looks like the same problem as in (4.8).

Here by considering the product $(y - \alpha x)(y + \alpha x) = y^2 - 2x^2$ we can give an *ad hoc* solution. With more terms we find

$$y = \alpha x + \frac{1}{4x} + \cdots, \tag{4.21}$$

and so $y^2 - 2x^2 = \alpha/2 + \cdots$. As $\alpha/2$ is not an integer, this is impossible for large x.

It is more elegant to eliminate α by noting that

$$2x(y^2 - 2x^2) = \alpha x + \cdots = y + \cdots$$

and indeed we find after some calculation that

$$2x(y^2 - 2x^2) - y = -\frac{1}{2x} + \cdots. \tag{4.22}$$

This leads to a contradiction for sufficiently large x.

How to eliminate the ad-hockery? The idea is to use the expansion (4.21) to find a second polynomial $G \neq 0$, also in $\mathbb{Z}[X, Y]$, such that the series for $G(x, y)$ involves only negative powers of x as in (4.22). This is like the first problem in Chapter 1, where we wanted only t^p for prime p. So we can hope to do this with an auxiliary polynomial and indeed G is it. Then provided the series is not identically zero (which rules out $G = F$ of course), we get a contradiction for sufficiently large $|x|$.

It turns out that we can make sure that the series for $G(x, y)$ is not identically zero simply by assuming $F(X, Y)$ irreducible over $\mathbb{Q}$ with degree say d in Y (as in the example (4.16) above), and restricting the degree of $G(X, Y)$ in Y to be strictly less than d. But then it is crucial that the α in the factors like $y - \alpha x$ has degree strictly less than d over $\mathbb{Q}$. Sadly this rules out interesting equations like the Thue equations $x^3 - 2y^3 = 1$ or $x^3 - 2y^3 = m$. This seems inevitable in principle. Indeed, if we could treat these, then we could presumably treat also $x^2 - 2y^2 = 1$. But this Pell equation is well-known to have infinitely many solutions.

Thue equations will be handled in Chapter 12 by more elaborate methods. However in the present chapter we can do for example $y(x^3 - 2y^3) = x$.

Let us work this through for

$$F(X, Y) = Y(Y^2 - 2X^2) - X \tag{4.23}$$

as in (4.19) (with $d = 3$), to see how something like

$$G(X, Y) = 2X(Y^2 - 2X^2) - Y \tag{4.24}$$

as in (4.22) arises more naturally.

We try

$$G(X, Y) = P(X) + Q(X)Y + R(X)Y^2$$

somewhat as in (2.21), with P, Q, R in $\mathbb{Z}[X]$ of degree at most L. We have of course $3(L + 1)$ unknowns to play with. With the series

$$\lambda(t) = \alpha t + \frac{1}{4t} + \cdots$$

(we now change over to t to avoid confusion with the integers x) of (4.21), there may appear in $G(t, \lambda(t))$ non-negative powers of t, but nothing more than $1, t, t^2, \ldots, t^{L+2}$. So the elimination of these involves $L + 3$ equations, easily less than $3(L + 1)$ for large L at our disposal (and even $L = 1$).

However the presence of $\alpha = \sqrt{2}$ means that the equations are not defined over $\mathbb{Q}$! For example with $L = 1$ and

$$P(X) = p_0 + p_1X, \quad Q(X) = q_0 + q_1X, \quad R(X) = r_0 + r_1X$$

and the more detailed expansion

$$\lambda(t) = \alpha t + \frac{1}{4t} + \frac{0}{t^2} + \cdots,$$

we find

$$\begin{aligned} p_0 + \frac{1}{4}q_1 + \frac{\alpha}{2}r_0 &= 0 \\ p_1 + q_0\alpha + \frac{\alpha}{2}r_1 &= 0 \\ \alpha q_1 + 2r_0 &= 0 \\ 2r_1 &= 0 \end{aligned}$$

using $\alpha^2 = 2$. As we want rational $p_0, p_1, q_0, q_1, r_0, r_1$ each equation, of the form $\Lambda_1 + \alpha\Lambda_2 = 0$, must be split into

$$\Lambda_1 = \Lambda_2 = 0. \tag{4.25}$$

For example the second equation splits as $p_1 = q_0 + r_1/2 = 0$. Thus we obtain twice as many equations, here eight, which is too many. And indeed one checks at once that the only solution is $p_0 = p_1 = q_0 = q_1 = r_0 = r_1 = 0$.

In general we will obtain $2(L+3)$ equations, which luckily is still less than $3(L+1)$ for sufficiently large L. In fact $L = 4$ will do, and we can obtain $L = 3$ in (4.24) only by extra luck (for example the last of the four equations above splits into just itself).

We are almost ready to state and prove the main result of this chapter generalizing (4.16). But first we must say something about the representation of large solutions of $F(x, y) = 0$ by convergent Laurent series. This is analysis not number theory; however we did not find a completely satisfactory discussion in the readily accessible literature, and so we supply full details. As some of the arguments involved will be used later in our proof of the Schneider–Lang Theorem in Chapter 19, as well as in Chapter 5 and the Appendix, the material is not out of place in this book.

Here is the analytic result on polynomials like (4.16). We write $\mathbb{C}[[z]]$ for the ring of formal power series $\sum_{k=0}^{\infty} a_k z^k$ in z with complex coefficients a_k. By itself this carries no connotations of convergence; for example we allow

$$\sum_{k=0}^{\infty} k! z^k = 1 + z + 2z^2 + 6z^3 + 24z^4 + \cdots \tag{4.26}$$

even though this converges only for $z = 0$ (still $\mathrm{ord}_{z=0}$ can be given a meaning). The polynomial ring $\mathbb{C}[z]$ can be regarded as a subset of $\mathbb{C}[[z]]$ in the obvious way. The easy proof that $\mathbb{C}[z]$ is an integral domain extends at once to show that $\mathbb{C}[[z]]$ is too.

Proposition 4.1 *Let F in $\mathbb{C}[X,Y]$ be a polynomial of the shape*

$$F(X,Y) = (Y-\alpha_1 X)\cdots(Y-\alpha_d X) + f(X,Y)$$

for different complex numbers $\alpha_1,\ldots,\alpha_d$ and a polynomial f of total degree strictly less than d. Then there are $\psi_1,\ldots,\psi_d$ in $z\mathbb{C}[[z]]$ and $R \geq 1$ with the following properties. For all $i=1,\ldots,d$ and $\lambda_i(X) = \alpha_i X + X\psi_i(1/X)$ we have $F(X,\lambda_i(X)) = 0$. Further let x,y be complex numbers with $|x| \geq R$ and $F(x,y)=0$. Then there is some $i=1,\ldots,d$ such that $\psi_i(1/x)$ converges and $y = \lambda_i(x)$.

Before we make a digression to prove this, we record some preliminary results. The first allows us to find all units in $\mathbb{C}[[z]]$. It is purely formal, so we state it for an arbitrary field $\mathcal{K}$. We are temporarily working near $(0,0)$, rather than (∞,∞) as in Proposition 4.1.

Lemma 4.2 *Let $\sigma(z) = c + \cdots$ be in $\mathcal{K}[[z]]$ with constant term $c \neq 0$. Then there is $\psi(z)$ in $\mathcal{K}[[z]]$ such that $\sigma(z)\psi(z) = 1$.*

Proof We substitute

$$\psi(z) = a_0 + a_1 z + a_2 z^2 + \cdots$$

formally into $\sigma(z)\psi(z) = 1$ and we find the a_k successively. To begin with, $ca_0 = 1$ so $a_0 = 1/c$ is in $\mathcal{K}$. And if we have determined $a_0,\ldots,a_{k-1}$ in $\mathcal{K}$ for some $k \geq 1$, then the coefficient of z^k in

$$(c+\cdots)(a_0+\cdots+a_k z^k+\cdots) = 1$$

is $ca_k + P$ for a fixed polynomial P, in $a_0,\ldots,a_{k-1}$ but not $a_k, a_{k+1},\ldots$, with coefficients in $\mathcal{K}$. Thus a_k in $\mathcal{K}$ is determined by induction from $a_0,\ldots,a_{k-1}$. □

The resulting ψ could be written as σ^{-1}; it is clearly unique. So $\mathcal{K}[[z]]$ has units $c+\cdots$ $(c \neq 0)$, and it is easy to see that there are no others.

A faster method in practice is to write $\sigma = c + \tau$ and then informally

$$\sigma^{-1} = c^{-1}\left(1+\frac{\tau}{c}\right)^{-1} = c^{-1}\left(1-\frac{\tau}{c}+\frac{\tau^2}{c^2}-+\cdots\right),$$

where τ^k has no term in $1,z,\ldots,z^{k-1}$. For example

$$\begin{aligned}\frac{1}{1-z-z^2} &= 1 + (z+z^2) + (z+z^2)^2 + \cdots \\ &= 1 + z + 2z^2 + 3z^3 + 5z^4 + 8z^5 + 13z^6 + \cdots\end{aligned}$$

is the generating function of the Fibonacci sequence.

For the next result we must be more careful with the uniqueness, and we take a big field $\mathcal{K}'$ with a subfield $\mathcal{K}$ and the corresponding rings $\mathcal{K}'[[z]], \mathcal{K}[[z]]$.

Lemma 4.3 *Let $S(z,w)$ be in $\mathcal{K}[z,w]$ with*

$$S(0,0)=0, \quad \frac{\partial S}{\partial w}(0,0)\neq 0$$

for the formal partial derivative. Then there is a unique $\psi(z)$ in $z\mathcal{K}'[[z]]$ such that $S(z,\psi(z))=0$; and in fact $\psi(z)$ lies in $z\mathcal{K}[[z]]$.

Proof Geometrically this means that the plane curve defined by $S=0$ is non-singular at $(0,0)$, and even that the tangent there is not "vertical". So things like $S(z,w)=z-w^2$ are ruled out, which would have led to $w=z^{1/2}$, not a power series. But $S(z,w)=z-w-w^2$ is alright.

We split off the linear part as

$$S(z,w)=bz+cw-T(z,w)$$

with $c=(\partial S/\partial w)(0,0)\neq 0$. We substitute

$$\psi(z)=a_1z+a_2z^2+\cdots$$

formally into $S(z,\psi(z))=0$ and we find the a_k successively as in the previous proof. To begin with, $(b+ca_1)z+\cdots=0$ so $a_1=-b/c$ is in $\mathcal{K}$. And if we have determined $a_1,\ldots,a_{k-1}$ in $\mathcal{K}$ for some $k\geq 2$, then

$$bz+c(a_1z+\cdots+a_kz^k+\cdots) \;=\; T(z,a_1z+\cdots+a_{k-1}z^{k-1}+\cdots),$$

where the coefficient of z^k on the right-hand side is a fixed polynomial, in $a_1,\ldots,a_{k-1}$ but not $a_k,a_{k+1},\ldots$, with coefficients in $\mathcal{K}$. Thus a_k in $\mathcal{K}$ is determined by induction from $a_1,\ldots,a_{k-1}$. □

The above example $S(z,w)=z-w-w^2$ can also be done informally with

$$w=z-w^2=z-(z-w^2)^2=z-(z-(z-w^2)^2)^2=\cdots,$$

which comes out in the end as

$$w=z-z^2+2z^3-5z^4+14z^5-42z^6+132z^7-429z^8+1430z^9+\cdots. \tag{4.27}$$

One can also solve the quadratic to get

$$w=\frac{-1+\sqrt{1+4z}}{2}=-\frac{1}{2}+\frac{1}{2}\sum_{k=0}^{\infty}\binom{1/2}{k}(4z)^k; \tag{4.28}$$

but this is a bit dodgy for characteristic two. In fact not much survives of (4.27), say in $\mathbb{F}_2[[z]]$; we see only $w=z+z^2+z^4+z^8+\cdots$. And indeed this is just the Mahler series of Chapter 3, because then $w^2 = z^2+z^4+z^8+z^{16}+\cdots$ so $w + w^2 = z$. For stranger things see Exercise 4.35 and especially Exercise 4.37.

All that was just formal. Now we want to rule out things like (4.26) turning up in this way. In fact for $\mathcal{K} = \mathbb{C}$ we want local convergence (near $z = 0$ of course), which for $\sum_{k=0}^{\infty} a_k z^k$ is equivalent to the existence of real A and $E \geq 0$ with

$$|a_k| \leq AE^k \quad (k = 0, 1, 2 \ldots). \tag{4.29}$$

Lemma 4.4 *Let ψ, χ, ω be in $\mathbb{C}[[z]]$ such that the formal derivatives ψ', χ', ω' lie in $\mathbb{C}[\psi, \chi, \omega]$. Then ψ, χ, ω are locally convergent.*

Proof We have

$$\psi' = P_1(\psi, \chi, \omega), \quad \chi' = Q_1(\psi, \chi, \omega), \quad \omega' = R_1(\psi, \chi, \omega) \tag{4.30}$$

with polynomials P_1, Q_1, R_1 in $\mathbb{C}[X, Y, Z]$. This is the first stage (I_1) of the following inductive situation.

(I_k): *There are polynomials P_k, Q_k, R_k with total degrees at most $a + bk$ and coefficients of absolute values at most $k!c^k$ such that*

$$\psi^{(k)} = P_k(\psi, \chi, \omega), \quad \chi^{(k)} = Q_k(\psi, \chi, \omega), \quad \omega^{(k)} = R_k(\psi, \chi, \omega)$$

for the higher derivatives, where a, b, c are independent of k.

To deduce (I_k) from (I_{k-1}) for $k \geq 2$, we consider

$$\psi^{(k)} = \frac{d}{dz} P_{k-1}(\psi, \chi, \omega) = P_k(\psi, \chi, \omega)$$

where

$$P_k = \frac{\partial P_{k-1}}{\partial X} P_1 + \frac{\partial P_{k-1}}{\partial Y} Q_1 + \frac{\partial P_{k-1}}{\partial Z} R_1. \tag{4.31}$$

Using $\mathcal{D}$ for total degree, we see that $\mathcal{D}(\partial P_{k-1}/\partial X) \leq \mathcal{D}(P_{k-1})$, and similarly for Y, Z. So

$$\mathcal{D}(P_k) \leq \mathcal{D}(P_{k-1}) + d_1$$

where

$$d_1 = \max\{\mathcal{D}(P_1), \mathcal{D}(Q_1), \mathcal{D}(R_1), 1\}.$$

This makes it clear that

$$\mathcal{D}(P_k) \leq d_1 k \quad (k = 2, 3, \ldots),$$

and even for $k = 1$. Similarly for $\mathcal{D}(Q_k), \mathcal{D}(R_k)$.

What about the coefficients?

We use $\mathcal{L}(A)$ for the sum of the absolute values of the coefficients of any A in $\mathbb{C}[X,Y,Z]$, sometimes called the length. It is easy to verify $\mathcal{L}(A+B) \leq \mathcal{L}(A)+\mathcal{L}(B)$ and also with a bit more care

$$\mathcal{L}(AB) \leq \mathcal{L}(A)\mathcal{L}(B) \tag{4.32}$$

conveniently without any multiplying factor, the main reason why we shall be using the length so much in this book. And further

$$\mathcal{L}(\partial A/\partial X) \leq \mathcal{D}(A)\mathcal{L}(A),$$

and similarly for Y, Z.

It follows from (4.31) that $\mathcal{L}(P_k) \leq 3d_{k-1}\mathcal{L}(P_{k-1})l_1$ with

$$d_{k-1} \;=\; \max\{\mathcal{D}(P_{k-1}), \mathcal{D}(Q_{k-1}), \mathcal{D}(R_{k-1}), 1\} \;\leq\; d_1 k,$$

for

$$l_1 = \max\{\mathcal{L}(P_1), \mathcal{L}(Q_1), \mathcal{L}(R_1)\}.$$

Now a simple induction gives

$$\mathcal{L}(P_k) \leq k!(3d_1 l_1)^k \quad (k = 2, 3, \ldots),$$

and even for $k = 1$. Similarly for $\mathcal{L}(Q_k), \mathcal{L}(R_k)$.

So (I_k) is proved for all k.

Finally with $\psi(z) = \sum_{k=0}^{\infty} a_k z^k$ we have

$$k!a_k = \psi^{(k)}(0) = P_k(\psi(0), \chi(0), \omega(0)).$$

There follows

$$k!|a_k| \;\leq\; k!(3d_1 l_1)^k (1 + |\psi(0)| + |\chi(0)| + |\omega(0)|)^{d_k}$$

and so (4.29). Similarly for χ, ω. □

There are other proofs, for example by majorization techniques (see Exercise 4.39). Some of the arguments here will be generalized in Chapter 19.

The present proof is actually quite delicate. For with the forbidden $\psi(z) = \sum_{k=0}^{\infty} k!z^k$ in (4.26) we have

$$(z\psi)' = \sum_{k=0}^{\infty}(k+1)!z^k = \frac{\psi - 1}{z}.$$

So $\psi' = (\psi - 1 - z\psi)/z^2$; but this is not polynomial in z. No trouble: just introduce $\omega = 1/z$. Then

$$\psi' = \omega^2(\psi - 1 - z\psi) = \omega^2(\psi - 1 - \chi\psi)$$

with $\chi = z$, as well as $\chi' = 1,\ \omega' = -\omega^2$.

So P_1, Q_1, R_1 exist in (4.30)! But of course ω is not in $\mathbb{C}[[z]]$.

It is not too difficult to prove that if $c + \cdots$ $(c \neq 0)$ is locally convergent, then so is its inverse as in Lemma 4.2 (see Exercise 4.9). But we don't need it here. The following is an analogue for Lemma 4.3.

Lemma 4.5 *Let $S(z, w)$ be in $\mathbb{C}[z, w]$ and $\psi(z)$ in $\mathbb{C}[[z]]$ as in Lemma 4.3 with $\mathcal{K}' = \mathcal{K} = \mathbb{C}$. Then ψ is locally convergent.*

Proof Formally differentiating $S(z, \psi(z)) = 0$ gives $S_z(z, \psi) + \psi' S_w(z, \psi) = 0$ for the partial derivatives $S_z = \partial S/\partial z$, $S_w = \partial S/\partial w$. Note that $S_w(z, \psi) \neq 0$ because its constant term $S_w(0, 0) \neq 0$ by hypothesis. So this time $\psi' = -S_z(z, \psi)/S_w(z, \psi)$. As with $\sum_{k=0}^{\infty} k! z^k$ above, take $\chi = z$ and $\omega = 1/S_w(z, \psi)$; by Lemma 4.2 this now does exist in $\mathbb{C}[[z]]$. Then ψ', χ' are clearly in $\mathbb{C}[\psi, \chi, \omega]$, and

$$\omega' = -\omega^2(S_{wz}(z, \psi) + \psi' S_{ww}(z, \psi))$$

(for the higher derivatives) also in $\mathbb{C}[\psi, \chi, \omega]$. Finally Lemma 4.4 gives what we want. □

Now we return from $(0, 0)$ back to (∞, ∞).

Lemma 4.6 *Let K be a subfield of $\mathbb{C}$, let α be in K, and let F in $K[X, Y]$ be a polynomial of the shape*

$$F(X, Y) = F_0(X, Y) + f(X, Y)$$

for $F_0(X, Y)$ homogeneous of degree d divisible by $Y - \alpha X$ but not $(Y - \alpha X)^2$ or X, and a polynomial f of total degree strictly less than d. Then there is a unique locally convergent $\psi(z)$ in $z\mathbb{C}[[z]]$ so that for

$$\lambda(X) = \alpha X + X\psi\left(\frac{1}{X}\right) = \alpha X + \lambda_0 + \frac{\lambda_1}{X} + \cdots$$

we have $F(X, \lambda(X)) = 0$ identically in X; and in fact $\psi(z)$ lies in $zK[[z]]$.

Proof We can assume

$$F(X, Y) = (Y - \alpha X)(Y - \alpha_2 X) \cdots (Y - \alpha_d X) + f(X, Y)$$

with all $\alpha_i \neq \alpha$. We define $S(z, w) = z^d F(1/z, (w + \alpha)/z)$ in $K[z, w]$, which is

$$S(z, w) = w(w + \alpha - \alpha_2) \cdots (w + \alpha - \alpha_d) + zT(z, w)$$

with T also in $K[z, w]$. Then $S(0, 0) = 0$ and $S_w(0, 0)$, the coefficient of w, is $(\alpha - \alpha_2) \cdots (\alpha - \alpha_d) \neq 0$. We are now in the situation of Lemma 4.3 with $\mathcal{K}' = \mathbb{C}, \mathcal{K} = K$; accordingly there is $\psi(z)$ in $zK[[z]]$, locally convergent by Lemma 4.5, with $S(z, \psi(z)) = 0$. By the definition of S this gives $F(1/z, (\psi(z) + \alpha)/z) = 0$, and now putting $X = 1/z$ we get what we want. □

We can now prove Proposition 4.1. By Lemma 4.6 there are certainly locally convergent $\psi_1, \ldots, \psi_d$ in $z\mathbb{C}[[z]]$ such that for $\lambda_i(X) = \alpha_i X + X\psi_i(1/X)$ we have $F(X, \lambda_i(X)) = 0$ for all $i = 1, \ldots, d$. But we can go much further. Consider the polynomial

$$z^d F\left(\frac{1}{z}, Y\right) - \prod_{i=1}^{d}(zY - \alpha_i - \psi_i(z))$$

in $Q[Y]$, where Q is the quotient field of the integral domain $\mathbb{C}[[z]]$. The coefficient of Y^d is zero; so the degree is strictly less than d. But the polynomial vanishes at each $(\alpha_i + \psi_i(z))/z$ in Q. These are all different because the α_i are. So the polynomial has more zeroes than its degree, and must therefore be identically zero. After the substitution $z = 1/X$ we find

$$F(X, Y) = \prod_{i=1}^{d}\left(Y - \alpha_i X - X\psi_i\left(\frac{1}{X}\right)\right) = \prod_{i=1}^{d}(Y - \lambda_i(X)) \qquad (4.33)$$

identically in X, Y.

Now choose $R \geq 1$ such that all $\psi_i(z)$ converge whenever $|z| < 1/R$. Let (x, y) in $\mathbb{C}^2$ be such that $F(x, y) = 0$ and $|x| > R$. We can substitute $(X, Y) = (x, y)$ into (4.33). We find

$$0 = F(x, y) = \prod_{i=1}^{d}(y - \lambda_i(x)).$$

There follows $y = \lambda_i(x)$ for some i, as required. This completes the proof of Proposition 4.1.

Here is an example for

$$F(X, Y) = Y(Y^2 - 2X^2) - X = Y(Y - \alpha X)(Y + \alpha X) - X$$

with $\alpha = \sqrt{2}$ corresponding to (4.19). We found

$$\lambda_1(X) = \alpha X + \frac{1}{4X} + \cdots$$

as in (4.21). By changing the sign of $\sqrt{2}$ (or applying the non-trivial automorphism of $\mathbb{Q}(\sqrt{2})$ as a field), we get also

$$\lambda_2(X) = -\alpha X + \frac{1}{4X} + \cdots.$$

There is also

$$\lambda_3(X) = -\frac{1}{2X} + \cdots.$$

So $Y(Y^2 - 2X^2) - X$ factorizes as

$$\left(Y - \left(\alpha X + \frac{1}{4X} + \cdots\right)\right)\left(Y - \left(-\alpha X + \frac{1}{4X} + \cdots\right)\right)\left(Y - \left(-\frac{1}{2X} + \cdots\right)\right).$$

After division by X^2 this becomes an identity in $\mathbb{C}[[1/X]][Y]$.

We pass now to the main diophantine result of this chapter.

Theorem 4.7 *Let F in $\mathbb{Q}[X, Y]$ be a polynomial, irreducible over $\mathbb{Q}$, of the shape*

$$F(X, Y) = F_0(X, Y) + f(X, Y)$$

for $F_0(X, Y)$ homogeneous of degree d, squarefree and reducible over $\mathbb{Q}$, and a polynomial f of total degree strictly less than d. Then there are at most finitely many (x, y) in $\mathbb{Z}^2$ with $F(x, y) = 0$.

It is plausible that the irreducibility of F is not much of a restriction; for example to solve $GH = 0$ we just have to solve $G = 0$ and $H = 0$. But the reducibility of F_0 is crucial for the proof; it ensures that we meet factors $y - \alpha x$ only with α of degree strictly less than d over $\mathbb{Q}$. The squarefree nature of F_0 is also crucial for the proof; but even for the validity of the theorem itself, as the equation $x^2 = y$ shows.

For the proof we need (besides Proposition 4.1) just one preliminary result, involving of course an auxiliary polynomial; this time just a zero of order at least 1 suffices. We shall have to deal with number fields K; these are simply fields containing $\mathbb{Q}$ such that the dimension $[K : \mathbb{Q}]$ of K as a vector space over $\mathbb{Q}$ is finite.

Lemma 4.8 *Let K be a number field with $[K : \mathbb{Q}] = e$, let α be in K, and let ψ be in $zK[[z]]$. Then for any $L \geq e^2$ in $\mathbb{Z}$ there is non-zero G in $\mathbb{Z}[X, Y]$ of degree at most L in X and of degree at most e in Y such that*

$$\phi(z) = G\left(\frac{1}{z}, \frac{\alpha}{z} + \frac{\psi(z)}{z}\right) \tag{4.34}$$

is in $zK[[z]]$.

Proof We write

$$G(X, Y) = \sum_{i=0}^{L}\sum_{j=0}^{e} p_{ij}X^iY^j.$$

The condition on ϕ tells us that the coefficients of $z^{-L-e}, \ldots, z^{-1}, 1$ in the Laurent expansion are zero. Thus we have $L + e + 1$ linear forms Λ in $(L + 1)(e + 1)$ unknowns. These forms are defined over K which has dimension e over $\mathbb{Q}$, so if $(\omega_1, \ldots, \omega_e)$ is a basis for K over $\mathbb{Q}$, we can write $\Lambda = \omega_1\Lambda_1 + \cdots + \omega_e\Lambda_e$ with linear forms $\Lambda_1, \ldots, \Lambda_e$ defined over $\mathbb{Q}$. We now try to solve the system

$$\Lambda_1 = \cdots = \Lambda_e = 0 \tag{4.35}$$

(as in (4.25) above) of $e(L+e+1)$ homogeneous linear equations, which is possible non-trivially if

$$(L+1)(e+1) > e(L+e+1),$$

that is, $L \geq e^2$ as given. □

We can now prove Theorem 4.7. We first assume that $F_0(X,Y)$ is not divisible by X, and right at the end we show with a simple trick how to eliminate this assumption.

After dividing by a leading coefficient, we can write

$$F_0(X,Y) = (Y - \alpha_1 X) \cdots (Y - \alpha_d X) \tag{4.36}$$

for distinct complex $\alpha_1, \ldots, \alpha_d$, of course algebraic over $\mathbb{Q}$.

Now let (x,y) in $\mathbb{Z}^2$ be any solution of $F(x,y) = 0$ with $|x| \geq R$ for the $R \geq 1$ supplied by Proposition 4.1. There is some $i = 1, \ldots, d$ and $\psi = \psi_i$ in $z\mathbb{C}[[z]]$ such that $\psi(1/x)$ converges and $y = \lambda(x)$ for $\lambda(X) = \alpha X + X\psi(1/X)$ and $\alpha = \alpha_i$; further $F(X, \lambda(X)) = 0$ identically.

Then by Lemma 4.6 with $K = \mathbb{Q}(\alpha)$ we see that $\psi(z)$ must lie in $zK[[z]]$.

As $F_0(X,Y)$ is reducible over $\mathbb{Q}$, we have $[K : \mathbb{Q}] = e < d$. Let G, ϕ be as in Lemma 4.8 for say $L = e^2$.

We pause to show that ϕ is not identically zero. Now $G(X,Y)$ must involve Y because a non-zero polynomial in $1/z$ alone could not possibly be in $zK[[z]]$. Thus the vanishing of ϕ would imply that $\lambda(X)$ would be of degree at most $e < d$ over $\mathbb{Q}(X)$. However from $F(X, \lambda(X)) = 0$ and the irreducibility of $F(X,Y) = Y^d + \cdots$ over $\mathbb{Q}$ we see that $\lambda(X)$ has degree d over $\mathbb{Q}(X)$.

We can therefore write $\phi(z) = cz^D + \cdots$ for some $D \geq 1$ and some $c \neq 0$. Since the integer $G(x,y) = \phi(1/x)$ and $|x| \geq R \geq 1$, it follows that

$$0 < |G(x,y)| \leq 2c|x|^{-D} \leq 2c|x|^{-1} < 1 \tag{4.37}$$

for sufficiently large $|x|$. By the Fundamental Theorem this is impossible, and we have therefore proved the finiteness of the set of integer solutions, because the shape (4.36) of $F_0 = Y^d + \cdots$ shows that any fixed (small) x determines y in a finite set.

But what if $F_0(X,Y)$ happens to be divisible by X? We note that Theorem 4.7 is invariant under $SL_2(\mathbb{Z})$; that is, if we replace X, Y by $pX+qY, rX+sY$ for integers p,q,r,s with $ps-qr=1$. Then $Y-\alpha_i X$ becomes

$$(rX+sY)-\alpha_i(pX+qY) \;=\; (r-\alpha_i p)X+(s-\alpha_i q)Y.$$

So all we need to do to get rid of any factor X is to make sure that $s-\alpha_i q \neq 0$ $(i=1,\ldots,d)$; for example with $p=1, r=0, s=1$ and $q\neq 0, q\neq 1/\alpha_i$ whenever $\alpha_i\neq 0$.

This completes the proof of Theorem 4.7.

It will be clear that the result is effective in the sense already mentioned: for any given F we could in principle make a complete list of all solutions (x,y) (see for example Exercises 4.13 and 4.14). Indeed one can find explicit bounds in the literature, notably in the Grytczuk and Schinzel (1991) paper (which incidentally uses a different method of proof due to Skolem). This remark is significant because our proof in Chapter 12 of the finiteness of the set of solutions of say $x^3-2y^3=m$ is not effective. See also Exercise 4.15 on $y(y^2-x^2)=mx$.

The method here goes back to Runge (1887) – he was famed also for his work in numerical analysis. Bombieri (1983) showed how it can be generalized and then interpreted in terms of heights (this not unrelated to the material of our Appendix – see for example Exercise A.27); concretely this allows one to solve equations not just in rational integers, but also integers from a number field or even algebraic integers of bounded degree (see for example Exercises 11.25, 11.26, 11.27), and even so-called S-integers (these need p-adic tools to do them justice – see for example Exercise 14.41). Further he showed that it can be adapted to handle problems in the topic of irreducibility (for this topic see the next chapter). He also gave an alternative approach using G-functions (see Chapter 22). As a result, the method now has rather general connotations; for example some parts of Mihăilescu's proof (in Mihăilescu (2004) – see Bilu et al. (2014) for a self-contained account) might be considered in this context.

Exercises

4.1 Find all x in $\mathbb{Z}$ with x^4+x+7 a square.

4.2 Let a,x be in $\mathbb{Z}$ with x^4+x+a a square. Show that $|x|\leq\max\{1,|a|\}$.

4.3 Let $f(a)$ be the cardinality $\#\{x\in\mathbb{Z};\ x^4+x+a \text{ a square}\}$. Show that there are constants C such that for all $a\neq 0$ in $\mathbb{Z}$ we have

(a) $f(a)\leq C|a|^{1/2}$ [*Hint*: $y=x^2+n$].

(b) $f(a)\leq C|a|^{1/3}$ [*Hint*: $|n|\leq N$ or not].

4.4 Show that $\#\{(x,y) \in \mathbb{Z}^2;\ x^5 - y^7 = 1,\ 7 \nmid x\} < \infty$.

4.5 (a) There is a power series $\alpha = \alpha(z) = a_0 + a_1 z + a_2 z^2 + a_3 z^3 + a_4 z^4 + \cdots$ such that $\alpha^3 + \alpha = z$. Find a_0, a_1, a_2, a_3, a_4.

(b) Find $p_0, p_1, q_0, q_1, r_0, r_1$ in $\mathbb{Z}$, not all 0, such that $\mathrm{ord}_{z=0}\,\phi \geq 5$ for

$$\phi = (p_0 + p_1 z) + (q_0 + q_1 z)\alpha + (r_0 + r_1 z)\alpha^2.$$

4.6 Let $\mathcal{K}$ be a field and let $T(z,w)$ be in $\mathcal{K}[z,w]$ without constant term and linear term.

(a) Show that there are polynomials $P_n(X_1, \ldots, X_{n-1})$ in $\mathcal{K}[X_1, \ldots, X_{n-1}]$, such that $P_n(a_1, \ldots, a_{n-1})$ is the coefficient of z^n in $T(z, \sum_{k=1}^{\infty} a_k z^k)$ $(n = 2, 3, \ldots)$.

(b) Let b, c be in $\mathcal{K}$ with $c \neq 0$, and let $S(z,w) = bz + cw - T(z,w)$. Define $a_1, \ldots$ by $a_1 = -b/c$ and

$$a_n = P_n(a_1, \ldots, a_{n-1})/c \quad (n = 2, 3, \ldots).$$

For $\psi = \sum_{k=1}^{\infty} a_k z^k$ show that $S(z, \psi) = 0$.

4.7 Show that $\mathbb{C}[[z]]$ is an integral domain.

4.8 Let g_2, g_3 be in $\mathbb{C}$, and let ψ be in $\mathbb{C}[[z]]$ with $\psi'^2 = 4\psi^3 - g_2\psi - g_3$ (the Weierstrass ψ function). Show that ψ is locally convergent.

4.9 Let $\sigma = c + \cdots$ $(c \neq 0)$ be locally convergent in $\mathbb{C}[[z]]$. Show that $\psi = \sigma^{-1}$ in $\mathbb{C}[[z]]$ is locally convergent [*Hint*: majorize – see also Exercise 4.39].

4.10 Let $\psi = \sum_{k=0}^{\infty} k! z^k$ be in $\mathbb{C}[[z]]$. Show that there is no polynomial $P(z,w)$ with $\psi' = P(z, \psi)$.

4.11 Let x be in $\mathbb{C}$ and let A be monic in $\mathbb{C}[X]$ with $A(x) = 0$. Show that $|x| \leq \mathcal{L}(A)$.

4.12 Let x be in $\mathbb{C}$ and let $A \neq 0$ be in $\mathbb{C}[X]$ with leading coefficient a and $A(x) = 0$. Show that $|x| \leq \mathcal{L}(A)/|a|$.

4.13 Find all integers x, y with $y(y^2 - x^2) = x$ [*Hint*: at least one of $|y|$, $|y - x|$, $|y + x|$ is at most $|x|^{1/3}$ – or try solving for x in terms of y].

4.14 Find all integers x, y with $y(y^2 - 2x^2) = x$.

4.15 Show that there is c such that $\max\{|x|, |y|\} \leq c|m|$ for all integers x, y with $y(y^2 - x^2) = mx$ $(m \neq 0)$ [*Hint*: $y = \mu y_1, x = \mu x_1$].

* * * * * * * * * * * * * * * * * * * *

4.16 Show that $\#\{x \in \mathbb{Z};\ 2x^2 + 1 \text{ a square}\} = \infty$.

4.17 Let $H \geq 1$. Let $P(X) = X^4 + aX^3 + bX^2 + cX + d$ be in $\mathbb{Z}[X]$, not a square, with $|P| = \max\{|a|, |b|, |c|, |d|\} \leq H$, and let x be in $\mathbb{Z}$ with $P(x)$ a square. Show that $|x| \leq 26H^3$ (Masser, 1980b).

4.18 Let $H \geq 1$. Show that there is $P(X) = X^4 + aX^3 + bX^2 + cX + d$ in $\mathbb{Z}[X]$, not a square, with $|P| \leq H$, and x in $\mathbb{Z}$ with $P(x)$ a square but $|x| \geq H^3/4096$ (Masser, 1980b).

4.19 For any $\epsilon > 0$ show that there is $C(\epsilon)$ such that $\#\{x \in \mathbb{Z};\ x^2 + a \text{ a square}\} \leq C(\epsilon)|a|^\epsilon$ for every $a \neq 0$ in $\mathbb{Z}$.

4.20 For any $\epsilon > 0$ show that there is $C(\epsilon)$ such that $\#\{x \in \mathbb{Z};\ x^2 + x + a \text{ a square}\} \leq C(\epsilon)|a|^\epsilon$ for every a in $\mathbb{Z}$.

4.21 Is there $C(\epsilon)$ for any $\epsilon > 0$ such that $\#\{x \in \mathbb{Z};\ x^4 + x + a \text{ a square}\} \leq C(\epsilon)|a|^\epsilon$ for every a in $\mathbb{Z}$? I don't know.

4.22 Show that $\#\{x \in \mathbb{Z};\ x^3 + a \text{ a square}\} < \infty$ for every $a \neq 0$ in $\mathbb{Z}$ (this was proved in Mordell (1923) – see Baker (1968b) for references and much more).

4.23 Show that there are C, κ such that $\#\{x \in \mathbb{Z};\ x^3 + a \text{ a square}\} \leq C|a|^\kappa$ for every $a \neq 0$ in $\mathbb{Z}$ (Evertse and Silverman, 1986).

4.24 Is there $C(\epsilon)$ for any $\epsilon > 0$ such that $\#\{x \in \mathbb{Z};\ x^3 + a \text{ a square}\} \leq C(\epsilon)|a|^\epsilon$ for every $a \neq 0$ in $\mathbb{Z}$? No-one knows.

4.25 For any $\epsilon > 0$ show that there is $C(\epsilon)$ such that $|x| \leq C(\epsilon)\exp(|a|^{1+\epsilon})$ for any x and $a \neq 0$ in $\mathbb{Z}$ with $x^3 + a$ a square (Stark, 1973).

4.26 Is there is C such that $|x| \leq C\exp(|a|)$ for any x and $a \neq 0$ in $\mathbb{Z}$ with $x^3 + a$ a square? No-one knows.

4.27 Show that $\mathcal{L}(AB) \leq \mathcal{L}(A)\mathcal{L}(B)$ for A, B in $\mathbb{C}[X_1, \cdots, X_n]$.

4.28 Define $\mathbb{C}[[z, w]]$; and if $S = bz + cw + \cdots$ is in $\mathbb{C}[[z, w]]$ with $c \neq 0$, show that there is ψ in $\mathbb{C}[[z]]$ with $S(z, \psi(z)) = 0$.

4.29 We consider the differential equation

$$f'^2 = f^3 - f. \tag{$*$}$$

(a) Given $a \neq 0, \pm 1$ in $\mathbb{C}$, show that there is a unique $f = a + \cdots$ in $\mathbb{C}[[z]]$ with $(*)$.

(b) Show that there is a unique $f = 4z^{-2} + \cdots$ in $z^{-2}\mathbb{C}[[z]]$ with $(*)$.

(c) Show that f in (a) is locally convergent [*Hint*: f, f', f'' – why is $f' \neq 0$?].

(d) Show that $z^2 f$ in (b) is locally convergent [*Hint*: calculate a differential equation for $-1/f$].

(e) What happens in (a) for $a = 0, \pm 1$?

4.30 Generalize Lemma 4.4 to $\psi_1, \ldots, \psi_n$.

4.31 For ψ in $z\mathbb{C}[[z]]$ define $\exp \psi$ in $\mathbb{C}[[z]]$; and if $\psi = \sum_{k=1}^{\infty} \frac{k^{k-1}}{k!} z^k$ show that $z \exp \psi = \psi$.

4.32 For $\psi = 1 + \cdots$ in $\mathbb{C}[[z]]$ define $\log \psi$ in $z\mathbb{C}[[z]]$; and if $\psi = 1 + \cdots$ is in $\mathbb{C}[[z]]$ with

$$\psi = 1 + z\psi^{\lambda} = 1 + z \exp(\lambda \log \psi) \quad (\lambda > 0),$$

show that

$$\psi^{\mu} = 1 + \mu \sum_{k=1}^{\infty} \binom{\lambda k + \mu - 1}{k - 1} \frac{z^k}{k} \quad (\mu > 0).$$

4.33 For any field $\mathcal{K}$ show that $\mathcal{K}[[z]]$ is a principal ideal domain, find all ideals and units and show that the quotient field is $\mathcal{K}((z)) = \bigcup_{k=1}^{\infty} z^{-k}\mathcal{K}[[z]] = \mathcal{K}[[z]][1/z]$.

4.34 Let $\sigma = c + \cdots$ $(c \neq 0)$ be in $\mathbb{C}[[z]]$. Show that there is ψ in $\mathbb{C}[[z]]$ with $\psi^2 = \sigma$.

4.35 Let $\sigma = 1 + z$ be in $\mathbb{F}_2[[z]]$. Show that there is no ψ in $\mathbb{F}_2[[z]]$ with $\psi^2 = \sigma$.

4.36 Let $S(z, w)$ be in $\mathbb{C}[z, w]$ with $\partial S/\partial w \neq 0$. Show that there are k in $\mathbb{Z}$, e in $\mathbb{N}$ and ψ in $z^k\mathbb{C}[[z]]$ with $S(z, \psi(z^{1/e})) = 0$ (Newton–Puiseux – see Eichler (1966) for a modern proof).

4.37 Show that there is no field $\mathbb{F}$ containing $\mathbb{F}_2$ and no k in $\mathbb{Z}$, e in $\mathbb{N}$ and ψ in $z^k\mathbb{F}[[z]]$ with $S(z, \psi(z^{1/e})) = 0$ for $S(z, w) = z(w^2 + w) - 1$.

4.38 In Exercise 4.37 show that formally $S(z, \Psi) = 0$ for

$$\Psi = \frac{1}{z^{1/2}} + \frac{1}{z^{1/4}} + \frac{1}{z^{1/8}} + \frac{1}{z^{1/16}} + \frac{1}{z^{1/32}} + \frac{1}{z^{1/64}} + \frac{1}{z^{1/128}} + \cdots$$

(compare Exercise 3.11).

4.39 (a) For $\Gamma \geq 0$ and $\gamma > 0$ (not necessarily Euler's constant but it might be) define

$$\theta(z) = \gamma \sqrt{1 - \frac{4\Gamma(\Gamma + \gamma)z}{\gamma^2(1 - z)}} = \gamma - \frac{2\Gamma(\Gamma + \gamma)}{\gamma} z + \cdots$$

and

$$\overline{\psi}(z) = \frac{\gamma - \theta(z)}{2(\Gamma + \gamma)} = \sum_{k=1}^{\infty} \overline{a_k} z^k.$$

Show that $\gamma\overline{\psi} = \Gamma z + \Gamma \sum_{i+j\geq 2} z^i \overline{\psi}^j$.

(b) Let S and $\psi(z) = \sum_{k=1}^{\infty} a_k z^k$ be as in Lemma 4.3 with $\mathcal{K} = \mathbb{C}$. Show that $|a_k| \leq |\overline{a_k}|$ $(k = 1, 2, 3, \ldots)$ for $\Gamma = \mathcal{L}(S)$ and $\gamma = |\partial S/\partial w(0,0)|$ [*Hint*: Exercise 4.6].

4.40 If A in Exercise 4.11 is divisible by $(X - x_1) \cdots (X - x_r)$, show that

$$\max\{1, |x_1|\} \cdots \max\{1, |x_r|\} \leq \mathcal{L}(A).$$

4.41 If $A(X) = X^d + a_1 X^{d-1} + a_2 X^{d-2} + \cdots + a_d$ in Exercise 4.11, show that $|x| \leq d \max\{1, |a_1|, |a_2|^{1/2}, \cdots |a_d|^{1/d}\}$.

4.42 (p-adic). Let p be prime and let $S(w)$ be in $\mathbb{Z}_p[w]$ with $S(0)$ in $p\mathbb{Z}_p$ but $S'(0)$ not in $p\mathbb{Z}_p$. Show that there is a unique α in $p\mathbb{Z}_p$ with $S(\alpha) = 0$.

4.43 (modulo p). Let p be prime and write $[n] = t^{p^n} - t$ in $\mathbb{F}_p(t)$, with

$$F_k = [k][k-1]^p[k-2]^{p^2} \cdots [1]^{p^{k-1}} \quad (k = 1, 2, \ldots)$$

and $F_0 = 1$. Show that the formal power series

$$\psi(z) = \sum_{k=0}^{\infty} \frac{(-1)^k}{F_k} z^{p^k}$$

satisfies $\psi(tz) = t\psi(z) - \psi(z)^p$ (proved by Carlitz (1935) – see also Goss (1996) for the modern theory – and the beginning of the eponymous modules developing into Drinfeld modules).

4.44 Let F be as in Theorem 4.7 but in $\mathbb{Z}[X, Y]$ and with total degree d. Show that there are $c = c(d)$ and $\kappa = \kappa(d)$ such that $\max\{|x|, |y|\} \leq c\mathcal{L}(F)^{\kappa}$ for all (x, y) in $\mathbb{Z} \times \mathbb{Z}$ with $F(x, y) = 0$ (Hilliker and Straus, 1983; Grytczuk and Schinzel, 1991).

5

Irreducibility

Theorem 4.7 of the previous chapter can be reformulated in terms of reducibility. It implies that there are at most finitely many y in $\mathbb{Z}$ for which there exists x in $\mathbb{Z}$ with $F(x, y) = 0$; that is, $F(X, y)$ has the factor $X - x$. If Y does not divide $F_0(X, Y)$, then it can be shown fairly easily using the concept of integrality (see Exercise 5.2) that this is almost equivalent to $F(X, y)$ having a linear factor over $\mathbb{Q}$. That may remind us of the Hilbert Irreducibility Theorem for unrestricted $\mathcal{F}(X, Y)$ in $\mathbb{Q}[X, Y]$ irreducible over $\mathbb{Q}$; namely that there exist infinitely many y in $\mathbb{Q}$ such that $\mathcal{F}(X, y)$ is irreducible over $\mathbb{Q}$. By pushing the techniques of the previous chapter a little further, we can prove the following stronger version of this result for our special F under slightly different hypotheses.

Theorem 5.1 *Let F in $\mathbb{Q}[X, Y]$ be a polynomial, irreducible over $\mathbb{Q}$, of the shape*

$$F(X, Y) = F_0(X, Y) + f(X, Y)$$

for $F_0(X, Y)$ homogeneous of degree d, with a non-repeated linear factor over $\mathbb{Q}$ but no factor Y, and a polynomial f of total degree strictly less than d. Then there are at most finitely many y in $\mathbb{Z}$ such that $F(X, y)$ in $\mathbb{Q}[X]$ is reducible over $\mathbb{Q}$.

Thus we get irreducibility for all but finitely many y in $\mathbb{Z}$, not just infinitely many y in $\mathbb{Q}$. But the F is restricted as in Theorem 4.7 together with the condition that F_0 has a linear factor $X - \alpha Y$ for α that is rational and not just algebraic of degree strictly less than d. Further F_0 no longer has to be squarefree, but only one such factor $X - \alpha Y$ is allowed. Also here the irreducibility of F is clearly needed. It can often be checked in practice by examining the various homogeneous parts.

An example is the irreducibility of

$$X(X^3 - 2y^3) - y = X^4 - 2y^3X - y$$

for all but finitely many y in $\mathbb{Z}$. The new feature is that the possibility of a factorization into two quadratics has now been ruled out. Here a factorization $F = X(X^3 - 2Y^3) - Y = GH$ would force for example $G = X + c$, but $F(-c, Y) \neq 0$.

It is interesting that the result becomes false if the rational factor of F_0 is Y, again because of Pell. There is an example

$$F(X,Y) = Y(Y^2 - 2X^2) - 3Y + 2 = -2YX^2 + (Y+2)(Y-1)^2$$

wherein $F(X, y)$ becomes reducible if $2y(y + 2)(y - 1)^2$ is a square. This will be the case if just $2y(y + 2)$ is a square u^2. The resulting equation reduces to $u^2 - 2v^2 = -2$ with $v = y + 1$. So $u = 2u_1$ is even; and now we see the Pell equation $v^2 - 2u_1^2 = 1$ with infinitely many solutions. Therefore $F(X, y)$ is reducible for infinitely many y in $\mathbb{Z}$. For example

$$F(X, 22619536) = -2(4756X - 76069496205)(4756X + 76069496205).$$

Even though both factors are linear, this does not give a counterexample to Theorem 4.7 because the rational x with $F(x, y) = 0$ are almost never integers; thus for $y = 22619536$ we get

$$x = \pm\frac{76069496205}{4756}.$$

Our proof of Theorem 5.1 would be quite fast with height theory, as hinted at the end of the previous chapter. To avoid this we need two new ingredients: first some inequalities, in more precise form associated with the name of Gelfond, for the length function $\mathcal{L}$ used in the proof of Lemma 4.4; and second resultants.

For the Gelfond-type inequalities we may as well extend $\mathcal{L}$ in the obvious way to any number of variables, so to polynomials in $\mathbb{C}[X_1, \ldots, X_n]$. The result goes in the direction opposite to that in (4.32).

Lemma 5.2 *For any integer $d \geq 0$ and $n \geq 1$ there is a positive constant $c = c(d, n)$ depending only on d, n such that*

$$\mathcal{L}(AB) \geq c\mathcal{L}(A)\mathcal{L}(B)$$

for any A, B in $\mathbb{C}[X_1, \ldots, X_n]$ with AB of total degree at most d.

Proof We may suppose $A \neq 0, B \neq 0$ and even $\mathcal{L}(A) = \mathcal{L}(B) = 1$ (by scaling). By listing the coefficients in some order, we can identify the set of polynomials of unit length in $\mathbb{C}[X_1, \ldots, X_n]$ of degree at most d with a compact subset $C(d) = C_n(d)$ of $\mathbb{C}^D$ (where $D = \binom{d+n}{n}$ the binomial coefficient). We may also suppose that A has degree at most r, and that B has degree at most s, for fixed r, s with $r+s = d$. Now because $AB \neq 0$ we see that $\pi(A, B) = \mathcal{L}(AB)$ defines a map π from the product $C(r) \times C(s)$ to the real interval $(0, \infty)$. This is easily seen to be continuous. Its value is therefore bounded below by some $c > 0$; and now we get what we want. □

We apologize for this proof, which does not readily supply an explicit value of c; what Gelfond did was to obtain such values that are essentially optimal in their dependence on the degree d (see Proposition 14.1(i) later).

We turn now to resultants, which are extremely classical but which do not turn up in every course in algebra (compare Cassels (1991) chapter 16 entitled "Remedial mathematics"). As we need them also in later chapters, we give a self-contained exposition.

Let $\mathcal{K}$ be a field, and let A, B be in $\mathcal{K}[X]$. We have already had occasion to consider their possible coprimeness. This can be tested by means of the euclidean algorithm, but that is more suited to computation with examples. An alternative method is Sylvester's "dialytic elimination". If they are not coprime, then they have a common factor which has a zero ξ in some extension of $\mathcal{K}$. It follows that $A(\xi) = B(\xi) = 0$. Writing

$$A(X) = a_0X^r + \cdots + a_r, \quad B(X) = b_0X^s + \cdots + b_s, \tag{5.1}$$

we get

$$a_0\xi^r + \cdots + a_r = 0, \quad b_0\xi^s + \cdots + b_s = 0$$

and we try to eliminate ξ.

Now we see a connexion with the second problem of Chapter 1; but this time we don't use auxiliary polynomials. From now on assume $r \geq 1, s \geq 1$. We multiply the first equation by $\xi^{s-1}, \ldots, \xi, 1$ and the second by $\xi^{r-1}, \ldots, \xi, 1$. We obtain $r + s$ homogeneous linear equations in the vector

$$\boldsymbol{\xi} = (\xi^{r+s-1}, \ldots, \xi, 1)^t \tag{5.2}$$

(transposition) with a certain matrix. This matrix can be viewed in any number of textbooks but it usually looks as if $r = s$ (and even in the definitive Gelfand et al. (1994) p. 400 – by the way, the reader is advised that they write their polynomials backwards). After some experiments at improvement, we began to suspect that a perfectly unambiguous typing is not possible, so we illustrate with specific r, s; for example with $r = 5, s = 8$ we see in all its glory

$$\begin{pmatrix} a_0 & a_1 & a_2 & a_3 & a_4 & a_5 & 0 & 0 & 0 & 0 & 0 & 0 & 0 \\ 0 & a_0 & a_1 & a_2 & a_3 & a_4 & a_5 & 0 & 0 & 0 & 0 & 0 & 0 \\ 0 & 0 & a_0 & a_1 & a_2 & a_3 & a_4 & a_5 & 0 & 0 & 0 & 0 & 0 \\ 0 & 0 & 0 & a_0 & a_1 & a_2 & a_3 & a_4 & a_5 & 0 & 0 & 0 & 0 \\ 0 & 0 & 0 & 0 & a_0 & a_1 & a_2 & a_3 & a_4 & a_5 & 0 & 0 & 0 \\ 0 & 0 & 0 & 0 & 0 & a_0 & a_1 & a_2 & a_3 & a_4 & a_5 & 0 & 0 \\ 0 & 0 & 0 & 0 & 0 & 0 & a_0 & a_1 & a_2 & a_3 & a_4 & a_5 & 0 \\ 0 & 0 & 0 & 0 & 0 & 0 & 0 & a_0 & a_1 & a_2 & a_3 & a_4 & a_5 \\ b_0 & b_1 & b_2 & b_3 & b_4 & b_5 & b_6 & b_7 & b_8 & 0 & 0 & 0 & 0 \\ 0 & b_0 & b_1 & b_2 & b_3 & b_4 & b_5 & b_6 & b_7 & b_8 & 0 & 0 & 0 \\ 0 & 0 & b_0 & b_1 & b_2 & b_3 & b_4 & b_5 & b_6 & b_7 & b_8 & 0 & 0 \\ 0 & 0 & 0 & b_0 & b_1 & b_2 & b_3 & b_4 & b_5 & b_6 & b_7 & b_8 & 0 \\ 0 & 0 & 0 & 0 & b_0 & b_1 & b_2 & b_3 & b_4 & b_5 & b_6 & b_7 & b_8 \end{pmatrix}. \quad (5.3)$$

The reader will note that the diagonal consisting of a_0 glides smoothly into the diagonal consisting of b_8. This happens in general with a_0 and b_s, when we get a square matrix of order $r+s$ which we will denote by $\mathrm{RES} = \mathrm{RES}(A, B)$.

So $\mathrm{RES}\,\boldsymbol{\xi} = 0$; and as $\boldsymbol{\xi} \neq 0$ this implies $\det \mathrm{RES} = 0$.

Conversely, if $\det \mathrm{RES} = 0$ then there is $\boldsymbol{\xi} \neq 0$ with $\mathrm{RES}\,\boldsymbol{\xi} = 0$; but as we do not yet know that $\boldsymbol{\xi}$ has the special form (5.2) we cannot reverse the argument to deduce that A, B have a common zero and so are not coprime. Indeed *a priori* $\boldsymbol{\xi}$ might be supposed to be defined over the original field $\mathcal{K}$ containing the entries of RES and not over some extension. We will show in a moment that the argument can indeed be reversed, at least provided $a_0 \neq 0$ or $b_0 \neq 0$. The latter condition is necessary; for example with $r = s = 1$ the polynomials $A(X) = B(X) = 0X + 1$ are coprime, but $\det \mathrm{RES} = 0$.

But first we define the resultant $\mathrm{res}(A, B)$ to be just this $\det \mathrm{RES}(A, B)$.

In fact there is a subtle ambiguity in the notation, because A and B do not determine r and s in (5.1). They do if we insist on both $a_0 \neq 0$ and $b_0 \neq 0$; but we prefer not to be so dogmatic, as it is too restrictive for some applications. A more pedantic notation would be for example $\mathrm{res}_{rs}(A, B)$, and this is helpful when some general resultant has to be specialized in the sense of algebraic geometry. However it is not really necessary in this book.

Here are some simple examples.

When $r = s = 1$ then

$$\mathrm{res}(A, B) = \begin{vmatrix} a_0 & a_1 \\ b_0 & b_1 \end{vmatrix} = a_0 b_1 - a_1 b_0.$$

When $r = 1, s = 2$ then

$$\mathrm{res}(A, B) = a_0^2 b_2 - a_0 a_1 b_1 + a_1^2 b_0,$$

which is $a_0^2 B(-a_1/a_0)$ if $a_0 \neq 0$.

When $r = s = 2$ then the formula is a bit long but may be nicely shortened to

$$\mathrm{res}(A,B) = \begin{vmatrix} a_0 & a_1 & a_2 & 0 \\ 0 & a_0 & a_1 & a_2 \\ b_0 & b_1 & b_2 & 0 \\ 0 & b_0 & b_1 & b_2 \end{vmatrix} = (a_0b_2 - a_2b_0)^2 - (a_0b_1 - a_1b_0)(a_1b_2 - a_2b_1).$$

Most of the formulae make sense over rings, so we assume that our $\mathcal{K}$ is the quotient field of some integral domain $\mathcal{R}$. Here is all we need about resultants.

Proposition 5.3 *Suppose A, B are in $\mathcal{R}[X]$ of degrees at most $r \geq 1, s \geq 1$ respectively.*

(i) There are P, Q in $\mathcal{R}[X]$, not both zero, with P of degree at most $s-1$ and Q of degree at most $r-1$, such that

$$PA + QB = \mathrm{res}(A,B). \tag{5.4}$$

Further if $\mathrm{res}(A,B) \neq 0$, *then the coefficients of P are maximal minors corresponding to the top s elements of the last column of* $\mathrm{RES}(A,B)$, *and the coefficients of Q are maximal minors corresponding to the bottom r elements of the last column of* $\mathrm{RES}(A,B)$.

(ii) Suppose the coefficients a_0, b_0 of A, B respectively are not both zero. Then A, B are coprime if and only if $\mathrm{res}(A,B) \neq 0$.

Proof For (i) we perform an operation on the columns κ of $\mathrm{RES}(A,B)$. Replace the last column κ_{r+s} by the linear combination

$$X^{r+s-1}\kappa_1 + \cdots + X\kappa_{r+s-1} + \kappa_{r+s}$$

(including $X^r\kappa_s$ and $X^s\kappa_r$). We get a new matrix, which with $r = 5, s = 8$ as above is

$$\begin{pmatrix} a_0 & a_1 & a_2 & a_3 & a_4 & a_5 & 0 & 0 & 0 & 0 & 0 & 0 & X^7A \\ 0 & a_0 & a_1 & a_2 & a_3 & a_4 & a_5 & 0 & 0 & 0 & 0 & 0 & X^6A \\ 0 & 0 & a_0 & a_1 & a_2 & a_3 & a_4 & a_5 & 0 & 0 & 0 & 0 & X^5A \\ 0 & 0 & 0 & a_0 & a_1 & a_2 & a_3 & a_4 & a_5 & 0 & 0 & 0 & X^4A \\ 0 & 0 & 0 & 0 & a_0 & a_1 & a_2 & a_3 & a_4 & a_5 & 0 & 0 & X^3A \\ 0 & 0 & 0 & 0 & 0 & a_0 & a_1 & a_2 & a_3 & a_4 & a_5 & 0 & X^2A \\ 0 & 0 & 0 & 0 & 0 & 0 & a_0 & a_1 & a_2 & a_3 & a_4 & a_5 & XA \\ 0 & 0 & 0 & 0 & 0 & 0 & 0 & a_0 & a_1 & a_2 & a_3 & a_4 & A \\ b_0 & b_1 & b_2 & b_3 & b_4 & b_5 & b_6 & b_7 & b_8 & 0 & 0 & 0 & X^4B \\ 0 & b_0 & b_1 & b_2 & b_3 & b_4 & b_5 & b_6 & b_7 & b_8 & 0 & 0 & X^3B \\ 0 & 0 & b_0 & b_1 & b_2 & b_3 & b_4 & b_5 & b_6 & b_7 & b_8 & 0 & X^2B \\ 0 & 0 & 0 & b_0 & b_1 & b_2 & b_3 & b_4 & b_5 & b_6 & b_7 & b_8 & XB \\ 0 & 0 & 0 & 0 & b_0 & b_1 & b_2 & b_3 & b_4 & b_5 & b_6 & b_7 & B \end{pmatrix}. \tag{5.5}$$

In general this gives a new matrix $\mathrm{RES}_X(A,B)$ but does not change its determinant, which delivers (5.4) with P,Q in $\mathcal{R}[X]$; it is then clear by developing the new determinant by this last column that the coefficients of P,Q are the required minors of $\mathrm{RES}(A,B)$. Further if $\mathrm{res}(A,B) \neq 0$, then it is trivial that P,Q are not both zero (or one can also see this non-trivially by looking at their coefficients). But why does this stay true if $\mathrm{res}(A,B) = 0$?

Well, in that case there is a non-trivial relation

$$p_1\rho_1 + \cdots + p_s\rho_s + q_1\rho_{s+1} + \cdots + q_r\rho_{s+r} = 0$$

between the rows ρ of $\mathrm{RES}(A,B)$, and we can assume that the coefficients lie in $\mathcal{R}$. The same relation persists for the rows of $\mathrm{RES}_X(A,B)$. So considering the last column of $\mathrm{RES}_X(A,B)$ (as in (5.5) for example) we deduce

$$p_1X^{s-1}A + \cdots + p_sA + q_1X^{r-1}B + \cdots + q_rB = 0.$$

And since $\mathrm{res}(A,B) = 0$ this delivers another (5.4), visibly with P,Q not both zero. This completes the proof of (*i*). In fact here too one can take the coefficients of P,Q as minors of $\mathrm{RES}(A,B)$ (see Exercise 5.6).

For (*ii*) we already saw in the preamble that if A,B are not coprime then the resultant is zero. So now suppose that A,B are coprime. If $\mathrm{res}(A,B) = 0$ then from (5.4) we get $PA = -QB$. Assume for example that $a_0 \neq 0$. Then $A \neq 0$. If $Q = 0$ then it would follow that $P = 0$ too, a contradiction. So $Q \neq 0$. Now A divides QB, and because A,B are coprime it follows that A divides Q. But now the degree of A is exactly r, while the degree of $Q \neq 0$ is at most $r-1$. This second contradiction completes the proof of (*ii*). □

Here is an example with $\mathcal{R} = \mathbb{Z}[x,y]$. We take

$$A(t) = t^3 + t - x, \quad B(t) = t^4 + t - y$$

and the resultant with respect to t is none other than the left-hand side of (1.4). The substitution $x = t^3 + t$, $y = t^4 + t$ in (5.4) gives the right-hand side of (1.4).

We are soon ready to prove Theorem 5.1; first a preliminary in the style of Lemma 4.8. This time we need a zero of higher order.

Lemma 5.4 *Let α be in $\mathbb{Q}$, and let ψ be in $z\mathbb{Q}[[z]]$. Then for any $L \geq 0, d \geq 2$ in $\mathbb{Z}$ there is non-zero G in $\mathbb{Z}[X,Y]$ of degree at most L in X and of degree at most $d-1$ in Y such that*

$$\phi(z) = G\left(\frac{1}{z}, \frac{\alpha}{z} + \frac{\psi(z)}{z}\right) \tag{5.6}$$

is in $z^{L(d-1)}\mathbb{Q}[[z]]$.

Proof Note that we can get divisibility by a full $z^{L(d-1)}$ instead of just z; this is thanks to the rationality of α.

We write

$$G(X,Y) = \sum_{i=0}^{L}\sum_{j=0}^{d-1} p_{ij}X^iY^j.$$

The condition on ϕ tells us that now the coefficients of

$$z^{-(L+d-1)}, \ldots, z^{-1}, 1, z, \ldots, z^{L(d-1)-1}$$

in the Laurent expansion are zero. Thus we have

$$(L+d-1) + L(d-1) \;=\; (L+1)d - 1$$

linear forms, now over $\mathbb{Q}$, in $(L+1)d$ unknowns. The corresponding system of homogeneous linear equations is therefore solvable non-trivially. □

We now prove Theorem 5.1. Of course it is trivial for $d = 1$, so we can assume $d \geq 2$. For easier comparison with the proof of Theorem 4.7 it is convenient to interchange the variables first; thus now $F_0(X,Y)$ has a linear factor over $\mathbb{Q}$ but no factor X, and we want to prove that there are at most finitely many x in $\mathbb{Z}$ such that $F(x,Y)$ in $\mathbb{Q}[Y]$ is reducible over $\mathbb{Q}$. Now $F_0(X,Y)$ is divisible by $Y - \alpha X$ for some α in $\mathbb{Q}$, but not by $(Y-\alpha X)^2$. We can clearly suppose that $F(X,Y)$ is in $\mathbb{Z}[X,Y]$.

By Lemma 4.6 with $K = \mathbb{Q}$, there is locally convergent $\psi(z)$ in $z\mathbb{Q}[[z]]$ such that for $\lambda(X) = \alpha X + X\psi(1/X)$ we have $F(X,\lambda(X)) = 0$. We choose $L \geq 1$ in $\mathbb{Z}$ sufficiently large (to be specified later). By Lemma 5.4 there is $G \neq 0$ in $\mathbb{Z}[X,Y]$ of degree at most L in X and of degree at most $d-1$ in Y such that ϕ defined by (5.6) is in $z^{L(d-1)}\mathbb{Q}[[z]]$.

We pause to show that ϕ is not identically zero. Now $G(X,Y)$ must involve Y because a non-zero polynomial in $1/z$ alone could not possibly be in $z^{L(d-1)}\mathbb{Q}[[z]]$. Thus the vanishing of ϕ would imply that $\lambda(z)$ would be of degree at most $d-1$ over $\mathbb{Q}(z)$. However from $F(z,\lambda(z)) = 0$ and the irreducibility of $F(X,Y) = aY^d + \cdots$ $(a \neq 0)$ over $\mathbb{Q}$, we see that $\lambda(z)$ has degree d over $\mathbb{Q}(z)$.

Let x be any sufficiently large complex number. Then for $y = \lambda(x)$ we have $G(x,y) = \phi(1/x)$ and so the by now standard estimate gives the bounds

$$0 < |G(x,y)| \leq c|x|^{-L(d-1)} \tag{5.7}$$

with c depending only on G so implicitly also on F and the yet-to-be-specified L (but not of course on x). This is much better than (4.37), and our immediate goal is to deduce a contradiction when x is integral. However $y = \lambda(x)$ is probably not rational let alone integral, so we cannot deduce a contradiction from the Fundamental Theorem; anyway we haven't yet used the hypothesis of reducibility of $F(x, Y)$.

In fact this hypothesis, now for x in $\mathbb{Z}$, implies $F(x, Y) = A_1 \cdots A_t$ with $A_1, \ldots, A_t$ in $\mathbb{Q}[Y]$, irreducible over $\mathbb{Q}$, of degrees at most $d-1$. Since $F(x, Y)$ is in $\mathbb{Z}[Y]$, Gauss's Lemma shows that we can take them in $\mathbb{Z}[Y]$. Now

$$0 = F(x, \lambda(x)) = F(x, y) = A_1(y) \cdots A_t(y).$$

It follows that there is some $A = A_i$ $(i = 1, \ldots, t)$, say of degree r with $1 \leq r \leq d-1$, with $A(y) = 0$. In particular y is an algebraic number of degree at most $d-1$. The expert will be able to use some form of Liouville inequality or height inequality or product formula (for all these see later) to deduce a contradiction from (5.7); we appeal to resultants.

More precisely we consider the resultant of $A(Y)$ and $B(Y) = G(x, Y)$, with B of degree at most s for some s with $1 \leq s \leq d-1$.

So the resultant N of $A(Y)$ and $B(Y)$ makes sense. It lies in $\mathbb{Z}$.

If $N = 0$ then by Proposition 5.3(*ii*) we see that A, B are not coprime. As A is irreducible this implies that A divides B. Because $A(y) = 0$ we get also $G(x, y) = B(y) = 0$, contradicting (5.7).

So we can assume $N \neq 0$; in this case we will obtain a contradiction if $|x|$ is sufficiently large.

By Proposition 5.3(*i*) there are P, Q in $\mathbb{Z}[Y]$, with Q of degree at most $d-2$, such that (5.4) holds; and the coefficients of Q are maximal minors corresponding to the bottom r elements of the last column of the matrix RES(A, B). Putting $Y = y$ in (5.4) gives $N = Q(y)B(y)$. We already know that $B(y) = G(x, y)$ is very small, but how big can $Q(y)$ be?

Well, as $y = \lambda(x) = \alpha x + x\psi(1/x)$ we see easily that $|y| \leq c|x|$ (and from here on, it is convenient to use c repeatedly for various possibly different constants as in (5.7) depending only on F and the soon-to-be-specified L but not on x). It follows that

$$|Q(y)| \leq \mathcal{L}(Q) \max\{1, |y|\}^{d-2} \leq c\mathcal{L}(Q)|x|^{d-2}.$$

A crude estimation of the coefficients of Q as subdeterminants gives

$$\mathcal{L}(Q) \leq (d-1)(r+s-1)!\mathcal{L}(A)^s\mathcal{L}(B)^{r-1} \leq c\mathcal{L}(A)^{d-1}\mathcal{L}(B)^{d-2}$$

because only $r-1$ rows involve the coefficients of B (and in fact the factorial counting the number of terms in a determinant can, as in (4.32), be eliminated). As $B(Y) = G(x, Y)$ we see from Lemma 5.4 that $\mathcal{L}(B) \leq c|x|^L$. And $F(x, Y) = A(Y)\tilde{A}(Y)$ for some $\tilde{A}(Y)$ in $\mathbb{Z}[Y]$, so the patiently waiting Lemma 5.2 shows that

$$\mathcal{L}(A) \le \mathcal{L}(A)\mathcal{L}(\tilde{A}) \le c\mathcal{L}(A\tilde{A}) \le c|x|^d$$

in this veritable frenzy of estimating. Putting all these together gives a grand total of

$$0 < |N| = |Q(y)||G(x,y)| \le c|x|^{L(d-2)+d^2-2}|G(x,y)|.$$

Finally the Fundamental Theorem on (5.7) will lead to a contradiction if $|x|$ is sufficiently large, provided the exponents satisfy

$$L(d-2)+d^2-2 < L(d-1)$$

for example $L = d^2 - 1$. This completes the proof of Theorem 5.1.

Results of this form were first given by Sprindzhuk; see Sprindzuk (1993) for instance. The following is a particularly amusing example. If the irreducible $\mathcal{F}$ in $\mathbb{Q}[X,Y]$ satisfies $\mathcal{F}(0,0) = 0$ and $\frac{\partial \mathcal{F}}{\partial X}(0,0) \neq 0$ (compare with Lemma 4.3), then there are at most finitely many prime numbers p such that $\mathcal{F}(X,p)$ in $\mathbb{Q}[X]$ is reducible over $\mathbb{Q}$ (see Bilu and Masser (2006) for a quick proof). This applies to $\mathcal{F} = X^2 + X - 2Y$; but note that now $\mathcal{F}(X,y) = (X-n)(X+n+1)$ is reducible for every triangular number $y = n(n+1)/2$, so the conclusion of Theorem 5.1 fails.

As mentioned before, Bombieri (1983) gave different proofs using heights or G-functions.

For more on Hilbert's Irreducibility Theorem see Lang (1983) as well as Schinzel (1982) and Schinzel (2000).

For more on resultants see the fascinating book of Gelfand et al. (1994); however their resultant is the same as ours only up to sign, due to backwards writing.

Exercises

5.1 Let K be a number field with $[K:\mathbb{Q}] = e$, let α be in K, and let ψ be in $zK[[z]]$. Show that for any $L \ge 0, M \ge 0, T \ge 0$ in $\mathbb{Z}$ with $(L+1)(M+1) > e(L+M+T)$, there is non-zero G in $\mathbb{Z}[X,Y]$ of degree at most L in X and of degree at most M in Y such that

$$\phi(z) = G\left(\frac{1}{z}, \frac{\alpha}{z} + \frac{\psi(z)}{z}\right)$$

is in $z^T K[[z]]$ (the case $M = e, T = 1$ is Lemma 4.8, and the case $e = 1, M = d-1, T = L(d-1)$ is Lemma 5.4).

5.2 (a) Let x be in $\mathbb{Q}$ and $a_0, \ldots, a_d$ in $\mathbb{Z}$ with $a_0x^d + \cdots + a_d = 0$. Show that a_0x is in $\mathbb{Z}$ [*Hint*: multiply by a_0^{d-1} and consider the integral closure $\mathbb{Z}_{\mathbb{C}}$ of $\mathbb{Z}$ in $\mathbb{C}$].

(b) Let F in $\mathbb{Z}[X, Y]$ be a polynomial of the shape $F(X, Y) = F_0(X, Y) + f(X, Y)$ for $F_0(X, Y)$ homogeneous of degree d with no factor Y, and a polynomial f of total degree strictly less than d, and let y be in $\mathbb{Z}$. If the coefficient of X^d is a_0, show that $F(X, y)$ has factor $X - x$ for some x in $\mathbb{Z}/a_0$ if and only if it has a linear factor over $\mathbb{Q}$.

5.3 (a) Let F be as in Theorem 4.7, but with F_0 not divisible by Y. Show that there are at most finitely many (x, y) in $\mathbb{Q} \times \mathbb{Z}$ with $F(x, y) = 0$.

(b) Find F as in Theorem 4.7 such that there are infinitely many (x, y) in $\mathbb{Q} \times \mathbb{Z}$ with $F(x, y) = 0$.

5.4 Let F be as in Theorem 4.7, but with F_0 not divisible by Y and with $d = 2$ or $d = 3$. Show that there are at most finitely many y in $\mathbb{Z}$ with $F(X, y)$ reducible in $\mathbb{Q}[X]$.

5.5 Let

$$\lambda(r, s) = \inf \frac{\mathcal{L}(AB)}{\mathcal{L}(A)\mathcal{L}(B)}$$

taken over all $A \neq 0, B \neq 0$ in $\mathbb{R}[X]$ with A of degree at most r and B of degree at most s.

(a) Show that $\lambda(1, 1) \geq 1/2$. [*Hint*: essentially $A = X + x, B = X - y$ with $x \geq y \geq 0$].

(b) Show that $\lambda(1, 1) = 1/2$.

(c) Show that $\lambda(r, r) \leq 1/2^r$.

5.6 Let $\mathcal{R}$ be a commutative ring, and let A, B be in $\mathcal{R}[X]$ with A of degree at most $r \geq 1$ and B of degree at most $s \geq 1$ and $\mathrm{res}(A, B) = 0$. Show that there are P, Q in $\mathcal{R}[X]$, not both zero, with P of degree at most $s - 1$ and Q of degree at most $r - 1$, and having minors of $\mathrm{RES}(A, B)$ as coefficients, such that $PA + QB = 0$ [*Hint*: minor size is matrix rank].

5.7 Let $P(X, Y)$ be the resultant of $A = t^{1948} + t^{666} + 1 - X$ and $B = t^{1291} + t^{163} + t - Y$ with respect to t.

(a) Show that $P(t^{1948} + t^{666} + 1, t^{1291} + t^{163} + t) = 0$.

(b) Show that P has degree at most 1291 in X and at most 1948 in Y.

(c) Show that $P \neq 0$ (compare with Exercise 1.11).

5.8 Let $r, s \geq 1$. Show that there is $c = c(r, s) > 0$ with the following property. For A in $\mathbb{Z}[X]$ irreducible of degree r, ξ in $\mathbb{C}$ with $A(\xi) = 0$, and B in $\mathbb{Z}[X]$ of degree s with $B(\xi) \neq 0$ we have

$$|B(\xi)| \geq c\mathcal{L}(A)^{-s}\mathcal{L}(B)^{-(r-1)} \max\{1, |\xi|\}^{-(r-1)}.$$

5.9 Let F, G be in $\mathbb{C}[X, Y]$ without common factor in $\mathbb{C}[X, Y] \setminus \mathbb{C}$. Show that there are at most finitely many (x, y) in $\mathbb{C}^2$ with $F(x, y) = G(x, y) = 0$. [*Hint*: $GL_2(\mathbb{C})$].

* * * * * * * * * * * * * * * * * * *

5.10 Does Exercise 5.4 hold for $d = 4$? I don't know.

5.11 Is $\lambda(1, 2) = 1/3$ in Exercise 5.5? I don't know.

5.12 Show that $\lambda(r, s) \geq 1/2^{r+s}$, so $\lambda(r, r) \geq 1/4^r$ [*Hint*: Exercise 4.40].

5.13 Show that

$$\lim_{r\to\infty} \frac{\log \lambda(r, r)}{r} = -\frac{4}{\pi} \int_0^{\pi/2} \log \left(2\cos\left(\frac{t}{2}\right)\right) dt,$$

so $\lambda(r, r)$ is about $1/\Lambda^r$ with $\Lambda = 3.2099123\ldots$ (Boyd, 1992).

5.14 Show that $\mathcal{L}(AB) \geq \frac{1}{2}\mathcal{L}(A)\mathcal{L}(B)$ for all A, B in $\mathbb{C}[X]$ with AB of degree at most 2.

5.15 Define $S(U, T_1, T_2, T_3, T_4)$ as

$$U^6 - T_2U^5 + (T_1T_3 - T_4)U^4 - (T_3^2 + T_1^2T_4 - 2T_2T_4)U^3$$
$$+(T_1T_3T_4 - T_4^2)U^2 - T_2T_4^2U + T_4^3.$$

Show that $X^4 - t_1X^3 + t_2X^2 - t_3X + t_4$ in $\mathbb{Q}[X]$ splits into two quadratic factors in $\mathbb{Q}[X]$ if and only if

$$S(U, t_1 + 4r, t_2 + 3rt_1 + 6r^2, t_3 + 2rt_2 + 3r^2t_1 + 4r^3, t_4 + rt_3 + r^2t_2 + r^3t_1 + r^4)$$

has a divisor in $\mathbb{Q}[U]$ of degree 1 for $r = 0, 1, 2, 3, 4, 5, 6$ (Oehler Master Thesis 2001).

5.16 Find all y in $\mathbb{Z}$ with $|y| < 100$ and $X^4 + X + y$ reducible in $\mathbb{Q}[X]$.

5.17 Show that $\text{res}(t + a, B(t)) = B(-a)$.

5.18 Calculate $\text{res}((t + a)(t + a'), B(t))$.

5.19 Let $\mathcal{R}$ be a commutative ring, and let A, B be in $\mathcal{R}[X]$ with A, B both of degree $d \geq 1$, with $r = \text{res}(A, B) \neq 0$.

(a) Show that there are unique P, Q in $\mathcal{R}[X]$, of degree at most $d - 1$, such that $PA + QB = r$.

(b) If P, Q both have degree $d - 1 \geq 1$ show that

$$\text{res}(P, B) = (-1)^d b_0^{2d-1} r^{d-1}, \quad \text{res}(A, Q) = a_0^{2d-1} r^{d-1}.$$

5.20 Try again Exercise 1.10.

5.21 In Exercise 5.8 show that $|B(\xi)| \geq c\mathcal{L}(A)^{-s}\mathcal{L}(B)^{-(r-1)}$.

5.22 Let $\mathcal{K}$ be a field, and let F, G be in $\mathcal{K}[X,Y]$, both involving Y, and with degrees at most r_X, s_X in X and r_Y, s_Y in Y. Show that the resultant with respect to Y is in $\mathcal{K}[X]$, and if non-zero it has degree at most $r_X s_Y + r_Y s_X$.

5.23 In Exercise 5.9 show that the set of (x,y) has at most $\mathcal{D}(F)\mathcal{D}(G)$ elements ($\mathcal{D}$ = total degree).

5.24 Let $F_1, \ldots, F_m$ be in $\mathbb{C}[X,Y]$ without common factor in $\mathbb{C}[X,Y]\backslash\mathbb{C}$. Show that there are at most finitely many (x,y) in $\mathbb{C}^2$ with

$$F_1(x,y) = \cdots = F_m(x,y) = 0.$$

5.25 (a) Show that the resultant of A, B in Gelfand et al. (1994) is

$$\mathrm{res}(B,A) = (-1)^{rs}\,\mathrm{res}(A,B).$$

(b) Show that

$$\mathrm{res}(A_1A_2,B) = \mathrm{res}(A_1,B)\,\mathrm{res}(A_2,B), \quad \mathrm{res}(A,B_1B_2) = \mathrm{res}(A,B_1)\,\mathrm{res}(A,B_2).$$

5.26 If $A(X) = \prod_{i=1}^{r}(X-\alpha_i)$ show that $\mathrm{res}(A,B) = \prod_{i=1}^{r} B(\alpha_i)$.

5.27 Show that

$$\mathrm{res}\left(\prod_{i=1}^{r}(X-\alpha_i),\ \prod_{j=1}^{s}(X-\beta_j)\right) = \prod_{i=1}^{r}\prod_{j=1}^{s}(\alpha_i-\beta_j).$$

5.28 Let A, B be in $\mathbb{C}[X] \setminus \mathbb{C}$, and let α, β be in $\mathbb{C}$ with $A(\alpha) = B(\beta) = 0$. Show that $C(\alpha+\beta) = 0$, where $C(X)$ is the resultant of $A(t)$ and $B(X-t)$ with respect to t.

5.29 In Exercise 5.28 find a resultant E with $E(\alpha\beta) = 0$.

5.30 In Exercise 5.28 find a resultant F with $F(B(\alpha)) = 0$.

5.31 (a) Suppose $A(\alpha) = 0$ for some A in $\mathbb{Q}[X]$ irreducible over $\mathbb{Q}$. Show that α is a root of unity if and only if some $\mathrm{res}(A(X), X^n - 1) = 0$ ($n = 1, 2, 3, \ldots$).

(b) Show that there is a polynomial F, homogeneous of degree six in three variables, with the property that if $a\alpha^2 + b\alpha + c = 0$ for $aX^2 + bX + c$ in $\mathbb{Q}[X]$ irreducible over $\mathbb{Q}$, then α is a root of unity if and only if $F(a,b,c) = 0$.

5.32 Let A, B be in $\mathbb{C}[X]$ with degree $m \geq 1$, and let

$$\frac{A(U)B(V) - A(V)B(U)}{U - V} = \sum_{i=0}^{m-1}\sum_{j=0}^{m-1} c_{ij}U^iV^j.$$

Show that $\text{res}(A, B) = (-1)^{m(m-1)/2} \det c_{ij}$ (Bézout–Cayley – see Gelfand et al. (1994) for more).

5.33 Suppose that R is the formal resultant $\det \text{RES}(A, B)$ of A and B (as in (5.3) for example), and that these have a common simple zero at ξ and no other common zeroes. Show that

$$(\xi^r, \xi^{r-1}, \ldots, 1) = \left(\frac{\partial R}{\partial a_0}, \frac{\partial R}{\partial a_1}, \ldots, \frac{\partial R}{\partial a_r}\right),$$

$$(\xi^s, \xi^{s-1}, \ldots, 1) = \left(\frac{\partial R}{\partial b_0}, \frac{\partial R}{\partial b_1}, \ldots, \frac{\partial R}{\partial b_s}\right)$$

when interpreted projectively.

5.34 Show that the homogeneous polynomials

$$a_0X^r + a_1X^{r-1}T + \cdots + a_rT^r, \quad b_0X^s + b_1X^{s-1}T + \cdots + b_sT^s$$

have a common zero (ξ, τ) in projective $\mathbb{P}_1$ if and only if the Sylvester determinant vanishes.

5.35 Let $A_i(X, Y) = a_iXY + b_iX + c_iY + d_i$ be in $\mathbb{C}[X, Y]$ with $a_i \neq 0$ $(i = 1, 2, 3)$. Show that there exists (ξ, η) in $\mathbb{C}^2$ with $A_i(\xi, \eta) = 0$ $(i = 1, 2, 3)$ if and only if

$$\begin{vmatrix} a_1 & b_1 & c_1 & d_1 & 0 & 0 \\ 0 & 0 & a_1 & b_1 & c_1 & d_1 \\ a_2 & b_2 & c_2 & d_2 & 0 & 0 \\ 0 & 0 & a_2 & b_2 & c_2 & d_2 \\ a_3 & b_3 & c_3 & d_3 & 0 & 0 \\ 0 & 0 & a_3 & b_3 & c_3 & d_3 \end{vmatrix} = 0$$

(Dixon, 1908).

5.36 Let

$$A_i(X, Y, T) = a_iX^2 + b_iXY + c_iY^2 + d_iXT + e_iYT + f_iT^2 \quad (i = 1, 2, 3)$$

be in $\mathbb{C}[X, Y, T]$. Show that there exists a polynomial in the coefficients, of degree 12, which vanishes if and only if the A_i $(i = 1, 2, 3)$ have a common zero in $\mathbb{P}_2$. It is too long to be exhibited here, but when $a_1 = c_2 = f_3 = 0$ it is

$$\begin{vmatrix}
b_1 & d_1 & 0 & 0 & 0 & 0 & c_1 & f_1 & 0 & e_1 & 0 & 0 \\
0 & 0 & b_2 & e_2 & 0 & 0 & a_2 & 0 & f_2 & 0 & d_2 & 0 \\
0 & 0 & 0 & 0 & d_3 & e_3 & 0 & a_3 & c_3 & 0 & 0 & b_3 \\
0 & 0 & 0 & c_1 & 0 & f_1 & 0 & 0 & e_1 & 0 & b_1 & d_1 \\
0 & 0 & 0 & 0 & 0 & f_2 & 0 & 0 & e_2 & a_2 & b_2 & d_2 \\
0 & 0 & 0 & c_3 & 0 & 0 & 0 & 0 & e_3 & a_3 & b_3 & d_3 \\
0 & 0 & 0 & 0 & f_1 & 0 & 0 & d_1 & 0 & b_1 & c_1 & e_1 \\
0 & a_2 & 0 & 0 & f_2 & 0 & 0 & d_2 & 0 & b_2 & 0 & e_2 \\
0 & a_3 & 0 & 0 & 0 & 0 & 0 & d_3 & 0 & b_3 & c_3 & e_3 \\
0 & 0 & c_1 & 0 & 0 & 0 & b_1 & 0 & 0 & d_1 & e_1 & f_1 \\
a_2 & 0 & 0 & 0 & 0 & 0 & b_2 & 0 & 0 & d_2 & e_2 & f_2 \\
a_3 & 0 & c_3 & 0 & 0 & 0 & b_3 & 0 & 0 & d_3 & e_3 & 0
\end{vmatrix}$$

(Dixon, 1908).

5.37 Let $F(X, Y)$ be as in Theorem 5.1. Let $c > 0$ and $\delta < 1/d$. Show that there are at most finitely many $y = r/s$ with r, s in $\mathbb{Z}$, $1 \leq s \leq c|r|^{\delta}$, and $F(X, y)$ reducible.

6

Elliptic curves – Stepanov's Method

We make another jump to count points on elliptic curves E over finite fields. A classical result of Hasse, from around 1930, is that over $\mathbb{F}_q$ their number differs from $q+1$ by at most $2\sqrt{q}$ (here the odd-looking $q+1$ comes from the zero of the group law). A modern proof (see for example Silverman (1992) chapter V) uses the separability of an endomorphism coming from the Frobenius map. Actually what we prove here, for simplicity with primes $q = p$, has nothing to do with elliptic curves.

Theorem 6.1 *Let C in $\mathbb{F}_p[X]$ be a cubic polynomial. Then the number N of (x, y) in $\mathbb{F}_p^2$ with $y^2 = C(x)$ satisfies $|N - p| \leq 8\sqrt{p}$.*

A completely trivial bound is $N \leq p^2$. This can immediately be improved by noting that for each x in $\mathbb{F}_p$ there are at most two solutions of $y^2 = C(x)$; thus $N \leq 2p$. As $N \geq 0$, we get only $|N - p| \leq p$, worse for all $p \geq 67$. The bound in Theorem 6.1 reflects the fact that $y^2 = w$ has either two or no solutions, and that on average there is exactly one.

In fact for $p \geq 67$ our Theorem 6.1 implies that $N > 0$, itself far from trivial. For example $N = 0$ with $y^2 = x^3 - x - 1$ over $\mathbb{F}_3$.

We warm-up by proving that $N > 0$ whenever $p \geq 5$. But it is instructive first to consider the analogous results for quadratic polynomials, even though these are easy. We need the following simple but crucial remark.

Lemma 6.2 *For $p \geq 3$ let w be in $\mathbb{F}_p$. Then $w^{(p-1)/2} = 0, 1, -1$ according to whether $w = 0$ or w is a non-zero square or w is a non-square.*

Proof This is familiar from most proofs of quadratic reciprocity. For example the polynomial $X^{(p-1)/2} - 1$ has at most $(p-1)/2$ different zeroes in $\mathbb{F}_p$, and these can be identified as the $x = m^2$ for $m = 1, 2, \ldots, (p-1)/2$; further if $x \neq 0$ is anything else then $(x^{(p-1)/2})^2 = 1$ and so $x^{(p-1)/2} = -1$. □

So let us prove that $N > 0$ with $y^2 = Q(x)$ for quadratic Q and $p \geq 3$; that is, there is always a solution. We write $r = (p-1)/2 \geq 1$. If there are no solutions, then by Lemma 6.2 we have

$$Q(x)^r = -1 \tag{6.1}$$

for all x in $\mathbb{F}_p$. So for $\phi(X) = Q(X)^r + 1$ we have $\phi(x) = 0$ for all x in $\mathbb{F}_p$. Thus $\phi(X)$ must be divisible by

$$\prod_{x \in \mathbb{F}_p} (X - x) = X(X-1)\cdots(X-p+1) = X^p - X.$$

But ϕ has degree $2r$ so is not zero; further $2r = p - 1 < p$, so we get a contradiction. The argument may remind us of the Chevalley Theorem on the existence of solutions of $F(x_1, \ldots, x_n) = 0$, in which the auxiliary equation $1 - F(x_1, \ldots, x_n)^{p-1} = (1 - x_1^{p-1}) \cdots (1 - x_n^{p-1})$ is considered.

The case $Q(X) = -X^2 - 1$ is the first step in many proofs of the Lagrange Theorem on four squares.

To prove $N > 0$ for $y^2 = C(x)$ with cubic C and $p \geq 5$, we must differentiate. Write E for the analogous quantity

$$E = C^r. \tag{6.2}$$

Then

$$E' = rC^{r-1}C' = r\frac{C'}{C}E = -\frac{1}{2}\frac{C'}{C}E, \tag{6.3}$$

a differential equation which will play here just as fundamental a role as $(e^z)' = e^z$ in the complex exponential situation.

Thus $E(x) = -1$ as in (6.1) implies

$$E'(x) = \frac{1}{2}\frac{C'(x)}{C(x)}. \tag{6.4}$$

So E' too behaves like a fixed rational function (of course it is not so identically, any more than E is identically -1).

We now consider

$$\Phi = E + 1 + (X^p - X)f \tag{6.5}$$

for undetermined f in $\mathbb{F}_p(X)$. If $N = 0$, then $E(x) = -1$ for all x in $\mathbb{F}_p$ and so certainly also $\Phi(x) = 0$ (if some denominator doesn't vanish). We will choose f so that even the derivative $\Phi'(x) = 0$. To do this, note that

$$\Phi' = E' + (X^p - X)f' - f$$

and so by (6.4)

$$\Phi'(x) = \frac{1}{2}\frac{C'(x)}{C(x)} - f(x).$$

Thus we simply choose $f(X) = C'(X)/2C(X)$.

To avoid possible nuisance with denominators and their vanishing, we consider instead the polynomial

$$\phi = 2C\Phi = 2(E+1)C + (X^p - X)C'. \tag{6.6}$$

We see that $\phi(x) = \phi'(x) = 0$ for all x with $E(x) = -1$; that is, all x in $\mathbb{F}_p$ by our assumption $N = 0$.

Over $\mathbb{C}$, of course $\phi(x) = \phi'(x) = 0$ means that $\phi(X)$ is divisible by $(X-x)^2$. If this persists over $\mathbb{F}_p$, then we could conclude that our $\phi(X)$ is divisible by $\prod_{x\in\mathbb{F}_p}(X-x)^2 = (X^p - X)^2$. This time ϕ has degree at most

$$\max\{3r+3, p+2\} = 3r+3 = \frac{3}{2}(p+1) < 2p \tag{6.7}$$

because $p \geq 5$. But a slight snag: we have to rule out

$$\phi = 0. \tag{6.8}$$

Well, in (6.6) the first term $2(E+1)C$ dominates $(X^p - X)C'$ with respect to degree, so the degree is exactly $3r + 3$ in (6.7).

We have therefore shown that if $p \geq 5$ then $y^2 = C(x)$ always has a solution. Even this is not trivial and actually a special case of a theorem of F.K. Schmidt.

However we must check the above $(X - x)^2$ business. This is easy. From just $\phi(x) = 0$ we get $\phi(X) = (X - x)\phi_1(X)$ for $\phi_1(X)$ in $\mathbb{F}_p[X]$; and now $0 = \phi'(x) = \phi_1(x)$ shows that $\phi_1(X)$ is divisible by a second $X - x$. We will soon generalize this sort of thing.

Actually the comfortable inequality in (6.7) enables us to prove much more than just $N > 0$. Let $\mathcal{S}$ be the set of x in $\mathbb{F}_p$ with $E(x) = -1$, with cardinality s. We know already that $s < p$. But with ϕ as in (6.6), our calculations show that $\phi(x) = \phi'(x) = 0$ on $\mathcal{S}$. So $\phi(X)$ is divisible by $\prod_{x\in\mathcal{S}}(X - x)^2$. It follows that $2s$ is at most the degree $(3/2)(p + 1)$ of ϕ; that is

$$s \leq \frac{3}{4}p + \frac{3}{4} \tag{6.9}$$

asymptotically smaller than p.

This means that N is not too small. By Lemma 6.2 each x in $\mathbb{F}_p$ has $E(x) = -1, 0, +1$. We have just considered $E(x) = -1$; write now s_{-1} in place of s. It provides no solution of $y^2 = C(x)$. Then $E(x) = 0$ means $C(x) = 0$, which has say $s_0 \leq 3$ solutions. These give s_0 solutions $(x, 0)$ of

$y^2 = C(x)$. And the set of x with $E(x) = +1$, say with cardinality s_{+1}, are such that $C(x)$ is a non-zero square. This gives exactly two solutions $(x, \pm y)$ of $y^2 = C(x)$. It follows that

$$p = s_{-1} + s_0 + s_{+1}, \quad N = s_0 + 2s_{+1} = 2p - s_0 - 2s_{-1}. \tag{6.10}$$

As $s_{-1} \leq (3/4)(p+1)$, we get

$$N \geq \frac{1}{2}(p-3) - s_0 \geq \frac{1}{2}(p-9) \tag{6.11}$$

considerably better than $N > 0$.

It gets even better for quadratic Q and $y^2 = Q(x)$. We can use exactly the same (6.6) (with C replaced throughout by Q). Now it is not quite so easy to see that ϕ is not identically zero, but we end up with

$$N \geq p - c \tag{6.12}$$

for some small c independent of p and Q, provided Q is squarefree (see also Exercise 6.5).

This is almost best possible: first, equations like $y^2 = ax^2$ for non-square a have only $N = 1$ solution $(0, 0)$, and second, $y^2 = x^2 - 1$ has $N = p - 1$ solutions

$$\left(\frac{1}{2}\left(\frac{1}{u} + u\right), \frac{1}{2}\left(\frac{1}{u} - u\right)\right)$$

if $p \geq 3$ (found by substituting $x - y = u, x + y = v = 1/u$). We pause here to note that the same result holds for any equation

$$y^2 = x^2 - z \tag{6.13}$$

for any non-zero z in $\mathbb{F}_p$ (see Exercise 6.6).

To improve (6.11) further, we have to use higher derivatives. For example by differentiating (6.3) we obtain

$$E'' = \frac{3C'^2 - 2CC''}{4C^2}E. \tag{6.14}$$

So E'' also behaves like a fixed rational function on the set $\mathcal{S}$ where $E = -1$. Now we try

$$\Phi = E + 1 + (X^p - X)f + (X^p - X)^2 g$$

in place of (6.5) so as to make $\Phi(x) = \Phi'(x) = \Phi''(x) = 0$. After clearing denominators, we end up with (after paying attention to some peculiarly awkward signs)

$$\phi = 8(E+1)C^2 + 4(X^p - X)CC' + (X^p - X)^2(2CC'' - C'^2) \tag{6.15}$$

of degree at most

$$\max\{3r+6, p+5, 2p+4\}$$

which is $2p+4$ for large p. Now $\phi(x) = \phi'(x) = \phi''(x) = 0$ thus ϕ should be divisible by $\prod_{x\in\mathcal{S}}(X-x)^3$, so if there are no snags, we can hope to get

$$3s \le 2p+4, \quad s \le \frac{2}{3}p + \frac{4}{3}$$

asymptotically slightly better than (6.9).

Incidentally here arises a new problem: for $p = 2$ the polynomial $\phi = X^2$ has $\phi(0) = \phi'(0) = \phi''(0) = 0$ but is not divisible by X^3 (and even worse – all its derivatives vanish at $X = 0$). However this turns out to be not too serious.

The way forward may seem clear. We might choose some fixed L and try

$$\Phi = E + 1 + \sum_{i=1}^{L}(X^p - X)^i f_i$$

to get $\phi(x) = \phi'(x) = \cdots = \phi^{(L)}(x) = 0$, so ϕ should be divisible by the product $\prod_{x\in\mathcal{S}}(X-x)^{L+1}$ and we might end up with

$$(L+1)s \le Lp + c, \quad s \le \frac{L}{L+1}p + c,$$

presumably with $c = c(L)$ now depending on L. Unfortunately this is getting worse, not better; as L gets large the coefficient of p is approaching 1, not 1/2 as would be needed to deduce $N \ge (1-\epsilon)p$ in (6.10) like the bound (6.12) for quadratics.

The key is to double the number of terms so as to include $(E+1)(X^p - X)^i$ as well as $(X^p - X)^i$. Thus for example

$$\Phi = \sum_{i=0}^{L}(X^p - X)^i f_i + (E+1)\sum_{i=0}^{L}(X^p - X)^i g_i. \tag{6.16}$$

With suitable $f_0, \ldots, f_L, g_0, \ldots, g_L$ we can hope to get twice as many zeroes, perhaps of order at least $2L$. There would result $2Ls \le Lp + c(L)$, and then the coefficient of p in the upper bound for s would approach 1/2 as L gets large. So indeed we could hope for $N \ge (1-\epsilon)p$ for p sufficiently large in terms of ϵ; exactly how large would depend on $c(L)$.

Actually it is just as easy to get equally good upper bounds for N. For example if x in $\mathbb{F}_p$ occurs as a solution of $y^2 = C(x)$, then $E(x) = 0$ or 1. So s_{+1} would be large (recall $s_0 \le 3$), and we could carry out the proof with $E - 1$ instead of $E + 1$. This would lead to $N \le (1-\epsilon)p$ and so $|N - p| \le \epsilon p$.

Up to now the snags are two-fold: first, with the divisibility by high powers of $X-x$ and second, with showing that $\phi \neq 0$. The first can be dealt with quite easily as follows.

Lemma 6.3 *For an integer l with $1 \leq l \leq p$, a polynomial ϕ in $\mathbb{F}_p[X]$, and x in $\mathbb{F}_p$ suppose $\phi(x) = \phi'(x) = \cdots = \phi^{(l-1)}(x) = 0$. Then ϕ is divisible by $(X-x)^l$.*

Proof The Taylor development about $X = x$ gives a finite sum

$$\phi(X) = \sum_{n\geq 0} a_n(X-x)^n$$

with $\phi^{(n)}(x) = n!a_n$. For $n \leq l-1 < p$ we have $\phi^{(n)}(x) = 0$ and $n\neq 0$ in $\mathbb{F}_p$, so $a_n = 0$. Thus $\phi(X) = \sum_{n\geq l} a_n(X-x)^n$ is indeed divisible by $(X-x)^l$. □

The above example $\phi = X^2$ for $p = 2, l = 3, x = 0$ shows that some upper bound on l here is necessary, and it generalizes to $\phi = X^p$ for $l = p+1$ to show that the upper bound $l \leq p$ is best possible. In particular the differential conditions on ϕ are not quite equivalent to the existence of a zero at $X = x$ of order at least l. This is why we are wary of using our $\text{ord}_{X=x}$.

The second snag $\phi = 0$ is a little more complicated. It is convenient to modify the scheme (6.16) slightly in order to exploit better the pleasant differentiation properties of X^p. We also clear denominators in the usual way, after which the f_i, g_i become polynomials. In order to guarantee $\phi \neq 0$ their degree should not become too large, and in fact an upper bound $r-2 = (p-5)/2$ is convenient as long as $p > 3$. Actually we assume $p \geq 7$ for safety.

Lemma 6.4 *For any $L \geq 1$ in $\mathbb{Z}$ and any $A_0, \ldots, A_L, B_0, \ldots, B_L$ in $\mathbb{F}_p[X]$, not all zero and of degree at most $r-2$, the polynomial*

$$\phi = \sum_{i=0}^{L} (A_i + EB_i)X^{ip}$$

is not identically zero.

Proof If this is false, then we have $A + EB = 0$ for $A = \sum_{i=0}^{L} A_i X^{ip}, B = \sum_{i=0}^{L} B_i X^{ip}$. Again we exploit the differential equation (6.3). Differentiating and multiplying by $2C$ we deduce also $2CA' + E(2CB' - C'B) = 0$. Eliminating E we find

$$2CAB' - C'AB - 2CA'B = 0. \tag{6.17}$$

Now if all $A_i = 0$ then $\sum_{i=0}^{L} B_i X^{ip} = 0$, and knowing just that the B_i have degree strictly less than p would suffice to imply that all the $B_i = 0$, a contradiction. So there is some $A_k \neq 0$ with k maximal. Similarly there is some

$B_h \neq 0$ with h maximal. Substituting $A = \sum_{i=0}^{k} A_i X^{ip}, B = \sum_{i=0}^{h} B_i X^{ip}$ into (6.17) gives

$$(2CA_kB_h' - C'A_kB_h - 2CA_k'B_h)X^{(k+h)p} = \sum_{l=0}^{k+h-1} D_l X^{lp}. \tag{6.18}$$

Here the D_l have degrees at most

$$\max\{3+(r-2)+(r-3), 2+(r-2)+(r-2), 3+(r-3)+(r-2)\} = p-3$$

and so the right-hand side of (6.18) has degree at most $(k+h)p-3$. It follows that the left-hand side must vanish, and so

$$2CA_kB_h' - C'A_kB_h - 2CA_k'B_h = 0.$$

Now if A_k has degree $a \leq r-2$ and B_h has degree $b \leq r-2$, we find that the coefficient of X^{a+b+2} above is a non-zero multiple of $2b-3-2a$. Hence $a-b = -3/2 = r-1$ in $\mathbb{F}_p$. But this is impossible since in $\mathbb{Z}$ we have $-(r-2) \leq a-b \leq r-2$. □

We are now ready to begin the proof of Theorem 6.1 with the help of a suitable auxiliary polynomial, this time (at least informally) having large order zeroes at several points. In line with the notation $s_{\pm 1}$, we use $\mathcal{S}_\sigma$ for the set of x in $\mathbb{F}_p$ with $E(x) = \sigma$.

Lemma 6.5 *Fix $\sigma = \pm 1$. Then for any $L \geq 1$ in $\mathbb{Z}$ there are $A_0, \ldots, A_L, B_0, \ldots,$ B_L in $\mathbb{F}_p[X]$, not all zero and of degree at most $r-2$, such that for*

$$\phi = \sum_{i=0}^{L} (A_i + EB_i)X^{ip} \tag{6.19}$$

we have $\phi(x) = \phi'(x) = \cdots = \phi^{(2L-1)}(x) = 0$ for all x in $\mathcal{S}_\sigma$ provided $6L^2 - 2L + 3 < p$.

Proof We will have to differentiate j times (for $0 \leq j < 2L$). Glancing at (6.3) and (6.14), we see that it is plausible to expect a denominator C^j. Accordingly we write

$$C^j \phi^{(j)} = \sum_{i=0}^{L} (A_{ij} + EB_{ij})X^{ip} \tag{6.20}$$

for the effect of formally differentiating (6.19) j times.

We claim that the A_{ij}, B_{ij} are in $\mathbb{F}_p[X]$ with degree at most $r-2+2j$.

This is trivial for $j = 0$. To prove it for all j we use induction, somewhat as in the proof of Lemma 4.4. Thus dividing (6.20) by C^j, replacing j with $j-1 \geq 0$ and differentiating using (6.3), we end up with

$$A_{ij} = -(j-1)C'A_{i,j-1} + CA'_{i,j-1},$$

$$B_{ij} = -(j-1-r)C'B_{i,j-1} + CB'_{i,j-1}.$$

Thus if $B_{i,j-1}$ has degree at most $d = r - 2 + 2(j-1)$, then B_{ij} has degree at most

$$\max\{2+d, 3+d-1\} = r-2+2j;$$

and similarly for A_{ij}. This proves the above claim.

Now we want the equations $\phi^{(j)}(x) = 0$ $(j = 0, \ldots, 2L-1)$ for all x in $\mathcal{S}_\sigma$. As then $C(x) \neq 0$ this follows from

$$\sum_{i=0}^{L}(A_{ij}(x) + \sigma B_{ij}(x))x^{ip} = 0 \quad (j = 0, \ldots, 2L-1)$$

and that in turn from

$$\sum_{i=0}^{L}(A_{ij}(x) + \sigma B_{ij}(x))x^{i} = 0 \quad (j = 0, \ldots, 2L-1).$$

In fact we can even afford

$$\sum_{i=0}^{L}(A_{ij}(X) + \sigma B_{ij}(X))X^{i} = 0 \quad (j = 0, \ldots, 2L-1) \tag{6.21}$$

identically in X. Why is this?

Well, in the original A_i, B_i in (6.19) we have

$$R = 2(L+1)(r-1) = Lp + p - 3L - 3$$

unknowns. The coefficients in (6.21) are linear forms in these over $\mathbb{F}_p$, and (6.21) for each j gives $r - 2 + 2j + L + 1$ homogeneous linear equations. So in all, we have

$$S = \sum_{j=0}^{2L-1}(r-2+2j+L+1) = 2L(r-2) + 6L^2 = Lp - 5L + 6L^2$$

equations. So $R > S$ provided $p > 6L^2 - 2L + 3$. This completes the proof. □

Next, by Lemma 6.3 and Lemma 6.5 we see that ϕ is divisible by the product $\prod_{x \in \mathcal{S}_\sigma}(X - x)^{2L}$ provided $2L \leq p$. So for the cardinality s_σ of $\mathcal{S}_\sigma$ we deduce from Lemma 6.4 that $2Ls_\sigma$ is at most the degree of ϕ, which is at most $r - 2 + 3r + pL = 2p - 4 + pL$. So

$$s_\sigma \leq \frac{p}{2} + \frac{p-2}{L}. \tag{6.22}$$

It follows from (6.10) that

$$N \le s_0 + 2\left(\frac{p}{2} + \frac{p-2}{L}\right) \le p + 3 + \frac{2p-4}{L}$$

as well as

$$N \ge 2p - s_0 - 2\left(\frac{p}{2} + \frac{p-2}{L}\right) \ge p - 3 - \frac{2p-4}{L};$$

so in combination

$$|N - p| \le 3 + \frac{2p-4}{L}. \tag{6.23}$$

We have accumulated the side conditions

$$p \ge 7, \ \ L \ge 1, \ \ 6L^2 - 2L + 3 < p, \ \ 2L \le p$$

for L. So for large p we could take L asymptotically about $\sqrt{p/6}$, giving about $2\sqrt{6}\sqrt{p}$ instead of $8\sqrt{p}$ in Theorem 6.1.

To get the full result is just light drudgery. We can assume $p \ge 67$, else as already remarked the trivial bound $|N - p| \le p$ suffices. Then we choose $L = [\sqrt{p/6}] \ge 2$ for the integer part. And

$$p \ge 6L^2 > 6L^2 - 2L + 3 > 4L$$

settles the side conditions. Finally $\sqrt{p} > \alpha/\beta = 7.28\ldots$ for $\alpha = (3/\sqrt{6}) + 8$, $\beta = (8/\sqrt{6}) - 2$, and now the inequality $\alpha\sqrt{p} - 7 < \beta p$ reduces just to

$$3 + \frac{2p-4}{\sqrt{p/6} - 1} < 8\sqrt{p},$$

which by (6.23) gives us what we want. This completes the proof of Theorem 6.1.

One interesting way of obtaining the better error bound $2\sqrt{p}$ (well-known to be essentially best possible) is to extend Theorem 6.1 to counting points over finite fields $\mathbb{F}_q$ with general $q = p^e$. One can prove the analogous bound

$$|N_e - q| \le 8\sqrt{q} \tag{6.24}$$

for the number N_e of points. The only obstacle is that the condition $2L \le p$ arising from Lemma 6.3 generally no longer holds when L is about $\sqrt{q}$. The solution is to consider "hyperderivatives" or "divided derivatives" or "Hasse derivatives" or "Hasse–Schmidt derivatives". In zero characteristic these amount to replacing $(\mathrm{d}/\mathrm{d}X)^n$ by $(\mathrm{d}/\mathrm{d}X)^n/n!$; then the condition $l \le p$ disappears in the analogue of Lemma 6.3 (see Exercise 6.10). Surprisingly such hyperderivatives are vital in zero characteristic as we will see in Chapters 9, 12, 17 and (indirectly) the Appendix.

Now, at least for elliptic curves, it is known (and is comparatively elementary) that there exist complex numbers μ, ν such that

$$N_e - q = -\mu^e - \nu^e. \tag{6.25}$$

From (6.24) as $e \to \infty$ it is easy to deduce $|\mu|, |\nu| \leq \sqrt{p}$. Then (6.25) for $e = 1$ gives $|N - p| \leq 2\sqrt{p}$.

These bounds can also be formulated in terms of characters. The Legendre symbol $(\frac{\cdot}{p})$ is a real character χ on $\mathbb{F}_p$ so that $\chi(w) = w^{(p-1)/2}$ (in $\mathbb{F}_p$) just as in Lemma 6.2. Thus $y^2 = w$ has $1 + \chi(w)$ solutions, and it follows at once that

$$N - p = \sum_{x \in \mathbb{F}_p} \chi(C(x)).$$

Thus for example

$$\left| \sum_{x \in \mathbb{F}_p} \chi(C(x)) \right| \leq 2\sqrt{p}. \tag{6.26}$$

This sort of phenomenon will be illustrated again in the next chapter.

Finally we should point out that Stepanov's techniques (developed in a series of papers from 1969 to 1974) are by no means restricted to $y^2 = C(x)$ for cubic C. Without much extra difficulty one can treat polynomials of general degree. And examples like (6.6), (6.15) remain useful, even though their usefulness decreases as the degree increases. Still, with (6.15) one can prove that the equation $y^2 = Q(x)$ for any quintic Q over $\mathbb{F}_p$ has a solution in $\mathbb{F}_p^2$ as long as $p \geq 17$ (see Exercise 6.1). The geometers will recognize this as a curve whose genus is probably two. We may note that the equation

$$y^2 = x^5 + 3x^4 + x^3 + 6x^2 + 6$$

over $\mathbb{F}_7$ has no solution in $\mathbb{F}_7^2$. I don't know any similar examples for $p = 11, 13$.

And then one can treat arbitrary algebraic curves $F(x, y) = 0$; an excellent account (with references) is given by Schmidt (1976). For more examples see Exercises 6.8 and 6.9. Bombieri gave a geometric version of the method. The Riemann Hypothesis as proved by Weil yields an error term $2g\sqrt{p}$ where g is the genus; but sometimes the method of auxiliary polynomials leads to improvements (see for example Exercise 7.15(d) for Fermat's $x^m + y^m = 1$, with $g = (m-1)(m-2)/2$ uncomfortably large). For more details, see the article Zannier (1998) and the notes to chapter 2 of Zannier (2012), as well as Corvaja and Zannier (2013). See also Stöhr and Voloch (1986) for another geometric method. And one can proceed to the following chapter for an apparently different sort of application of Stepanov's Method.

Exercises

6.1 Let $p \geq 17$ be a prime, and let $Q(X)$ be in $\mathbb{F}_p[X]$ of degree 5. Show that there are x, y in $\mathbb{F}_p$ with $y^2 = Q(x)$ [*Hint*: $P = 8(E+1)Q^2 + 4FQQ' + F^2(2QQ'' - Q'^2)$ with $F = X^p - X$, $E = Q^r$, $r = (p-1)/2$ and P, P', P''. Careful with $P = 0$!].

6.2 Let $p \geq 17$ be a prime, and let $Q(X) = aX^4 + bX^3 + cX^2 + dX + e$ be in $\mathbb{F}_p[X]$ with $a \neq 0$.

(a) When a is a square in $\mathbb{F}_p$ show that there are x, y in $\mathbb{F}_p$ with $y^2 = Q(x)$. [*Hint*: coefficient of X^{4r+8}].

(b) When a is not a square in $\mathbb{F}_p$ show that there are no x, y in $\mathbb{F}_p$ with $y^2 = a(x^2 - a)^2$.

6.3 Let $p \geq 3$ be a prime, and let $Q(X) = aX^2 + bX + c$ be in $\mathbb{F}_p[X]$ with $0 \neq a$ a square. Show that there are at least $p - 3$ solutions (x, y) in $\mathbb{F}_p^2$ of $y^2 = Q(x)$.

****

6.4 In Exercise 6.3 show that there are at least $p - 1$ solutions (x, y) in $\mathbb{F}_p^2$ of $y^2 = Q(x)$.

6.5 Let $p \geq 3$ be a prime and let $Q(X)$ be squarefree in $\mathbb{F}_p[X]$ of degree 2. Show that there are at least $p - 1$ solutions (x, y) in $\mathbb{F}_p^2$ of $y^2 = Q(x)$.

6.6 Show that for any $z \neq 0$ in $\mathbb{F}_p$ there are exactly $p - 1$ solutions (x, y) in $\mathbb{F}_p^2$ of $y^2 = x^2 - z$.

6.7 Let $p \geq 3$ be a prime and let $Q(X)$ be in $\mathbb{F}_p[X]$ of degree 2. Without using auxiliary polynomials determine the exact number of solutions (x, y) in $\mathbb{F}_p^2$ of $y^2 = Q(x)$.

6.8 Show that there is k with the following property. Let $p \geq k$ be a prime and let $Q(X)$ be in $\mathbb{F}_p[X]$ of degree 4, not of the shape $a(D(X))^2$ ($a \in \mathbb{F}_p$, $D(X) \in \mathbb{F}_p[X]$). Then there are x, y in $\mathbb{F}_p$ with $y^2 = Q(x)$ (according to Porcaro's 1998 Master Thesis $k = 17$ suffices).

6.9 Show that there is k with the following property. Let $p \geq k$ be a prime and let $C(X)$ be in $\mathbb{F}_p[X]$ of degree 3, not of the shape $a(D(X))^3$ ($a \in \mathbb{F}_p$, $D(X) \in \mathbb{F}_p[X]$). Then there are x, y in $\mathbb{F}_p$ with $y^3 = C(x)$ (according to Porcaro's 1998 Master Thesis $k = 47$ suffices).

6.10 Let $\mathcal{K}$ be a field. Define $\mathcal{K}$-linear maps δ_i from $\mathcal{K}[X]$ to $\mathcal{K}[X]$ ($i = 0, 1, 2, \ldots$) by $\delta_i X^j = \binom{j}{i} X^{j-i}$ ($j \geq i$) with binomial coefficients and otherwise

$\delta_i X^j = 0$. Let P be in $\mathcal{K}[X]$, let x be in $\mathcal{K}$, and let $l \geq 1$. Show that $\delta_i P(x) = 0$ $(i = 0, 1, \ldots, l-1)$ if and only if $(X - x)^l$ divides P (compare Lemma 6.3).

6.11 Let $q = p^e$, let $C(X)$ be in $\mathbb{F}_p[X]$ of degree 3 and let N be the cardinality $\#\{(x, y) \in \mathbb{F}_q^2;\ y^2 = C(x)\}$. Show that $|N - q| \leq \gamma\sqrt{q}$ with γ independent of p, q, C.

6.12 Let $E/\mathbb{F}_p$ be an elliptic curve. It is known that there are α, β in $\mathbb{C}$ such that $\#E(\mathbb{F}_q) = q + 1 - \alpha^e - \beta^e$ for all $q = p^e$. Show that $|\alpha|, |\beta| \leq \sqrt{p}$. Calculate α, β for $y^2 = x^3 - 2$ and $p = 5$.

6.13 Is there $Q(X)$ in $\mathbb{F}_{11}[X]$ of degree 5 such that there are no x, y in $\mathbb{F}_{11}$ with $y^2 = Q(x)$? I don't know.

6.14 With δ_i as in Exercise 6.10 show that $\delta_i(PQ) = \sum_{a+b=i} \delta_a(P)\delta_b(Q)$.

6.15 With δ_i as in Exercise 6.10 show that

$$\delta_i(P^n) = P^{n-i} \sum_{\mathbf{a}} C(\mathbf{a}) \delta_0(P)^{a_0} \delta_1(P)^{a_1} \cdots \delta_i(P)^{a_i}$$

where the sum is taken over all $\mathbf{a} = (a_0, a_1, \ldots, a_i)$ with non-negative coordinates satisfying

$$a_0 + a_1 + \cdots + a_i = i, \quad a_1 + \cdots + i a_i = i$$

and the $C(\mathbf{a})$ are non-negative integers. Some version for undivided derivatives is attributed to the Blessed Francesco Faà di Bruno (who even has $\phi(P)$ instead of P^n).

7
Exponential sums

The best-known of these is due to Gauss, namely

$$G = \sum_{k=1}^{p} \exp\left(\frac{2\pi i k^2}{p}\right) = \sum_{x \in \mathbb{F}_p} \exp\left(\frac{2\pi i x^2}{p}\right) \tag{7.1}$$

with a prime $p \neq 2$, for which it is fairly easy to show

$$|G| = \sqrt{p}. \tag{7.2}$$

This resembles (6.26) in that we have a sum of p roots of unity, for which the trivial estimate p is greatly improved. As the proof of (7.2) will be imitated soon, we give it here.

In fact with the complex conjugate $\overline{G}$ we have

$$|G|^2 = G\overline{G} = \sum_{x \in \mathbb{F}_p} \exp\left(\frac{2\pi i x^2}{p}\right) \sum_{y \in \mathbb{F}_p} \exp\left(-\frac{2\pi i y^2}{p}\right),$$

which is

$$\sum_{z \in \mathbb{F}_p} N_z \exp\left(\frac{2\pi i z}{p}\right),$$

where N_z is the number of solutions of $x^2 - y^2 = z$ in $\mathbb{F}_p^2$. By the discussion around (6.13) with $x - y = u, x + y = v$ we know that $N_z = p - 1$ for $z \neq 0$ (see also Exercise 6.6); and clearly $N_0 = 2p - 1$. Thus

$$|G|^2 \;=\; 2p - 1 + (p-1) \sum_{k=1}^{p-1} \exp\left(\frac{2\pi i k}{p}\right) \;=\; 2p - 1 - (p-1) \;=\; p.$$

Similar things are known for sums of the form $\sum_{k=1}^{p} \exp(2\pi i f(k)/p)$, where f is a polynomial or even a rational function. In the middle of the last century, Heilbronn introduced a rather different kind of sum

$$H = \sum_{k=1}^{p} \exp\left(\frac{2\pi i k^p}{p^2}\right),$$

for which no non-trivial upper bound was known until 1996, when Heath-Brown showed that it has order of magnitude at most $p^{11/12}$. His remarkable proof (parts of which were anticipated by Mitkin) used several elements of Stepanov's Method. Here we prove the following explicit estimate.

Theorem 7.1 *We have* $|H| \le 4p^{11/12}$.

In fact not even the switch to $\mathbb{F}_p$ as in (7.1) is obvious. We should remark that if k, k' are congruent modulo p then $k' = k + ap$ for a in $\mathbb{Z}$, and so

$$k'^p = (k+ap)^p = k^p + \binom{p}{1} k^{p-1}(ap) + E$$

with E divisible by p^2. Thus k^p, k'^p are congruent modulo p^2 and indeed it makes sense to write

$$H = \sum_{x \in \mathbb{F}_p} \exp\left(\frac{2\pi i x^p}{p^2}\right).$$

We can see at least a superficial resemblance to elements of the previous chapter by calculating $|H|^2$ as for G above. Only now we put $x - y = u, x = uv$ for $x \neq y$. Then

$$x^p - y^p = u^p(v^p - (v-1)^p).$$

To evaluate this modulo p^2, we note that for each v in $\mathbb{F}_p$ there is a unique $D(v)$ in $\mathbb{F}_p$ with

$$v^p - (v-1)^p \equiv 1 - pD(v) \qquad \text{mod } p^2.$$

This holds even when $p = 2$, but for simplicity we continue to assume $p \neq 2$. Taking $x = y$ then $x \neq y$ we get

$$|H|^2 = p + \sum_{z \in \mathbb{F}_p} s_z H^*(1 - pz), \tag{7.3}$$

where s_z is the number of solutions of $D(v) = z$ and

$$H^*(w) = \sum_{0 \neq u \in \mathbb{F}_p} \exp\left(\frac{2\pi i w u^p}{p^2}\right).$$

This equation $D(v) = z$ will be analogous to the $E(x) = \sigma$ of the previous chapter. For D like E is a polynomial. In fact

$$X^p - (X-1)^p - 1 = -\sum_{l=1}^{p-1} (-1)^{p-l} \binom{p}{l} X^l$$

in $\mathbb{Z}[X]$ and

$$\binom{p}{l} = \frac{p}{l}\left(\frac{p-1}{1}\right)\cdots\left(\frac{p-l+1}{l-1}\right) \equiv (-1)^{l-1}\frac{p}{l} \qquad \bmod p^2.$$

We find indeed the polynomial

$$D(X) = \sum_{l=1}^{p-1} \frac{X^l}{l} = X + \frac{X^2}{2} + \cdots + \frac{X^{p-1}}{p-1} \tag{7.4}$$

in $\mathbb{F}_p[X]$. What is more, D like E has a differential equation

$$D'(X) = \sum_{l=1}^{p-1} X^{l-1} = \frac{X^p - X}{X(X-1)} \tag{7.5}$$

even simpler than (6.3) in the sense that D' behaves like the zero function on all of $\mathbb{F}_p$ except at 0 and 1.

Thus we may hope to get bounds for the s_z in (7.3) as we did for the s_σ in the previous chapter.

In fact upper bounds will suffice as in (6.22) there. For Cauchy–Schwarz on (7.3) gives $|H|^2 \le p + \sqrt{W}$, where

$$W = \sum_{z\in\mathbb{F}_p} s_z^2 \sum_{z\in\mathbb{F}_p} |H^*(1-pz)|^2.$$

Here

$$|H^*(1-pz)|^2 = \sum_{0\neq u\in\mathbb{F}_p}\sum_{0\neq t\in\mathbb{F}_p} \exp\left(\frac{2\pi i(u^p - t^p)}{p^2}\right)\exp\left(-\frac{2\pi iz(u^p - t^p)}{p}\right).$$

Now summing the far right factor over z gives zero unless $u^p = t^p$, which in $\mathbb{F}_p$ is $u = t$. It follows that

$$\sum_{z\in\mathbb{F}_p} |H^*(1-pz)|^2 = p(p-1).$$

Also

$$\sum_{z\in\mathbb{F}_p} s_z^2 \le (\max_{z\in\mathbb{F}_p} s_z)\sum_{z\in\mathbb{F}_p} s_z = p\max_{z\in\mathbb{F}_p} s_z.$$

We end up with

$$|H| \le 2p^{3/4}(\max_{z\in\mathbb{F}_p} s_z)^{1/4}. \tag{7.6}$$

Thus indeed we need only upper bounds for s_z. The trivial bound is p, and this gives something slightly worse than the trivial $|H| \le p$.

And we see now from (7.6) that Theorem 7.1 is an immediate consequence of the following result.

Proposition 7.2 *For any z in $\mathbb{F}_p$ we have $s_z \leq 9p^{2/3}$.*

The analogous estimate for s_σ in the previous chapter was obtained using an auxiliary polynomial in X, E, X^p as in Lemma 6.5. There the degree in E was restricted to 1; in fact $CE^2 = C^p$ is simply a cubic polynomial in X^p and would have brought nothing new. We have D in place of E, and the right-hand side of (7.4) strongly resembles the Taylor expansion of $-\log(1 - X)$ about $X = 0$; in fact one stops just before the term X^p/p, which makes no sense in characteristic p. Now over $\mathbb{C}$ all the powers of $-\log(1 - X)$ are linearly independent over $\mathbb{C}(X)$, which amounts to an algebraic independence result as in Lemma 3.2. This suggests that we may be able to use higher powers of D in our auxiliary polynomial. We don't have a blank cheque, because of course $X, D(X)$ are certainly not algebraically independent over $\mathbb{F}_p$.

Here is an analogue of Lemma 6.4.

Lemma 7.3 *Suppose $P(X, Y, Z)$ in $\mathbb{F}_p[X, Y, Z]$ is not identically zero, of degree at most L in X and of degree at most M in Y. Then if*

$$L < p, \quad (2L + 1)M < p, \tag{7.7}$$

the polynomial

$$\phi = P(X, D(X), X^p)$$

is not identically zero.

Proof We remark that this is slightly weaker than the original result of Heath-Brown, where (7.7) is replaced by $(L+1)(M+1) \leq p$ in our notation, which as he remarked is best possible in some sense. Our result still suffices for Theorem 7.1; moreover our proof is along lines already laid out in Chapter 5, at least if a certain resultant does not vanish. See also similar uses of the resultant in Proposition 8.1 and the discussion in the last part of Chapter 22.

We use the field $C = \mathbb{F}_p(X^p)$. Clearly $\gamma' = 0$ for any γ in C; and it is easy to prove the converse. For example writing $\gamma = f/g$ for coprime f, g in $\mathbb{F}_p[X]$, we deduce $f'g = fg'$ from $\gamma' = 0$ and so f divides f', which by degree considerations forces $f' = 0$. If $f = \sum_i a_i X^i$ we see that $a_i = 0$ for all i not divisible by p, so f is in $\mathbb{F}_p[X^p]$; similarly for g, and so γ is in C. We also need to know that X has degree p over C; but this is easy because of $(T - X)^p = T^p - X^p$ in $C[T]$, which is irreducible in $C[T]$ because $(T - X)^r = T^r - rXT^{r-1} + \cdots$ is not in $C[T]$ for $r = 1, \ldots, p - 1$.

It suffices to deduce a contradiction from

$$F(X, D(X)) = 0 \tag{7.8}$$

with non-zero $F(U,V)$ in $C[U,V]$ having degree at most L in U and degree at most M in V, also with (7.7). We have used new variables here for clarity later. Here we are entitled to assume that the counterexample has been chosen with the degree in V minimal, say m for $0 \leq m \leq M$.

Now $m = 0$ is impossible by $L < p = [C(X) : C]$.

Similarly we may assume that F has no factor in $C[U]$ of positive degree.

This implies that F is irreducible in $C[U,V]$, for any factor would have smaller degree in V and also satisfy the analogue of (7.7).

Differentiating (7.8) and using (7.5) gives

$$G(X, D(X)) = 0 \tag{7.9}$$

with

$$G(U,V) \;=\; U(U-1)F_U(U,V) + (X^p - U)F_V(U,V) \tag{7.10}$$

and the partial derivatives F_U, F_V; here we recall that the coefficients of F can be regarded as constants for this step. These equations would of course also be valid with U^p instead of X^p; however, our X^p can be smuggled into the coefficient field C, so that the degree of G in U is at most $L+1$ not $L+p$.

We would like to take the resultant of $F(U,V), G(U,V)$ with respect to V; that is, considered as polynomials in V with coefficients in $\mathcal{R} = C[U]$. Note that they both have degrees at most $m \geq 1$ in V.

Thus the resultant K exists, in $C[U]$. If it is non-zero, we find that its degree in U is at most

$$r_U s_V + r_V s_U \;\leq\; LM + (L+1)M \;=\; (2L+1)M \;<\; p$$

with r_U, r_V as the partial degrees of F and s_U, s_V the partial degrees of G (for example take a look at (5.3) or consult Exercise 5.22). In this case we are home at once. For (5.4) supplies Q, R in $C[U,V]$, not both zero and of degrees at most $m-1$ in V, with

$$Q(U,V)F(U,V) + R(U,V)G(U,V) = K(U). \tag{7.11}$$

This together with (7.8) and (7.9) implies $K(X) = 0$. But X has degree p over C.

The fun begins if the resultant is zero. Then (7.11) implies that F divides RG in $C[U,V]$. By degrees F does not divide R unless $R = 0$; but then (7.11) would imply $Q = 0$. As F is irreducible, it must divide G. By (7.10) we have $s_V \leq m = r_V$ and $s_U \leq 1 + r_U$. It follows that there are α, β in C such that

$$G(U,V) = (\alpha U + \beta)F(U,V). \tag{7.12}$$

We note now that $F(X, D(X)+1) \neq 0$. Otherwise, considering the difference $F(U, V+1) - F(U, V)$ and the minimality of m would imply that $F(U, V+1) = F(U, V)$. But it is easy to see that this implies $p \leq m \leq M$ contradicting (7.7). For example iteration would show that $F(U, V+T) - F(U, V)$ vanishes at $T = t$ for all t in $\mathbb{F}_p$, so all its coefficients in $C[T][U, V]$ must be divisible by $T^p - T$, forcing $m \geq p$.

The next step is odd. Pick any γ in C, and consider

$$\psi(X) = \psi_\gamma(X) = F(X, D(X) + \gamma). \tag{7.13}$$

As $D(X), D(X) + \gamma$ satisfy the same differential equation (7.5) we see that

$$X(X-1)\psi'(X) = G(X, D(X) + \gamma),$$

which by (7.12) is

$$(\alpha X + \beta)F(X, D(X) + \gamma) = (\alpha X + \beta)\psi(X).$$

So we get the strange-looking differential equation

$$\frac{\psi'(X)}{\psi(X)} = \frac{\alpha X + \beta}{X(X-1)}; \tag{7.14}$$

even stranger, the right-hand side is independent of the choice of γ in (7.13)! Provided $\psi \neq 0$, of course; and clearly this holds for sufficiently general γ and certainly infinitely many. In fact we don't use the special form of the numerator in (7.14), but it's nice to look at.

In particular, we noted above that $\psi \neq 0$ holds with $\gamma = 1$. Thus for the quotient

$$\chi(X) = \frac{F(X, D(X) + \gamma)}{F(X, D(X) + 1)}$$

we get $\chi'(X)/\chi(X) = 0$. Consequently $\chi(X) = \delta$ lies in C.

This is getting odder and odder. We have shown that for each of infinitely many γ in C there is δ in C such that

$$F(X, D(X) + \gamma) - \delta F(X, D(X) + 1) = 0.$$

If we introduce the polynomial

$$F(U, V + \gamma) - \delta F(U, V + 1) - (1 - \delta)F(U, V),$$

we see that it vanishes at $(X, D(X))$ but has degree in V smaller than that of F. So by minimality it must vanish identically in $C[U, V]$. We can now find the mysterious δ by looking at the second highest power of V in

$$F(U, V + \gamma) - F(U, V) = \delta(F(U, V + 1) - F(U, V));$$

as $m \leq M < p$ we find simply $\delta = \gamma$.

Thus for infinitely many γ in C we have

$$F(U,V+\gamma) \;=\; \gamma F(U,V+1)+(1-\gamma)F(U,V).$$

This implies that

$$F(U,V+T) \;=\; TF(U,V+1)+(1-T)F(U,V)$$

in $C[U,V,T]$, which in turn clearly implies that $m \le 1$ (e.g. put $V=0$).

We can therefore write $F(U,V) = A + BV$ for A,B in $C[U]$, both clearly non-zero. Now the vanishing of the resultant leads to

$$U(U-1)(AB'-A'B) = (X^p - U)B^2.$$

Let A,B have degrees a,b respectively. If $a \neq b$, then the leading term on the left-hand side is a multiple of $(b-a)U^{a+b+1}$ by something non-zero in C, while the leading term on the right-hand side is a multiple of U^{2b+1} by something non-zero in C. Hence $a=b$, and so this holds unconditionally. But now the left-hand side has degree at most $2b$ in U while the right-hand side still has degree $2b+1$, a contradiction. □

In fact we are now ready to begin the proof of Proposition 7.2 with the usual auxiliary polynomial. We use $\mathcal{S}_z$ for the set of x in $\mathbb{F}_p$ with $D(x)=z$; however it is convenient to remove 0, 1 from these giving $\mathcal{S}_z^*$.

Lemma 7.4 *Fix any z in $\mathbb{F}_p$. Then for any $L,M,N \ge 0$ in $\mathbb{Z}$ and $J \ge 1$ in $\mathbb{Z}$ there is $P(X,Y,Z)$ in $\mathbb{F}_p[X,Y,Z]$, not identically zero, of degree at most L in X, of degree at most M in Y and of degree at most N in Z, such that for*

$$\phi = P(X,D(X),X^p) \tag{7.15}$$

we have $\phi(x)=\phi'(x)=\cdots=\phi^{(J-1)}(x)=0$ for all x in $\mathcal{S}_z^*$ provided

$$(L+1)(M+1)(N+1) > J(L+N+J). \tag{7.16}$$

Proof We follow rather closely the technique of the proof of the analogous Lemma 6.5. We will have to differentiate j times (for $j=0,\ldots,J-1$). Glancing at (7.5), we see that it is plausible to expect $X^j(X-1)^j$ arising as a denominator. Accordingly we write

$$X^j(X-1)^j\phi^{(j)} = P_j(X,D(X),X^p) \tag{7.17}$$

for the effect of formally differentiating (7.15) j times.

We claim that the P_j are in $\mathbb{F}_p[X,Y,Z]$ with degrees at most $L+j, M, N+j$ in X,Y,Z respectively.

This is trivial for $j = 0$. To prove it for all j we use induction, again somewhat as in the proof of Lemma 4.4. Thus dividing (7.17) by $X^j(X-1)^j$, replacing j with $j-1 \geq 0$ and differentiating using (7.5), we end up with

$$P_j = X(X-1)\frac{\partial P_{j-1}}{\partial X} + (Z-X)\frac{\partial P_{j-1}}{\partial Y} - (j-1)(2X-1)P_{j-1}.$$

Thus the degrees in X and Z increase by at most 1 with each differentiation, and the degree in Y doesn't increase. The above claim is immediate.

Now we want the equations $\phi^{(j)}(x) = 0$ $(j = 0, \ldots, J-1)$ for all x in $\mathcal{S}_z^*$. As then $x \neq 0, 1$ this follows from

$$P_j(x, z, x^p) = 0 \quad (j = 0, \ldots, J-1)$$

and that in turn from

$$P_j(x, z, x) = 0 \quad (j = 0, \ldots, J-1).$$

In fact we can even afford

$$P_j(X, z, X) = 0 \quad (j = 0, \ldots, J-1) \tag{7.18}$$

identically in X. Why is this?

Well, in the original P in (7.15) we have

$$R = (L+1)(M+1)(N+1)$$

unknowns. The coefficients in (7.18) are linear forms in these over $\mathbb{F}_p$, and (7.18) for each j gives $(L+j)+(N+j)+1$ homogeneous linear equations. So in all we have

$$S = \sum_{j=0}^{J-1}(L+N+2j+1) = LJ + NJ + J^2$$

equations. And now $R > S$ gives the condition of the lemma. □

The results above, together with Lemma 6.3, show that the cardinality of $\mathcal{S}_z^*$ multiplied by the multiplicity J is bounded above by the degree of ϕ, provided $J \leq p$. Adding 2 for $x = 0, 1$ we deduce

$$s_z \leq 2 + \frac{L + (p-1)M + pN}{J} \tag{7.19}$$

subject also to (7.16) and (7.7).

The optimal choices of L, M, N, J as functions of p are not so clear as that for L in the preceding chapter. But a bit of playing around (for example the left-hand side of (7.16) could be thought of as LMN and thus this must be compared separately with JL, JN, J^2) shows that L, J are of order $p^{2/3}$ and

M, N of order $p^{1/3}$. Going further, we find that with L, M, N, J asymptotically $\lambda p^{2/3}, \mu p^{1/3}, \nu p^{1/3}, \kappa p^{2/3}$ and $\lambda = \theta^{-2}, \mu = \theta^{-1}, \nu = \theta^2, \kappa = \theta^{-2}$, we get an asymptotic bound of $3\theta p^{2/3}$ for s_z, where $\theta = 2^{1/3}$. This leads via (7.6) to the asymptotic bound of $cp^{11/12}$ for $|H|$, where

$$c = 2(3\theta)^{1/4} = 2.7886\ldots.$$

For Proposition 7.2 we need more light drudgery. We use

$$L = \left[\frac{1}{2}p^{2/3}\right] - 1, \quad M = N = [p^{1/3}], \quad J = \left[\frac{1}{4}p^{2/3}\right] \leq p$$

with $L \geq 1$ and $J \geq 1$ if $p \geq 11$, for which (7.7) is evident. As for (7.16), we divide it into three equal parts.

First

$$(L+1)(M+1)(N+1) \;>\; Lp^{2/3} \;>\; 3JL;$$

and *a fortiori* for $3JN$ because

$$L - N \;>\; \frac{1}{2}p^{2/3} - 2 - p^{1/3} \;=\; \frac{1}{2}(p^{1/3} - 1)^2 - \frac{5}{2} \;>\; 0$$

if $p \geq 37$. And in that case

$$(L+1)(M+1)(N+1) \;>\; \left(\frac{1}{2}p^{2/3} - 1\right)p^{2/3} \;>\; \frac{1}{4}p^{4/3} \;>\; 3J^2.$$

So now (7.19) becomes

$$s_z \;\leq\; 2 + \frac{(1/2)p^{2/3} - 1 + (p-1)p^{1/3} + p^{4/3}}{(1/4)p^{2/3} - 1} \;=\; S(p^{1/3})$$

for a rational function S. One checks that

$$9X^2 - S(X) = \frac{X^4 - 40X^2 + 4X + 12}{X^2 - 4}$$

and that the largest real zero of the numerator is less than 7. It follows that $s_z \leq 9p^{2/3}$ when $p > 7^3 = 343$. And if $p \leq 343$ then the trivial bound is $p \leq 7p^{2/3}$. This completes the proof of Proposition 7.2 and so also of Theorem 7.1.

Heath-Brown's original proof is in Heath-Brown (1996). Our explicit estimate happens to coincide with one given by Kowalski (2010). There he uses a simplification due to Yu (1999) of Heath-Brown's proof for his original version of our Lemma 7.3. See also Lemmas 4 and 5 of Mitkin (1992).

Heath-Brown and Konyagin (2000) have since improved the exponent to 7/8 by using an averaged version of Proposition 7.2 (see Exercise 7.18), and also applied the same method to more conventional Gauss sums like

$$\sum_{k=1}^{p} \exp\left(2\pi i k^m/p\right),$$

for m large. See Exercise 7.15 for a special case of their method, which has been extended recently by Bourgain, Gamburd and Sarnak in some investigations of Markoff numbers.

At the time of writing, the record exponent for Heilbronn's sum is any $\eta > 5/6$ due to Shkredov (2014).

The method has been further extended by Ghosh and Ward (2015) to get analogues of Proposition 7.2 for polynomials more general than (7.4) but still satisfying differential equations. See also Exercise 7.19.

Exercises

Here $H(w) = \sum_{k=1}^{p} \exp\left(2\pi i w k^p/p^2\right)$ and $G_m(w) = \sum_{k=1}^{p} \exp\left(2\pi i w k^m/p\right)$ for $p \geq 3$.

7.1 Show that $\sum_{x\in\mathbb{F}_p} \left(\frac{x^2-1}{p}\right) = -1$ (Legendre symbol).

7.2 When $p \geq 5$ show that $|G_3(1)| \leq 4p^{3/4}$ [*Hint*: If $x^3 - y^3 = z \neq 0$, deduce $v^2 = (zu^3-2)/6$ with $u = 2/(x-y)$, $v = (x+y)/(x-y)$. Careful with $u = 0$].

7.3 Show that $|H(w)| \leq 4p^{11/12}$ if w is not divisible by p.

* * * * * * * * * * * * * * * * * * * *

7.4 Show that there are exactly $\frac{1}{4}(p - (\frac{-1}{p}))$ elements x in $\mathbb{F}_p$ such that $x-1, x+1$ are both squares.

7.5 When $p \equiv 2 \bmod 3$ show that $G_3(1) = 0$.

7.6 Let χ be a character on $\mathbb{F}_p^*$, and let $G(\chi) = \sum_{k=1}^{p} \chi(k)\exp(2\pi i k/p)$. When $\chi \neq \chi_0$ show that $|G(\chi)| = p^{1/2}$.

7.7 When $p \equiv 1 \bmod 3$ show that $|G_3(1)| \leq 2p^{1/2}$ [*Hint*: $\chi^3 = \chi_0$].

7.8 Show that $G_3(1)$ is an algebraic number of degree at most 3.

7.9 When $p = 1291$ show that $\gamma^3 - 3873\gamma - 86497 = 0$ for $\gamma = G_3(1)$.

7.10 When $p \equiv 1 \bmod 3$ show that $\gamma^3 - 3p\gamma + ap = 0$ for $\gamma = G_3(1)$, with a in $\mathbb{Z}$ and $4p = a^2 + 27b^2$ with b in $\mathbb{Z}$.

7.11 Show that $H(w) = 0$ if w is divisible by p but not by p^2.

7.12 Show that the $H(w)$ for w not divisible by p with $0 < w < p^2$ consist of a set of p conjugate algebraic integers repeated $p-1$ times.

7.13 For $p = 7$ show that $H(1), H(2), H(3), H(4), H(6), H(9), H(12)$ are the different zeroes of $X^7 - 7X^6 + 49X^4 - 98X^2 - 49X + 7$.

7.14 Let $Q \neq 0$ be in $\mathbb{F}_p[X]$ of degree less than p and having T terms. Show that $Q, Q', \ldots, Q^{(T-1)}$ cannot all vanish at $X = 1$.

7.15 For a positive integer n let $\mathcal{S}(n)$ be the set of x in $\mathbb{F}_p$ with $x^n = (1-x)^n = 1$.

(a) For any $L, M, N \geq 0$ in $\mathbb{Z}$ show that there is $P(X, Y, Z)$ in $\mathbb{F}_p[X, Y, Z]$, not identically zero, of degree at most L in X, of degree at most M in Y and of degree at most N in Z, such that for $\phi = P(X, X^n, (1-X)^n)$ we have $\phi(x) = \phi'(x) = \cdots = \phi^{(J-1)}(x) = 0$ for all x in $\mathcal{S}(n)$, provided $(L+1)(M+1)(N+1) > J(L+J)$.

(b) Show that $\phi \neq 0$ provided $L + nM < p$ and $n \geq (L+1)(M+1)$ [*Hint*: Exercise 7.14].

(c) Deduce that there is c such that the cardinality $\#\mathcal{S}(n) \leq c(n^{2/3} + n^2/p)$ for any p, n [*Hint*: assume first $n \leq p-1$ and take $N = M$; but the choice of L, M, N, J now depends on the relationship of n with $p^{3/4}$].

(d) Deduce that there is c such that the equation $u^m + v^m = 1$ has at most $c(p+m^{4/3}p^{2/3})$ solutions (u, v) in $\mathbb{F}_p^2$ for p and any $m \geq 1$ dividing $p-1$ (Garcia and Voloch (1988), which uses a different method). The Riemann Hypothesis would give $p + cm^2p^{1/2}$. See also the notes to chapter 2 of Zannier (2012), as well as Corvaja and Zannier (2013).

7.16 (a) Extend Exercise 7.15(d) to $N_z = \#\{(u, v) \in \mathbb{F}_p^2;\ u^m + v^m = z\}$.

(b) Show that

$$\sum_{w \in \mathbb{F}_p} |G_m(w)|^4 = p \sum_{z \in \mathbb{F}_p} N_z^2 \leq p(m^2p^2 + S),$$

where

$$S = \sum_{0 \neq z \in \mathbb{F}_p} N_z^2 \leq N \sum_{z \in \mathbb{F}_p} N_z = p^2 N$$

with $N = \max_{0 \neq z \in \mathbb{F}_p} N_z$.

(c) Deduce that there is c such that

$$\left| \sum_{k=1}^{p} \exp\left(\frac{2\pi i k^m}{p}\right) \right| \leq c(m^{1/4}p^{3/4} + m^{7/12}p^{2/3})$$

for any p and any $m \geq 1$ dividing $p-1$ (Shparlinsky, 1991) [*Hint*: $G_m(w) = G_m(wl^m)$].

7.17 (a) With $G(\chi)$ as in Exercise 7.6 and $m \geq 1$ dividing $p-1$, show that

$$G_m(w) = \sum_{\chi^m=\chi_0,\ \chi\neq\chi_0} \overline{\chi(w)}G(\chi)$$

and deduce that $|G_m(w)| \leq (m-1)p^{1/2}$.

(b) Conclude that

$$\left|\sum_{k=1}^{p} \exp\left(\frac{2\pi i k^m}{p}\right)\right| \leq cm^{7/12}p^{2/3}$$

in Exercise 7.16(c).

(c) Show that the inequality in (b) holds for any positive integer m, whether dividing $p-1$ or not.

7.18 With s_z as in Proposition 7.2, show that there is an absolute constant c such that $\sum_{z\in\mathcal{Z}} s_z \leq c(p|\mathcal{Z}|)^{2/3}$ for any set $\mathcal{Z}$ in $\mathbb{F}_p$ with cardinality $|\mathcal{Z}| \geq 1$ (that is how Heath-Brown and Konyagin get their exponent $7/8$).

7.19 (a) Analogously to (7.4) let

$$E(X) = \sum_{l=0}^{p-1} \frac{X^l}{l!} = 1 + X + \frac{X^2}{2} + \cdots + \frac{X^{p-1}}{(p-1)!}$$

be the "truncated exponential function" in $\mathbb{F}_p[X]$. Show that $E'(X) = E(X) + X^{p-1}$.

(b) Suppose $P \neq 0$ is in $\mathbb{F}_p[X, Y, Z]$, of degree at most $L \geq 1$ in X and of degree at most $M \geq 1$ in Y. If $(2L+1)M < p$ show that the polynomial $P(X, E(X), X^p)$ is not identically zero [*Hint*: take $\psi(X) = F(X, E(X) + \gamma T)$ in the polynomial ring $\mathbb{F}_p[X, T]$ with derivation defined by $\mathrm{d}T/\mathrm{d}X = T$].

(c) Deduce that there is an absolute constant c such that for every z in $\mathbb{F}_p$ there are at most $cp^{2/3}$ values of x in $\mathbb{F}_p$ with $E(x) = z$ (Mitkin, 1992). Compare Proposition 7.2.

7.20 Let t_z be the number of x in Exercise 7.19(c). Show that there is an absolute constant c such that $\sum_{z\in\mathcal{Z}} t_z \leq c(p|\mathcal{Z}|)^{2/3}$ for any set $\mathcal{Z}$ in $\mathbb{F}_p$ with cardinality $|\mathcal{Z}| \geq 1$ (Seppi Master Thesis 2015).

8

Irrationality measures I – Mahler

Let θ be an irrational number, so that for any integers r and $s \geq 1$ we have $\theta - r/s \neq 0$ (one usually takes p, q here, but these letters have been used up in earlier chapters). It is now natural to ask how small $|\theta - r/s| > 0$ can be, and this can be answered by an inequality

$$\left|\theta - \frac{r}{s}\right| \geq \epsilon(r, s) \tag{8.1}$$

for all such r and s, where $\epsilon(r, s) > 0$ is an easily calculated function only mildly dependent (if at all) on the numerical value of θ. Such a function is usually called an irrationality measure (see for example Exercise 2.5 with $\theta = e$ and $\epsilon(r, s) = \frac{1}{2s(4s)!}$). We will see a very important example in Chapter 12.

If θ is not real, say with imaginary part $\Im\theta = \eta \neq 0$, then trivially (8.1) holds with $\epsilon(r, s) = |\eta|$ independent of r, s; so it is no great restriction to suppose that θ is real.

We can eliminate r by considering $\inf_{r\in\mathbb{Z}} |\theta - r/s|$, which is still positive, for example because $|\theta - r/s| < 1/2s$ for exactly one value of r, namely the integer nearest $s\theta$. Thus we can replace $\epsilon(r, s)$ by a function $\epsilon(s) > 0$ of s alone in (8.1).

A standard application of the Box or Pigeon-hole Principle shows that there are infinitely many s with $\epsilon(s) < 1/s^2$ for such a function, even for rational θ. More precisely for any integer $S > 1$ consider the $\theta_m = m\theta - [m\theta]$ ($m = 0, 1, \ldots, S - 1$) together with $\theta_S = 1$, where $[x]$ is the usual integer part. These $S + 1$ numbers lie in the real interval $[0, 1]$, and this can be divided into subintervals $[(l - 1)/S, l/S)$ ($l = 1, 2, \ldots, S - 1$) together with $[(S - 1)/S, 1]$, making S intervals in all. So there are m, n with $0 \leq m < n \leq S$ and θ_m, θ_n in the same subinterval. This leads to

$$|s\theta - r| \leq \frac{1}{S} \tag{8.2}$$

where $s = n - m$ satisfies

$$0 < s < S \tag{8.3}$$

(note that we cannot have both $m = 0$ and $n = S$) and r in $\mathbb{Z}$. Thus $\epsilon(s) \leq 1/(sS) < 1/s^2$, so that

$$\left|\theta - \frac{r}{s}\right| < \frac{1}{s^2}. \tag{8.4}$$

Now if θ is irrational, then as S tends to infinity we get infinitely many r/s (choose some S, then get r, s, then choose new $S > 1/|\theta - r/s|$ and so on), all with s determining r so infinitely many s; and if $\theta = a/b$ is rational, then trivially $s = b, 2b, \ldots$ will do.

It is sometimes useful to have (8.2), (8.3) also when $S > 1$ is not an integer; but this follows at once by applying it to $[S] + 1$.

We postpone to Chapter 12 a discussion of numbers like $\theta = e$ whose irrationality is proved by simple truncation arguments. In fact the main subjects of that chapter are algebraic numbers like $\theta = \sqrt{2}, \sqrt[3]{2}$ whose irrationality is rather trivial. At the moment we haven't yet proved the irrationality of π, but we cannot resist quoting Mahler's famous

$$\left|\pi - \frac{r}{s}\right| \geq \frac{1}{s^{42}} \quad (s \geq 2)$$

in Mahler (1953). We also skip over the values of the function considered in Chapter 2, and proceed to the situation of Chapter 3.

In Theorem 3.1 we saw that the values θ of the function

$$f(z) = \sum_{k=0}^{\infty} z^{2^k}$$

at all rational α with $0 < |\alpha| < 1$ are irrational. To begin with, we take here $\theta = f(2/3) = 1.349\ldots$, which required the auxiliary polynomial

$$P(X, Y) = (3 - 2X)Y^2 + (1 - 4X)Y - X \tag{8.5}$$

and the corresponding auxiliary function

$$\phi(z) \;=\; P(z, f(z)) \;=\; 2z^6 + \cdots \tag{8.6}$$

in (3.7), with a zero of order 6, for the proof of its irrationality. We first show that there exists $c > 0$ such that

$$\left|\theta - \frac{r}{s}\right| \geq \frac{c}{s^{77}} \tag{8.7}$$

for every $s \geq 1$.

With $\alpha = 2/3$ we consider as in (3.14) the numbers

$$\phi_n = \phi(\alpha^{2^{n+1}}) = P(\alpha^{2^{n+1}}, \theta - \alpha - \cdots - \alpha^{2^n})$$

satisfying by (8.6)

$$|\phi_n| \leq 3\alpha^{6.2^{n+1}} \tag{8.8}$$

for large enough n. Then they were rational but now they are probably not. However if θ is close to r/s then they are close to

$$f_n = P\left(\alpha^{2^{n+1}}, \frac{r}{s} - \alpha - \cdots - \alpha^{2^n}\right), \tag{8.9}$$

which are rational. More precisely $\phi_n = P_n(\theta), f_n = P_n(r/s)$ for $P_n(Y) = P(\alpha^{2^{n+1}}, Y - \alpha - \cdots - \alpha^{2^n})$, so $\phi_n - f_n = P'_n(\theta_n)(\theta - r/s)$ for the derivative and some θ_n between θ and r/s. Assuming $\epsilon = |\theta - r/s| < 1/2$, we find $|\theta_n| \leq 2$ as well as the length $\mathcal{L}(P_n) \leq 5L_n^2 + 5L_n + 1$ for any $n \geq 0$, where

$$L_n = \mathcal{L}(Y - \alpha - \cdots - \alpha^{2^n}) \leq 3.$$

Thus

$$|P'_n(\theta_n)| \leq 2\mathcal{L}(P'_n) \leq 4\mathcal{L}(P_n) \leq 250.$$

We end up with $|\phi_n - f_n| \leq 250\epsilon$. Multiplying up by the usual denominator $D_n s^2$ of f_n with $D_n = 3^{2^{n+1}}3^{2.2^n}$, and using (8.8), we deduce

$$250 D_n s^2 \epsilon \geq D_n s^2 |f_n| - 3D_n s^2 \alpha^{6.2^{n+1}}.$$

Assuming

$$f_n \neq 0 \tag{8.10}$$

we get

$$250 D_n s^2 \epsilon \geq 1 - 3\left(\frac{81}{64}\right)^{-2^{n+1}} s^2. \tag{8.11}$$

Thus to get a positive lower bound for ϵ we should certainly choose n with $(81/64)^{2^{n+1}} > 3s^2$; and for safety

$$(81/64)^{2^{n+1}} \geq 6s^2 \tag{8.12}$$

leads to

$$\epsilon \geq \frac{1}{500 D_n s^2} = \frac{1}{500.9^{2^{n+1}} s^2}. \tag{8.13}$$

At first sight the s^2 here looks agreeably suited to (8.4); but it is accompanied by a less pleasant $9^{2^{n+1}}$. This should be humoured by choosing n as small as possible subject to (8.12).

However the main snag is (8.10), just as in Chapter 3. We could sneak through analytically, by obtaining a lower bound for ϕ_n and using the fact that it is near f_n. But we prefer the algebraic way; and we will show in a moment that the equations

$$f_n = f_{n+1} = 0 \tag{8.14}$$

are impossible for any $n \geq 0$, similar to (2.2).

Assuming this, we can finish the proof of (8.7) as follows. Let m be the smallest n satisfying (8.12). Then $m \geq 2$ and $(81/64)^{2^m} < 6s^2$. We choose $n = m$ or $n = m + 1$ to make $f_n \neq 0$. Then (8.13) gives

$$\epsilon \geq \frac{1}{500.9^{2^{m+2}}s^2} > \frac{1}{500(6s^2)^\lambda s^2}$$

with $\lambda = \log(9^4)/\log(81/64) < 75/2$. This gives what we want, at least if n satisfies (8.8). But this is guaranteed by s sufficiently large; and for the remaining s we simply adjust c so that (8.7) continues to hold.

How do we prove the impossibility of (8.14)? By (8.9) we have $f_n = P(\zeta)$ for $\zeta = (\alpha^{2^{n+1}}, r/s - \alpha - \cdots - \alpha^{2^n})$. It follows that $f_{n+1} = Q(\zeta)$ for the new polynomial

$$Q(X, Y) = P(X^2, Y - X). \tag{8.15}$$

Thus $R(\alpha^{2^{n+1}}) = 0$ for the resultant $R(X)$ of $P(X, Y)$ and $Q(X, Y)$ with respect to Y. We find

$$R(X) = 4X^7(4X^3 - 12X^2 + 9X - 4).$$

By good luck the cubic factor has only one real zero, at $X = 2.177\ldots$; and we are done. Without the luck we could still have concluded the impossibility of (8.14) for large enough n independent of r, s; and this would have sufficed for (8.7).

In Chapter 12 we will prove results like (8.7) where the constant c is not known to be effective; that is, no explicit value can be given, even in principle. So it may be instructive here to convince ourselves that our present c is effective.

In fact this is fairly clear from the proof; however it may be tedious to calculate an explicit value. The trouble is caused by the requirements that n be sufficiently large. This occurs first in (8.8), where it looks like we have to estimate all the Taylor coefficients of $\phi(z)$. Such an infinite task can be avoided by an application of the maximum modulus principle, which will be used very often in later chapters under the popular name of the Schwarz Lemma. Namely, the function $\tilde{\phi}(z) = \phi(z)/z^6$ has a removable singularity at $z = 0$, and so for any complex z with $|z| \le 2/3$ we have $|\tilde{\phi}(z)| \le \sup_{|w|=2/3} |\tilde{\phi}(w)|$. The right-hand side here is $(2/3)^{-6}M$, where

$$M = \sup_{|w|=2/3} |\phi(w)| \le \mathcal{L}(P) \max\{1, \sup_{|w|=2/3} |f(w)|\}^2 < 4\mathcal{L}(P) = 44 \qquad (8.16)$$

(the second supremum being $f(2/3) < 2$). We conclude

$$|\phi(z)| \le 44(3/2)^6 |z|^6 \le 502|z|^6,$$

which for $z = \alpha^{2^{n+1}}$ gives a weaker version of (8.8) that is valid for all $n \ge 0$.

Then we were a bit lucky with $R(X)$, but we can avoid the numerical inspection of zeroes by assuming

$$f_n = f_{n+1} = f_{n+2} = 0. \qquad (8.17)$$

This leads not only to $R(\alpha^{2^{n+1}}) = 0$ but also to $S(\alpha^{2^{n+1}}) = 0$, where S stands in the same relation to R as Q did to P; namely $S(X) = R(X^2)$. The bad news is that the resultant of R and S turns out to be zero; the good news is that this is due to the large common powers of X, and in fact after depriving R of X^7 and S of X^{14} we get a resultant 6708252377088, quite definitely not zero. Thus (8.17) is impossible for any $n \ge 0$. This increases the exponent 77 in (8.7) to 152; but now the resulting c can be calculated without much trouble.

So much for irrationality measures for $f(2/3)$. What happens for general $f(a/b)$?

Recall from Chapter 3 that the irrationality of $f(a/b)$ was established using an auxiliary polynomial P whose degree depends now on a, b. The reader may convince herself that this makes no essential difference to the arguments, and that something like (8.7) can be established with an explicit exponent κ in place of 77, now also depending on a, b; at least as long as (8.10) holds, and then a suitable value would come from the analogue of (8.11). But the above trouble caused by having to take n sufficiently large is a bit more serious. The discussion around the Schwarz Lemma goes through without change provided we can estimate $\mathcal{L}(P)$ as in (8.16), and we deal with this at the end of the chapter.

However (8.10) may fail, and have to be replaced by considerations involving (8.14) or even (8.17). The resultant business is not all that clear, and one must struggle against the unfortunate tendency of the wretched things to vanish (just as in earlier chapters, especially in the proof of Lemma 7.3).

We sketch now a different way of disproving things like (8.17), only the reader will have to take a couple of things on trust; these will be explained in later chapters.

Suppose that we know upper bounds for the degree of P and its length, and we have bad luck in the shape of

$$f_n = f_{n+1} = \cdots = f_{n+K} = 0 \tag{8.18}$$

for (8.9) and some large K. For $\epsilon = |\theta - r/s|$ and $\theta = f(\alpha), \alpha = a/b$, and as usual $\phi(z) = P(z, f(z))$, we deduce that the

$$P(\alpha^{2^{m+1}}, f(\alpha) - \alpha - \cdots - \alpha^{2^m}) = \phi(\alpha^{2^{m+1}}) \quad (m = n, n+1, \ldots, n+K)$$

are small in terms of P, n, K and ϵ.

Using interpolation (for example Lagrange, or even Newton – see Chapter 9) we can find a polynomial $Q(X)$ of degree at most K, with similarly small coefficients, such that

$$Q(\alpha^{2^{m+1}}) = \phi(\alpha^{2^{m+1}}) \quad (m = n, n+1, \ldots, n+K).$$

So $\phi - Q$ has many zeroes.

Using an extension of the Schwarz Lemma (see also Chapter 9) we can deduce that $\phi(z) - Q(z)$ is small whenever $|z|$ is sufficiently small.

It follows that $\phi(z)$ is similarly small. Using Cauchy integrals (see Chapter 12) we can deduce that many derivatives $\phi^{(t)}(0)$ $(t = 0, 1, \ldots, T)$ are small.

As these are integers, they will be zero as soon as they are sufficiently small. It turns out that for any T we can secure this, provided K is large enough and then ϵ small enough (see Exercise 12.48 for more details). In that case the order of vanishing $\mathrm{ord}_{z=0} P(z, f(z)) > T$. But we can then derive a contradiction from the following result.

Proposition 8.1 *For any $L \geq 0$ in $\mathbb{Z}$ and any non-zero polynomial F in $\mathbb{C}[X, Y]$ of degree at most L in each variable we have* $\mathrm{ord}_{z=0} F(z, f(z)) \leq 12L^2$.

This is our first example of a "zero estimate", in which the main feature is its independence of the coefficients of the polynomial (now arbitrary complex numbers). As such things are very important for more advanced aspects, even if one does not insist on measures, we pause to give a full proof here.

We need a preliminary observation, but first some notation. The step from P to Q in (8.15) corresponds to an endomorphism σ on $\mathbb{C}[X, Y]$ defined by $\sigma X = X^2, \sigma Y = Y - X$; thus $Q = \sigma P$.

Lemma 8.2 *Suppose G in $\mathbb{C}[X, Y]$ is such that G divides σG. Then G is independent of Y.*

Proof Oddly enough this will be analytic; we will use the function f. This somehow reminds me of Littlewood's slogan: it can pay to find out what is the worst enemy of what you want to prove, and then induce him to change sides. We suppose that G involves Y and deduce a contradiction.

We can clearly find z_0 with $0 < |z_0| < 1$ such that $G(z_0, Y)$ involves Y. Then $G(z_0, W + f(z_0))$ involves W, and we can find w_0 with $G(z_0, w_0 + f(z_0)) = 0$. As G divides σG we get

$$0 = \sigma G(z_0, w_0 + f(z_0)) = G(z_0^2, w_0 + f(z_0) - z_0);$$

that is, $G(z_0^2, w_0 + f(z_0^2)) = 0$. Iteration gives $G(z_0^{2^n}, w_0 + f(z_0^{2^n})) = 0$ and so the function $\chi(z) = G(z, w_0 + f(z))$ has infinitely many zeroes $z = z_0^{2^n}$ ($n = 0, 1, 2, \ldots$). Well-known properties of analytic functions show that $\chi(z)$ must be identically zero. But $G \neq 0$ so this contradicts the algebraic independence of $z, f(z)$ proved in Lemma 3.2. □

We now prove Proposition 8.1.

Suppose first that F is irreducible over $\mathbb{C}$. If $\psi(z) = F(z, f(z))$ has a zero of order T at $z = 0$, then $\psi(z^2)$ has a zero of order $2T$ there. This is $Q(z, f(z))$ with $Q = \sigma F$. If $F(X, Y)$ does not involve Y then clearly $T \leq L \leq 3L^2$. We claim $T \leq 3L^2$ even if $F(X, Y)$ does involve Y. In that case also $Q(X, Y)$ involves Y. We may therefore construct the resultant of F and Q with respect to Y; it is a polynomial $R(X)$ in $\mathbb{C}[X]$, of the form $AF + BQ$ with A, B in $\mathbb{C}[X, Y]$. It follows that

$$R(z) = A(z, f(z))\psi(z) + B(z, f(z))\psi(z^2)$$

has a zero of order at least T at $z = 0$. If $R \neq 0$ then it has degree at most $3L^2$ (see for example Exercise 5.22), so we deduce $T \leq 3L^2$ as claimed. If $R = 0$ then F divides σF; however this is ruled out by Lemma 8.2.

Now in the arguments about irrationality or irrationality measure, our F arises from solving linear equations and no-one knows a way to guarantee its irreducibility. So the full proof needs some extra considerations. Generally we have $F = F_1 \cdots F_r$ for irreducible $F_1, \ldots, F_r$, say of total degrees $D_1, \ldots, D_r$ respectively, so that $\sum_{i=1}^{r} D_i \leq 2L$. Then

$$\mathrm{ord}_{z=0} F(z, f(z)) = \sum_{i=1}^{r} \mathrm{ord}_{z=0} F_i(z, f(z)) \leq 3 \sum_{i=1}^{r} D_i^2 \leq 12L^2$$

and the proof is complete.

Note that the original construction in Lemma 3.3 (for $M = L$) shows that $12L^2$ cannot be replaced by anything less than $T = L^2 + 2L$. Indeed this construction gives a possibility to get T as an upper bound. For if there is a zero of bigger order, then the vanishing of the first $T + 1$ Taylor coefficients now leads to a system of $T + 1$ homogeneous linear equations in $T + 1$ unknowns.

So there is a determinant, and one would be surprised if this was not a large integer, so rather probably non-zero. In that case there is no non-trivial solution and we get the upper bound T. At least one can check this for any given L, and for $L = 1, 2, 3, 4, 5, 6$ the determinants are (up to sign, which is anyway dependent on various orderings)

$$1,\ 4,\ 1600 = 2^6 5^2,\ 213909504 = 2^{22}.3.17,\ 2^{36}.7.658967,$$
$$2^{65}.3.7.44163068083$$

respectively (but no obvious pattern!). And for $L = 10$ it is bigger than 10^{161}.

However this method is not fool-proof; for example the "bad luck" already encountered in (3.7) shows that for $L = 1, M = 2$ the analogous $LM + L + M$ is not an upper bound when the degree in Y is at most M (the corresponding determinant then being zero). See also Exercises 20.92, 20.93, 20.94, 20.95, and 13.18 with different functions.

For $M = L$ it seems a rather basic problem to improve the upper bound in Proposition 8.1 to something asymptotic to L^2 for large L, but such things look extremely difficult (see however Exercise 13.17). We will see later (for example in Lemma 18.1) that the mere existence of bounds independent of the coefficients is a much easier matter. Similarly it can be shown that there is an upper bound for K in (8.18) depending only on the degree of P. Indeed by applying resultants as above, but directly to (8.18), one can prove $K \leq 25L^3$, where $L \geq 0$ in $\mathbb{Z}$ is an upper bound for the degree of P in each variable (compare Exercise 8.8).

Finally we promised to say something about the size of the coefficients in the auxiliary polynomials used here. This will become a major theme later. It concerns small non-trivial solutions of homogeneous linear equations, and is popularly known as Siegel's Lemma, even though the principle was first used earlier by Thue. For simplicity we restrict for the moment to the field $\mathbb{Q}$ and the ring $\mathbb{Z}$.

Proposition 8.3 *For positive integers M, N with $M < N$ and real $U \geq 1$ let u_{mn} ($m = 1, \ldots, M$; $n = 1, \ldots, N$) be integers with absolute values at most U. Then there are integers $x_1, \ldots, x_N$, not all zero, with*

$$\sum_{n=1}^{N} u_{mn} x_n = 0 \qquad (m = 1, \ldots, M)$$

and

$$|x_n| \leq (NU)^{\frac{M}{N-M}} \qquad (n = 1, \ldots, N).$$

Proof We have to solve $u(\mathbf{x}) = 0$, where u is the linear map from $\mathbb{R}^N$ to $\mathbb{R}^M$ defined by

$$u(\mathbf{x}) = \left(\sum_{n=1}^{N} u_{1n}x_n, \ldots, \sum_{n=1}^{N} u_{Mn}x_n\right)$$

for $\mathbf{x} = (x_1, \ldots, x_N)$. But non-trivially, and in $\mathbb{Z}^N$!

For real $\mathcal{X} > 0$ to be chosen shortly, denote by $\mathcal{S}$ the set of all $\mathbf{x}$ in the "cube" in $\mathbb{Z}^N$ with $0 \le x_1, \ldots, x_N \le [\mathcal{X}]$. The cardinality is $|\mathcal{S}| = ([\mathcal{X}]+1)^N$. The set $u(\mathcal{S})$ is also finite, and we can nicely estimate its cardinality by going to the cube centre to see that $|x_n - [\mathcal{X}]/2| \le [\mathcal{X}]/2$ $(n = 1, \ldots, N)$. It follows that $\left|\sum_{n=1}^N u_{mn}x_n - \xi_m\right| \le NU[\mathcal{X}]/2$ for $\xi_m = \sum_{n=1}^N u_{mn}[\mathcal{X}]/2$ $(m = 1, \ldots, M)$. Thus the mth coordinate of $u(\mathbf{x})$ lies in an interval of length at most $NU[\mathcal{X}]$, and so there are at most $NU[\mathcal{X}] + 1 \le NU([\mathcal{X}]+1)$ possibilities for it. Therefore $u(\mathcal{S})$ has cardinality $|u(\mathcal{S})| \le (NU)^M([\mathcal{X}]+1)^M$. We now choose $\mathcal{X}$ so that

$$|u(\mathcal{S})| < |\mathcal{S}|. \tag{8.19}$$

This is ensured by $(NU)^M([\mathcal{X}]+1)^M < ([\mathcal{X}]+1)^N$; that is, $[\mathcal{X}]+1 > (NU)^{\frac{M}{N-M}}$; and so $\mathcal{X} = (NU)^{\frac{M}{N-M}}$ is just fine.

From (8.19) it follows that u on $\mathcal{S}$ is not injective. Consequently there are distinct $\mathbf{x}', \mathbf{x}''$ in $\mathcal{S}$ with $u(\mathbf{x}') = u(\mathbf{x}'')$. Thus $u(\mathbf{x}) = 0$ for $\mathbf{x} = \mathbf{x}' - \mathbf{x}'' \neq 0$ with components x_n satisfying $|x_n| \le [\mathcal{X}] \le \mathcal{X}$ $(n = 1, \ldots, N)$. □

A somewhat popular error is to say only $U \ge 0$ in the statement of Siegel's Lemma. Indeed already for $U < 1/N$ the conclusion would imply all $x_n = 0$ an absurdity even though now all $u_{mn} = 0$ too. This error has been committed in Siegel's presence.

See also Exercise 8.15 for another proof using Minkowski's First Theorem in the Geometry of Numbers.

Up to now we usually had the situation $N = M+1$ at the limit of solvability (but occasionally M and N depended on other parameters which prevented this). In that situation the bound is $(NU)^{N-1}$, which is compatible with writing down an explicit solution by Cramér's Rule. For example with $M = 2, N = 3$ the system

$$u_{11}x_1 + u_{12}x_2 + u_{13}x_3 = 0$$

$$u_{21}x_1 + u_{22}x_2 + u_{23}x_3 = 0$$

is solved by

$$x_1 = \begin{vmatrix} u_{12} & u_{13} \\ u_{22} & u_{23} \end{vmatrix}, \quad x_2 = -\begin{vmatrix} u_{11} & u_{13} \\ u_{21} & u_{23} \end{vmatrix}, \quad x_3 = \begin{vmatrix} u_{11} & u_{12} \\ u_{21} & u_{22} \end{vmatrix}$$

with $|x_n| \le 2U^2$ (Proposition 8.3 gives $9U^2$). However this may be the trivial solution! In that case the rank of the corresponding matrix is at most one, and we may omit one of the equations, say the second. There are still explicit solutions like

$$x_1 = u_{12},\ x_2 = -u_{11},\ x_3 = 0$$

(which could also be trivial) but these usually do not satisfy our bound $(3U)^{1/2}$ in the situation $M = 1, N = 3$.

In all of what is to follow, it will not suffice to take $N = M+1$ (matrix almost square). Often one takes $N = 2M$ (matrix of domino proportions), when the bound becomes just NU, a great deal smaller. Even this is not good enough for some applications, where N/M has to be made large (matrix long and thin) and the bound is of order U^ϵ for small ϵ.

An extreme situation is obtained by making N tend to infinity. Sometimes the bound tends to 1, and as soon as it is strictly less than 2 we deduce that $x_n = -1, 0, 1$. We would have liked to illustrate this with the two problems in the Prologue, but it seems not to be possible (however see Exercises 8.13 and 8.14); and so we use instead the problem of finding a non-zero polynomial in $\mathbb{Z}[X]$ with a zero of order at least 200 at $X = 1$.

The obvious choice $P = (X-1)^{200}$ has a coefficient

$$90548514656103281165404177077484163874504589675413336841320.$$

But trying $P = \sum_{i=0}^{L} p_i X^i$ gives equations

$$\sum_{i=0}^{L} p_i i(i-1)\cdots(i-t+1) = 0 \quad (t = 0, \ldots, 199)$$

with $M = 200,\ N = L+1$ and $U = L^{199}$, say. So the upper bound is

$$((L+1)L^{199})^{199/(L-199)}.$$

Indeed as L tends to infinity this tends to 1. We conclude that there is a non-zero polynomial in $\mathbb{Z}[X]$ with coefficients ± 1 having a zero of order at least 200 at $X = 1$. However the smallest value of L making the bound strictly less than 2 is 779147. An example with much bigger degree

$$1606938044258990275541962092341162602522202993782792835301375$$

is

$$(X-1)(X^2-1)(X^4-1)(X^8-1)(X^{16}-1)\cdots(X^{2^{199}}-1).$$

Here is an example involving even larger numbers. At the end of Chapter 3 we saw that degree $L = 22126$ is needed just to prove the Mahler value $f(1947/1948)$ irrational. There we secured $\mathrm{ord}_{z=0} P(z, f(z)) \geq L^2 + 2L$, which are $M = L^2 + 2L$ equations in $N = (L+1)^2$ unknowns. So the coefficient bound is

$$(NU)^{M/(N-M)} = (NU)^{L^2+2L} \geq 2^{L^2} \geq 10^{100000000}.$$

We remark that the "Siegel Exponent" $M/(N-M)$ (sometimes called the Dirichlet Exponent) in Proposition 8.3 arises naturally out of the argument, and indeed it cannot be improved for any choice of M, N (see Exercise 8.17).

As for the term NU, it can be improved to $\sqrt{N}U$; and Beck (personal communication, 2015) has shown that there is an absolute constant $c > 0$ such that the result with $c\sqrt{N}U$ would be false for every N (even with $U = 1$).

Siegel was the first to formulate Proposition 8.3 in his Great Paper Siegel (1929) (see p. 213 and also the translation in Zannier (2014b) p. 6), but earlier Thue used the principle frequently for example in Thue (1909). An unconventional application is due to Strassen (1974), who used it to construct polynomials in $\mathbb{Q}[X]$ that are "hard to compute" from the point of view of complexity theory; for example

$$\sum_{i=0}^{L} 2^{2^{L^3 i}} X^i.$$

In a few applications one needs not just one non-zero solution, but (if $M < N-1$) two linearly independent solutions, or even some sort of basis. However these require Minkowski's Second Theorem (see Exercise 8.16).

Early irrationality measures (and much more) for values of Mahler's $f(z)$ (and many more) were obtained by Galochkin (1980) and Miller (1982). However their work gives no explicit exponents of the denominator s.

Exercises

Here $f(z) = \sum_{k=0}^{\infty} z^{2^k}$.

8.1 Let $\theta = f(2/3)$. Find $c > 0$ such that $|\theta - r/s| \geq c/s^{77}$ for all r and $s \geq 1$ in $\mathbb{Z}$.

8.2 Let α, η be in $\mathbb{C}$ with $0 < |\alpha| < 1$ and let $P(X, Y)$ be either zero or irreducible in $\mathbb{C}[X, Y]$ of degree at most $L \geq 1$ in X and degree at most $M \geq 1$ in Y. If $f_n = \cdots = f_{n+K} = 0$ for $f_k = P(\alpha^{2^{k+1}}, \eta - \alpha - \cdots - \alpha^{2^k})$ with $K = 3LM + M^2 + 1$, show that $P = 0$.

8.3 (a) Let b be in $\mathbb{N}$, and let x_1, x_2, x_3 be in $\mathbb{Z}$, not all zero, with $x_1 + bx_2 + b^2 x_3 = 0$. Show that $\max\{|x_1|, |x_2|, |x_3|\} \geq b$.

(b) Conclude that for each C and each $\varsigma < 1/2$, "Siegel over $\mathbb{Q}$" would be false for $M = 1, N = 3$ with $C(CU)^{\varsigma}$ instead of $(3U)^{1/2}$.

8.4 Find C with the following property. Let M, N be in $\mathbb{N}$ with $M < N$, let Ω be in $\mathbb{R}$ with $\Omega \geq 1$, and let ω_{mn} be in $R = \mathbb{Z} + \mathbb{Z}i$ with $|\omega_{mn}| \leq \Omega$. Then there are ξ_n in R, not all zero, with $\sum_{n=1}^{N} \omega_{mn}\xi_n = 0$ $(m = 1, \ldots, M)$ and $|\xi_n| \leq C(CN\Omega)^{M/(N-M)}$ $(n = 1, \ldots, N)$ (thus an explicit "Siegel over $K = \mathbb{Q}(i)$").

* * * * * * * * * * * * * * * * * * *

8.5 Let $\theta = f(1947/1948)$. Find κ such that $|\theta - r/s| \geq c/s^{\kappa}$ $(c > 0)$ for all r and $s \geq 1$ in $\mathbb{Z}$.

8.6 Find C such that for any $L \geq 1, M \geq 1$ in $\mathbb{Z}$ and any non-zero polynomial F in $\mathbb{C}[X, Y]$ of degree at most L in X and of degree at most M in Y we have $\mathrm{ord}_{z=0} F(z, f(z)) \leq CLM$.

8.7 Suppose Q in $\mathbb{C}[X, Y]$ and $t > 0$ are such that Q divides $\sigma^t Q$, with σ as in Lemma 8.2 and σ^t the iterate. Show that Q is independent of Y and has the form $cX^e \prod_{\zeta}(X - \zeta)^{e(\zeta)}$ for roots of unity ζ.

8.8 Let α, η be complex numbers with $0 < |\alpha| < 1$ and let $P(X, Y)$ in $\mathbb{C}[X, Y]$ be a polynomial of degree at most $L \geq 0$ in each variable. If the quantities $f_n = P(\alpha^{2^{n+1}}, \eta - \alpha - \cdots - \alpha^{2^n})$ satisfy $f_n = f_{n+1} = \cdots = f_{n+K} = 0$ for some $n \geq 0$ and $K = CL^3$ for some sufficiently large absolute constant C, show that $P = 0$ [*Hint*: Write $P_1, \ldots, P_m$ for the distinct irreducible factors of P involving Y, and P_0 for the biggest factor of P not involving Y, and find many zeroes of $P_0 \prod \mathrm{res}(P_i, \sigma P_j)$ where the resultants are with respect to Y, and the product is taken over all positive i, j for which the resultant somehow contrives not to vanish].

8.9 Prove the assertion of Exercise 8.8 with $|\alpha| > 1$ instead of $0 < |\alpha| < 1$, or with $|\alpha| = 1$ and α not a root of unity.

8.10 Show that the assertion of Exercise 8.8 would be false for $\alpha = 0$ or every root of unity α.

8.11 Can CL^3 be improved to CL^2 in Exercise 8.8? I don't know.

8.12 Find C such that for every $M \geq 2$ there is $P(X) \neq 0$ in $\mathbb{Z}[X]$, divisible by $(X - 1)^M$, with coefficients $0, \pm 1$ and degree at most $CM^2 \log M$.

8.13 Show that there is $P \neq 0$ in $\mathbb{Z}[X, Y]$, with coefficients of absolute value at most 9^{3239}, such that $P(t^{1948} + t^{666} + 1, t^{1291} + t^{163} + t) = 0$.

8.14 Is there $P \neq 0$ in $\mathbb{Z}[X, Y]$, with coefficients $-1, 0, +1$, such that $P(t^{1948} + t^{666} + 1, t^{1291} + t^{163} + t) = 0$? I don't know.

8.15 Let M, N be in $\mathbb{N}$ with $M < N$, and let u_{mn} be in $\mathbb{Z}$ with coefficient matrix of rank M. Using Minkowski's First Theorem show that there are $x_1, \ldots, x_N$ in $\mathbb{Z}$, not all zero, with $\sum_{n=1}^{N} u_{mn}x_n = 0$ $(m = 1, \ldots, M)$ and $|x_n| \leq N\Delta^{\frac{1}{N-M}}$ $(n = 1, \ldots, N)$, where $\Delta = \sqrt{\sum_\delta \delta^2}$ with the sum over all $M \times M$ subdeterminants δ of the matrix [*Hint*: the space defined by the equations intersects $\mathbb{Z}^N$ in a lattice whose determinant is Δ divided by the highest common factor of the δ].

8.16 Let M, N, u_{mn}, Δ be as in Exercise 8.15, and let $L = N - M$. Using Minkowski's Second Theorem show that there are $(x_{1l}, \ldots, x_{Nl})$ $(l = 1, \ldots, L)$ in $\mathbb{Z}^N$, linearly independent over $\mathbb{Q}$, with

$$\sum_{n=1}^{N} u_{mn}x_{nl} = 0 \qquad (m = 1, \ldots, M;\ l = 1, \ldots, L)$$

and

$$\prod_{l=1}^{L} \max\{|x_{1l}|, \ldots, |x_{Nl}|\} \leq N^L \Delta.$$

8.17 Let M, N be in $\mathbb{N}$ with $M < N$, let $L = N - M$, and let $U \geq 1$ be in $\mathbb{R}$. Let p_{ml} be different primes with $(U/2)^{1/L} \leq |p_{ml}| \leq U^{1/L}$. Let $x_1, \ldots, x_N$ be in $\mathbb{Z}$, not all zero, with

$$p_{m1} \cdots p_{mL} \left(\sum_{l=1}^{L} \frac{x_l}{p_{ml}} - x_{L+m} \right) = 0 \ (m = 1, \ldots, M).$$

Show that $\max\{|x_1|, \ldots, |x_N|\} \geq (U/2)^{M/(N-M)}$. Here $|u_{mn}| \leq U$ – thus the Siegel Exponent $M/(N - M)$ is best possible (Schmidt, 1991).

8.18 Let $\theta = f(1947/1948)$. Find $c > 0$ and κ such that $|\theta - r/s| \geq c/s^\kappa$ for all r and $s \geq 1$ in $\mathbb{Z}$.

8.19 Are there C and κ such that for any $L \geq 1$ in $\mathbb{Z}$ and any non-zero polynomial F in $\mathbb{C}[X, Y]$ of degree at most L in each variable we have $\mathrm{ord}_{z=0} F(z, \tilde{f}(z)) \leq CL^\kappa$ for $\tilde{f}(z) = \sum_{k=1}^{\infty} z^{k!}$? No-one knows.

8.20 Show that there are no C and κ such that for any $L \geq 1$ in $\mathbb{Z}$ and any non-zero polynomial F in $\mathbb{C}[X, Y]$ of degree at most L in each variable we have $\mathrm{ord}_{z=0} F(z, \tilde{\tilde{f}}(z)) \leq CL^\kappa$ for $\tilde{\tilde{f}}(z) = \sum_{k=1}^{\infty} z^{2^{k!}}$.

9

Integer-valued entire functions I – Pólya

Are there entire functions f such that $f(0), f(1), f(2), \ldots$ are rational integers? Of course: any polynomial with integer coefficients. But also $f(z) = z(z-1)/2$, which for any z in $\mathbb{Z}$ is a binomial coefficient. And more generally

$$\binom{z}{k} = \frac{z(z-1)\cdots(z-k+1)}{k!} \tag{9.1}$$

(which incidentally comes from a hyperderivative $(1/k!)(\mathrm{d}/\mathrm{d}X)^k X^z$ at $X = 1$). Are there any f not polynomials? Of course: for example $f(z) = \sin(\pi z)$. This is harmlessly bounded on $\mathbb{R}$ but on its rightful domain $\mathbb{C}$ it grows quite violently; for example $f(iy) = (e^{-\pi y} - e^{\pi y})/(2i)$ grows exponentially as the real y goes to infinity. So there are arbitrarily large complex z with $|f(z)| > e^{\pi|z|}/3$. Growing rather more slowly is $f(z) = 2^z$, with $|f(z)| = 2^{|z|}$ for real $z \geq 0$. More cunning looks $(-1)^z$, which could be defined as $\exp(z \log(-1))$, for example $e^{\pm \pi i z}$; but any choice of the logarithm comes back to at least $e^{\pi|z|}$.

Pólya proved that there is not much between polynomials and 2^z. More precisely: if f is entire with $f(0), f(1), f(2), \ldots$ in $\mathbb{Z}$ and $|f(z)| \leq cC^{|z|}$ for some real c and some real $C < 2$, then f is a polynomial. His proof goes as follows. Any entire f has a formal development by Newton Interpolation as

$$f(z) = a_0 + a_1 z + a_2 \frac{z(z-1)}{2} + \cdots = \sum_{k=0}^{\infty} a_k \binom{z}{k}, \tag{9.2}$$

where the

$$a_k = f(k) - \binom{k}{1} f(k-1) + \binom{k}{2} f(k-2) - + \cdots + (-1)^k f(0) \tag{9.3}$$

are the so-called forward differences. The series converges to $f(z)$ everywhere under the above growth condition, a fact not mentioned in every book on numerical analysis. To see this, note that for the special function $1/(w-z)$ and any $n \geq 1$, there is a finite version

$$\frac{1}{w-z}=\sum_{k=0}^{n-1}\frac{k!}{w(w-1)\cdots(w-k)}\binom{z}{k}+\frac{n!}{w(w-1)\cdots(w-n+1)(w-z)}\binom{z}{n}$$

easily verified by induction. Then multiplying by $f(w)/(2\pi i)$ and integrating over an anticlockwise circle defined by $|w| = 2n \geq |z|+1$, we get by Cauchy's Theorem

$$f(z)=\sum_{k=0}^{n-1} a_k\binom{z}{k}+a_n(z)\binom{z}{n} \tag{9.4}$$

with

$$a_k = \frac{k!}{2\pi i}\int\frac{f(w)dw}{w(w-1)\cdots(w-k)},$$

by partial fractions nothing other than (9.3), and

$$a_n(z) = \frac{n!}{2\pi i}\int\frac{f(w)dw}{w(w-1)\cdots(w-n+1)(w-z)}.$$

Clearly

$$|a_n(z)| \leq \frac{2cnC^{2n}}{\binom{2n}{n}} \leq 2cn(2n+1)(C/2)^{2n}$$

which tends to zero with n tending to infinity, and for $m = [|z|]$ we have

$$\left|\binom{z}{n}\right| \leq \binom{m+n}{n} = \binom{m+n}{m} \leq (|z|+n)^m$$

so that also the remainder in (9.4) tends to zero.

Similarly we get, on changing the radius to $2k$, the inequality $|a_k| \leq 2c(2k+1)(C/2)^{2k}$. Now under the integrality hypothesis a_k is a rational integer. It follows from the ubiquitous Fundamental Theorem that $a_k = 0$ for large enough k; thus from (9.2) we see that f must be a polynomial (and in fact an integer linear combination of binomial coefficients).

It is amusing to note that the formal Newton series for 2^z itself is $\sum_{k=0}^{\infty}\binom{z}{k}$, which is certainly 2^z for $z = 0, 1, 2, \ldots$. But already for $z = -1$ the sum is $1-1+1-1+-\cdots$, which doesn't converge to 2^{-1} even though it would very much like to.

Pólya's result can be turned round to assert that under suitable hypotheses at least one of $f(0), f(1), f(2), \ldots$ is not an integer. This is a very weak analogue of the results of Chapters 2 and 3, where a specific $f(\alpha)$ was proved irrational by means of an auxiliary polynomial in z and $f(z)$. It was observed by Waldschmidt that a weaker form of Pólya's result, with $C < \exp(1/283)$ according to Escassut and Mignotte, could be also proved with such a polynomial. It is instructive to present his proof, for at least three reasons: first, the use of auxiliary functions is vital for the beautiful result of Gramain in the next chapter, and second, the use of binomial coefficients (essentially the divided derivatives of Chapter 6 – see also Exercise 2.14) is crucial for Thue's Theorem later on, and third, it is likely that Gelfond's solution of Hilbert's Seventh Problem (see Chapter 19) was influenced by such considerations.

In Chapters 2 and 3 (and others) we constructed $\phi(z) = P(z, f(z))$ with a high order zero at $z = 0$. This is here not possible, as we know nothing about the corresponding Taylor series of $f(z)$ with coefficients involving $f^{(k)}(0)$ for $k = 0, 1, 2, \ldots$. We do know something about the Newton series with coefficients involving $f(s)$ for $s = 0, 1, 2, \ldots$, so a reasonable substitute would be to secure instead $\phi(s) = 0$ for many values of s, say $0, 1, 2, \ldots, S-1$. Just as a high order zero tends to imply that nearby function values are small (see Lemmas 2.4 and 3.4), so do many simple zeroes, and to see this we will use the Schwarz Lemma considerations of the previous chapter. In fact we can show that $\phi(S)$ is small, and then by the Fundamental Theorem it must be zero. We have thus acquired an extra zero of ϕ, which could be the start of a wonderful induction.

This strategy will not work quite so easily. The reason is that even a function $\phi(z) = z^k$, with a high order zero at $z = 0$, is not small enough at $z = 1$. The quantity analogous to M in (8.16) measuring the growth of ϕ is simply too big. Fortunately the solution is at hand: we use the binomial polynomials (eh?) $\binom{z}{k}$ instead, which are divided by factorials but stay integer-valued. Thus we are looking at linear combinations of $\binom{z}{i} f(z)^j$.

We now give the details of the proof, remaining informal and especially relaxed about the bound on C (which we bury at the end); after all, we just gave the full proof that $C < 2$ suffices. Thus we assume that f is integer-valued, with growth which we now write in the form

$$|f(z)| \leq c e^{|z|/\mathcal{X}} \tag{9.5}$$

and we reserve the right to take the absolute constant $\mathcal{X}$ (that is, not depending on anything) as big as we like so that the proof works, leading to $C = \exp(1/\mathcal{X})$. We can clearly assume $c \geq 1$.

We take $L \geq 0$ in $\mathbb{Z}$, for convenience supposed odd, and also $M \geq 0$ in $\mathbb{Z}$, and choose integer coefficients p_{ij}, not all zero, such that the function

$$\phi(z) = \sum_{i=0}^{L} \sum_{j=0}^{M} p_{ij} \binom{z}{i} f(z)^j \tag{9.6}$$

satisfies

$$\phi(0) = \phi(1) = \cdots = \phi(S-1) = 0 \tag{9.7}$$

for a suitable integer S. We have S equations and $(L+1)(M+1)$ unknowns, and we already saw that the Siegel Proposition 8.3 gives agreeable estimates when there are twice as many unknowns as equations. So for a quiet life we take (the integer)

$$S = \frac{(L+1)(M+1)}{2}. \tag{9.8}$$

The bound for the p_{ij} is then just $(L+1)(M+1)U$, where $U \geq 1$ is an upper bound for the coefficients of the p_{ij} in (9.6) with $z = s$ for $s = 0, 1, \ldots, S-1$.

We have for $i = 0, 1, \ldots, L$

$$\left|\binom{z}{i}\right| \leq \frac{|z|+1}{1}\frac{|z|+2}{2}\cdots\frac{|z|+i}{i} \leq \frac{|z|+1}{1}\frac{|z|+2}{2}\cdots\frac{|z|+L}{L},$$

which by Stirling's $L! > L^L e^{-L}$ gives

$$\left|\binom{z}{i}\right| \leq e^L \left(\frac{|z|+L}{L}\right)^L \quad (i = 0, 1, \ldots, L). \tag{9.9}$$

In particular for $z = s$ in (9.7) we get $e^L(M+2)^L$ in (9.9). Using (9.5) we end up with

$$|p_{ij}| \leq (L+1)(M+1)e^L(M+2)^L c^M e^{SM/\mathcal{X}}. \tag{9.10}$$

We next aim to show that $\phi(S)$ is small using the ideas of the Schwarz Lemma. As in the previous chapter we divide out by zeroes, here the factors $z, z-1, \ldots, z-(S-1)$ arising from (9.7); thus

$$\tilde{\phi}(z) = \frac{\phi(z)}{z(z-1)\cdots(z-S+1)} \tag{9.11}$$

is entire, and we have

$$|\tilde{\phi}(S)| \leq \sup_{|w|=S} |\tilde{\phi}(w)| \leq \sup_{|w|=3S} |\tilde{\phi}(w)|. \tag{9.12}$$

The far right-hand side here is at most $(2S)^{-S} \sup_{|w|=3S} |\phi(w)|$ and the far left-hand side is at least $S^{-S}|\phi(S)|$; at the peril of a large $\mathcal{X}$ in (9.5) coming out, we will not bother with the finer factorial estimates in the Pólya proof. We deduce $|\phi(S)| \leq 2^{-S} \sup_{|w|=3S} |\phi(w)|$, a considerable improvement on the maximum modulus principle for ϕ itself. Now the supremum can be estimated as we did for the application of the Siegel Lemma, and in addition to the right-hand side of (9.10) we get an extra

$$(L+1)(M+1)e^L(3M+4)^L c^M e^{3SM/\mathcal{X}}.$$

Here the $(L+1)(M+1)$ comes from the number of terms; this sort of thing will cause some trouble much later on, but here it is harmless. Anyway with a slight double exposure effect (or "déjà vu all over again" according to Yogi Berra) we get

$$|\phi(S)| \leq 2^{-S}(L+1)^2(M+1)^2 e^{2L}(M+2)^L(3M+4)^L c^{2M} e^{4SM/\mathcal{X}}. \quad (9.13)$$

Here the plucky little 2^{-S} has to beat a helluva lot of stuff, but still the lower bound $S \geq \frac{1}{2}L(M+1)$ will suffice. Now we can estimate from above the right-hand side as

$$(L+1)^2(M+1)^2 c^{2M}\left(\frac{e^2(M+2)(3M+4)}{2^{(M+1)/2}}\right)^L e^{4SM/\mathcal{X}}$$

hoping that the $e^{4SM/\mathcal{X}}$ can be beaten simply by choosing $\mathcal{X}$ sufficiently large.

We first fix M to make the big expression in brackets say at most $1/2$; in fact $M = 30$ will do. Now using $L+1 \leq 2L$ in S in the last exponential we get just

$$|\phi(S)| \leq 961c^{60}(L+1)^2 2^{-L} e^{3720L/\mathcal{X}}.$$

Finally for any fixed $\mathcal{X} > 3720/\log 2$, say $\mathcal{X} = 5367$, the right-hand side tends to zero as L tends to infinity, and we will be able to welcome the extra zero $\phi(S) = 0$.

Indeed this is the start of an induction, and we will soon prove that $\phi(s) = 0$ for all non-negative s in $\mathbb{Z}$. Waldschmidt went further by showing that $\sup_{|w|\leq R}|\phi(w)|$ actually tends to zero as R tends to infinity. This enabled him to conclude $\phi = 0$ identically.

We can reach this conclusion more simply with the Newton series for ϕ, which now formally vanishes. However $\phi(z)$ has a growth of order at most $C_1^{|z|}$ with any $C_1 > \exp(M/\mathcal{X}) = 1.0056\ldots$ and so this series converges to ϕ.

Whatever, the polynomial $\sum_{i=0}^{L}\sum_{j=0}^{M} p_{ij}\binom{z}{i}Y^j$ must involve Y because a non-zero polynomial of degree at most L in z cannot have $S = \frac{31}{2}(L+1)$ different zeroes. Thus $f(z)$ is algebraic over $\mathbb{C}(z)$. However the only entire algebraic functions are polynomials (see Exercise 9.4 below), and we will get our version of Pólya's Theorem as soon as we have carried out the above induction.

So suppose for some integer $S_1 \geq S$ we have shown that

$$\phi(0) = \phi(1) = \cdots = \phi(S_1 - 1) = 0.$$

Using the device of writing $S_1 = \frac{1}{2}(L+1)(M_1+1)$ for rational $M_1 \geq M$, we can easily check that

$$|\phi(S_1)| \leq 2^{-S_1}(L+1)^2(M+1)^2 e^{2L}(M+2)^L(3M_1+4)^L c^{2M} e^{4S_1 M/\mathcal{X}}$$

as in (9.13). And just as there, we get the upper bound

$$(L+1)^2(M+1)^2 c^{2M}\left(\frac{e^2(M+2)(3M_1+4)}{2^{(M_1+1)/2}}\right)^L e^{4S_1 M/\mathcal{X}},$$

itself at most

$$(L+1)^2(M+1)^2 c^{2M}\theta(M_1)^L \tag{9.14}$$

with

$$\theta(\mu) = \frac{e^2(M+2)(3\mu+4)e^{4M(\mu+1)/\mathcal{X}}}{2^{(\mu+1)/2}}.$$

Now $\theta(M) = .959\ldots < 1$ by our above choice of $M = 30$ and then $\mathcal{X} = 5367$. It is easy to check that $\theta(\mu)$ decreases as the real $\mu \geq M$ increases; for example

$$\frac{\theta'(\mu)}{\theta(\mu)} = \frac{3}{3\mu+4} + \frac{4M}{\mathcal{X}} - \frac{\log 2}{2}$$

decreases as $\mu \geq M$ increases, with negative value $-.292\ldots$ at $\mu = M$. Thus if L is chosen sufficiently large independently of M_1 (in fact we see that L depends only on the constant c) we can indeed deduce $\phi(S_1) = 0$. This achieves the desired induction step.

Actually it was not quite cricket to appeal to the Newton series just now, because the original Pólya Theorem follows so quickly from it. However the above upper bounds for $|\phi(S)|$ hold equally well for $\Phi = \sup_{|z|\leq 1}|\phi(z)|$ because then $|z(z-1)\cdots(z-S+1)| \leq S^S$ in the denominator of (9.11), just as for $z = S$. And similarly with S_1; in fact we get the bound (9.14) for $|\Phi|$ with each $S_1 \geq 1$ (now L is fixed as well as M). As $e^{4M/\mathcal{X}} < \sqrt{2}$, we see that $\theta(\mu) \to 0$ as $\mu \to \infty$. Thus we conclude $\Phi = 0$, and so $\phi = 0$.

This completes the proof of our version of Pólya's Theorem, with the atrocious

$$C = \exp(1/5367) < 1.00019. \tag{9.15}$$

However such atrocities have a respectable pedigree, as we will see in the next chapter; and certainly we have been somewhat cavalier in many of our estimates.

One way of doing better is with so-called Blaschke factors. Let F be entire, and from now on abbreviate $\sup_{|w|=R}|F(w)|$ to $|F|_R$. We have often used the Maximum Modulus Principle in the form

$$|F|_r \leq |F|_R \tag{9.16}$$

when $0 \leq r \leq R$. If $F(0) = 0$ then by applying this to $F(z)/z$ we obtain the improvement $|F|_r \leq (r/R)|F|_R$. If $F(z_0) = 0$ for some z_0 with $|z_0| < R$ instead of $F(0) = 0$, then application to $F(z)/(z - z_0)$ gives by estimating crudely $|F|_r \leq (r + |z_0|)/(R - |z_0|)|F|_R$. However if $z_0 \neq 0$ this is not necessarily an improvement of (9.16), especially if $|z_0|$ is close to R. To fix this we should divide F by

$$\beta(z) = \frac{R(z - z_0)}{R^2 - \bar{z}_0 z} \tag{9.17}$$

instead of $z - z_0$ and note that if $|w| = R$ then

$$|R^2 - \bar{z}_0 w| = |\bar{w}w - \bar{z}_0 w| = R|w - z_0|$$

so that $|\beta|_R = 1$; it follows that $|F|_r \leq |\beta|_r |F|_R$ with certainly $|\beta|_r \leq 1$ (because β is analytic for $|z| \leq R$).

In fact one can show that

$$|\beta|_r = \frac{R(r + |z_0|)}{R^2 + r|z_0|} \tag{9.18}$$

(see Exercise 9.14). These Blaschke factors (9.17) will be crucial in the next chapter; however we will not use (9.18) directly there.

Another way of getting a larger C in (9.15) is to replace the $f(z)^j$ in (9.6) by $f(z + j)$, faintly reminiscent of (9.3). This will also be crucial in the next chapter.

Pólya's original result is in Pólya (1915), and Waldschmidt's proof in Waldschmidt (1978). The result has been extended in many ways. For example if

$$C < \exp\sqrt{\left(\frac{\log 3}{2}\right)^2 + \left(\frac{\pi}{6}\right)^2} = 2.1358743377\ldots$$

then $f(z) = P(z) + 2^z Q(z)$ for polynomials P, Q in $\mathbb{Q}[z]$. An earlier result had just $C < 2e^{1/1500} = 2.00133\ldots$.

A natural problem is to do something similar for entire functions taking all of $\mathbb{Z}$ into $\mathbb{Z}$. Here the weaker growth condition $|f(z)| \leq cC^{|z|}$ for some real $C < (3 + \sqrt{5})/2 = 2.618\ldots$ ensures that f is a polynomial. But now the Fibonacci-style extremal example

$$f(z) = \left(\frac{3 + \sqrt{5}}{2}\right)^z + \left(\frac{3 - \sqrt{5}}{2}\right)^z$$

is not so obvious. Note also that the plausible contender

$$f(z) = \left(\frac{1+\sqrt{5}}{2}\right)^z + \left(\frac{1-\sqrt{5}}{2}\right)^z$$

grows more slowly on $\mathbb{Z}$ but in view of the second term (compare with $(-1)^z = e^{\pm\pi iz}$ above) needs a whole

$$C \geq \exp\sqrt{\log\left(\frac{1+\sqrt{5}}{2}\right)^2 + \pi^2} = 24.00431639\ldots$$

on $\mathbb{C}$.

We may also mention a variation by Gelfond: if $q \geq 2$ is an integer and $f(q), f(q^2), f(q^3)\ldots$ are integers, then $|f(z)| \leq cC^{(\log|z|)^2}$ $(|z| \geq 3)$ for $C < \exp(4/\log q)$ implies that f is a polynomial, also best possible.

For more on these problems see Buck (1948) and a bit more recently Gramain (1978).

There are also results about integer values of f and f' (and f'' and so on); but see Chapter 19 for the use of higher derivatives to prove things that are much more interesting.

Another variation concerns functions f mapping $\mathbb{N}$ to $\mathbb{Z}$, but defined only on $\mathbb{N}$ and satisfying the congruence conditions $f(n+m) \equiv f(n)$ modulo m for each m in $\mathbb{N}$ (and n). Hall (1971) and Rusza (1971) independently showed that if there are c and $C < e - 1$ such that $|f(n)| \leq cC^n$ for all n, then f is a polynomial. They used differences like (9.3). The C was then improved by Perelli and Zannier (1984), Zannier (1996) using instead various auxiliary functions. Rusza had conjectured that $C < e$ suffices. One can even combine the two problems by assuming f entire (with $|f(z)| \leq cC^{|z|}$); in that case they had improved C further to any $C < e + 1 = 3.718\ldots$ (Perelli and Zannier, 1981) – see Exercise 9.13.

Exercises

9.1 Show that

$$\frac{1}{w-z} = \sum_{k=0}^{n-1} \frac{k!}{w(w-1)\cdots(w-k)}\binom{z}{k} + \frac{n!}{w(w-1)\cdots(w-n+1)(w-z)}\binom{z}{n}$$

for $n = 1, 2, 3, \ldots$.

9.2 Let f be entire with $f(0), f(1), f(2), \ldots$ in $\mathbb{Z}$ and $|f|_R \le 1000(3/2)^R$ for all R. Using Newton show that f is a polynomial of degree at most 19, with length

$$\mathcal{L}(f) < 2000 \sum_{k=0}^{\infty} (2k+1) \left(\frac{3}{4}\right)^{2k} = \frac{800000}{49} < 16327.$$

9.3 (a) Let S be in $\mathbb{N}$. If ϕ is entire with $\phi(0) = \phi(1) = \cdots = \phi(S-1) = 0$ show that $|\phi|_1 \le 2^{-S}|\phi|_{3S}$.

(b) If ϕ is entire with $\phi(0) = \phi(1) = \cdots = 0$ and $|\phi|_R \le cC^R$ with $C < 2^{1/3}$ show **not** using Newton, that $\phi = 0$ identically.

9.4 If f is entire and $P(z, f(z)) = 0$ with $0 \ne P(z, w)$ in $\mathbb{C}[z, w]$, show that f is in $\mathbb{C}[z]$ [*Hint*: Exercise 4.12 and Liouville].

* * * * * * * * * * * * * * * * * * * *

9.5 Let P be in $\mathbb{C}[z]$ of degree L. Show that $|P|_R \le R^L|P|_1$ for any $R \ge 1$ [*Hint*: $w^L P(1/w)$].

9.6 For any $z_0, \ldots, z_{n-1}$ in $\mathbb{C}$ show that

$$\frac{1}{w-z} = \frac{1}{w-z_0} + \sum_{k=1}^{n-1} \frac{1}{w-z_k} \prod_{l=0}^{k-1} \frac{z-z_l}{w-z_l} + \frac{1}{w-z} \prod_{l=0}^{n-1} \frac{z-z_l}{w-z_l}$$

(Exercise 9.1 is the case $z_k = k$).

9.7 Let $z_0, \ldots, z_d$ in $\mathbb{C}$ be different, and let P be in $\mathbb{C}[z]$ with degree at most d. Show that

$$P(z) = P(z_0) + \sum_{k=1}^{d} \left(\sum_{i=0}^{k} \frac{P(z_i)}{\prod_{j=0, j\ne i}^{k} (z_i - z_j)} \right) (z-z_0) \cdots (z-z_{k-1}).$$

9.8 Let $\mathcal{Z} = \{z_0, z_1, \ldots\} \subset \mathbb{C}$ be countably infinite, and let $\mathcal{W} \subset \mathbb{C}$ be dense. Show that there is an entire function f, not a polynomial, with $f(z_i)$ in $\mathcal{W}$ ($i = 0, 1, \ldots$); in particular there is such an f with $f(\alpha)$ algebraic for every algebraic α (due to Weierstrass in 1885 and Stäckel in 1995 – see Mahler (1976) for more details).

9.9 Show that there is an entire function f, not a polynomial, with all derivatives $f^{(t)}(\alpha)$ ($t = 0, 1, 2, \ldots$) algebraic for every algebraic α (due to Faber in 1904 – again see Mahler (1976) for more details).

9.10 For $Z \ge 0$ show that there are at most 9Z complex numbers z with $|z| \le Z$ such that $e^z = z$.

9.11 For sufficiently large Z show that there are at least $Z/4$ complex numbers z with $|z| \leq Z$ such that $e^z = z$.

9.12 If f is entire with $f(0) = f(1) = f(2) = \cdots = 0$ and $|f|_R \leq cC^R$ with some c and some $C < e^\pi$, show that $f = 0$ identically (due to Carlson in 1914 – see Boas (1954) for more details).

9.13 Let f be entire with $|f|_R \leq cC^R$ with some c and some $C < e+1$, and suppose further that $f(n+m) \equiv f(n)$ modulo m for every m in $\mathbb{N}$. Show that f is a polynomial (Perelli and Zannier, 1981) [*Hint*: prove that $\prod_{p \leq k} p$ divides a_k in (9.3) above].

9.14 Let w be in $\mathbb{C}$ and R in $\mathbb{R}$ with $|w| \leq R$, and let $\beta(z) = \frac{R(z-w)}{R^2-\overline{w}z}$ (Blaschke). Let F be entire with $F(w) = 0$. Show that $|F|_r \leq |\beta|_r |F|_R$ and $|\beta|_r = \frac{R(r+|w|)}{R^2+r|w|} \leq 1$ $(0 \leq r \leq R)$.

9.15 Let f be entire with $|f|_R \leq cC^{R^{3/2}}$ mapping $\mathbb{Z}+i\mathbb{Z}$ into $\mathbb{Z}+i\mathbb{Z}$. Show that f is in $\mathbb{C}[z]$.

10

Integer-valued entire functions II – Gramain

With regard to the previous chapter, it is even arguable that the analogous problem for entire functions mapping $\mathbb{Z} + i\mathbb{Z}$ into $\mathbb{Z} + i\mathbb{Z}$ is more natural, and the only extremal example that easily springs to mind is the Weierstrass sigma function

$$z \prod_{\gamma} \left(1 - \frac{z}{\gamma}\right) \exp\left\{\frac{z}{\gamma} + \frac{1}{2}\left(\frac{z}{\gamma}\right)^2\right\} \tag{10.1}$$

with the product taken over all $\gamma \neq 0$ in $\mathbb{G} = \mathbb{Z} + i\mathbb{Z}$. This vanishes on $\mathbb{G}$ and grows as fast as $cC^{|z|^2}$ with $C = \exp(\pi/2)$ (see Exercise 10.4 and Exercise 20.98(b) with $\mu = 0$). It could be considered an analogue of

$$\sin(\pi z) = z \prod_{0 \neq n \in \mathbb{Z}} \left(1 - \frac{z}{n}\right) \exp\left(\frac{z}{n}\right) = z \prod_{n \in \mathbb{N}} \left(1 - \frac{z^2}{n^2}\right)$$

and so we might expect that such a growth with sufficiently small C might imply that f is a polynomial. That was proved by Gelfond, who found that

$$C < \exp\left\{\frac{\pi}{2(1 + \exp(164/\pi))^2}\right\} < 1 + 10^{-45}$$

suffices. This makes the atrocity (9.15) in Chapter 9 seem respectable. By way of compensation, we will exhibit a worse one in Chapter 17 and an absolutely terrible one in the Appendix.

Already this sort of result has consequences not far off transcendence or irrationality. For example, an analogue for f mapping the positive quadrant of $\mathbb{G}$ (or even the $a + ib$ with just $a \geq 0$) to $\mathbb{G}$ would imply that e^{π} is not in $\mathbb{Z}$, because of $f(z) = e^{\pi z}$ and $f(a+ib) = (e^{\pi})^a(-1)^b$ (of course $e^{\pi} = 23.14\ldots$ is a somewhat simpler proof). But even that every $e^{n\pi}$ $(n = 1, 2, \ldots)$ is not in $\mathbb{Z}$. And an analogue for f mapping $\mathbb{Z} + i\sqrt{163}\mathbb{Z}$ into $\mathbb{G}$ would imply that

$$e^{\pi\sqrt{163}} = 262537412640768743.99999999999925007259719 81\ldots \quad (10.2)$$

is not in $\mathbb{Z}$, not quite so obvious numerically (see Chapter 19 for more on this).

Much later Gruman found the much better upper bound $\exp(.0396\ldots) = 1.040\ldots$ for C, and I had already found a value involving the limit of a complicated sequence (a kind of two-dimensional version of Euler's constant – see Exercise 10.10 below); later it turned out that this gave $\exp(.167) = 1.181\ldots$ (the limit could meanwhile be evaluated after much computation to $1.819\ldots$, but the next decimal place is still unknown). At the same time I improved on the sigma function by using Newton Interpolation as in (9.2), but with the sequence $\{0, 1, 2, \ldots\}$ replaced by suitably ordered elements of $\mathbb{G}$. This led to growth at most $cC^{|z|^2}$ with any $C > C_0$ for the curious constant

$$C_0 = \exp\left(\frac{\pi}{2e}\right) = 1.7822269447490036742598008003959032\ldots \quad (10.3)$$

I also conjectured that this was best possible. In a beautiful paper Gramain proved it.

Theorem 10.1 *Let f be an entire function mapping $\mathbb{G}$ into $\mathbb{G}$. If there are constants c, C with $|f(z)| \leq cC^{|z|^2}$ and $C < C_0$ then f is a polynomial.*

Already Waldschmidt had noted that his method of the previous chapter leads to a somewhat weaker result; but since this plays a role in Gramain's proof, we illustrate the argument with the following version (which already turned up as Exercise 9.15).

Lemma 10.2 *Let f be an entire function mapping $\mathbb{G}$ into $\mathbb{G}$. If there are constants c, C with $|f(z)| \leq cC^{|z|^{3/2}}$ then f is a polynomial.*

Proof We use an auxiliary function like (9.6) but with z^i in place of $\binom{z}{i}$; indeed there seems to be no natural Gaussian analogue of these "binomial polynomials" (see however Exercises 10.3 and 10.7). Thus we take $L \geq 0, M \geq 0$ in $\mathbb{Z}$ and choose for example rational integer coefficients p_{ij}, not all zero, such that the function

$$\phi(z) = \sum_{i=0}^{L}\sum_{j=0}^{M} p_{ij} z^i f(z)^j \quad (10.4)$$

satisfies some analogue of (9.7). A natural condition would be

$$\phi(\gamma) = 0 \quad (10.5)$$

for all γ with $|\gamma| \leq S$ and a suitable integer S. Now the number N of such γ and so of equations is not so simple to count exactly, but for the moment this is not crucial, and by putting the disc in a square we see that

$$N \leq (2S+1)^2. \tag{10.6}$$

Each equation splits into two according to real and imaginary parts, as in (4.25). We have $(L+1)(M+1)$ unknowns, and we know that the Siegel Lemma gives agreeable estimates when there are twice as many unknowns as equations. So we need

$$(L+1)(M+1) \;\geq\; 4(2S+1)^2 \tag{10.7}$$

with a mental note that the two sides should be roughly equal. The bound for the p_{ij} is then just $(L+1)(M+1)U$, where $U \geq 1$ is an upper bound for the coefficients of the p_{ij} in (10.4) with $z = \gamma$. By the growth hypothesis we end up with

$$|p_{ij}| \;\leq\; (L+1)(M+1)S^L c^M C^{S^{3/2}M}, \tag{10.8}$$

since $|\Re w|$ and $|\Im w|$ (involving real and imaginary parts) are at most $|w|$.

In the previous chapter we went in a somewhat gingerly manner from zeroes at $0, 1, \ldots, S-1$ to a zero at S; but now we have more confidence and we will jump from $|\gamma| \leq S$ to $|\gamma'| \leq 2S$. As yet there is no need for Blaschke refinements, and so we simply divide ϕ by $\prod_{|\gamma| \leq S}(z-\gamma)$ to get $\tilde{\phi}$ as in (9.11). Now it is appropriate to use $|\tilde{\phi}|_{2S} \leq |\tilde{\phi}|_{7S}$ (in our maximum modulus notation), which leads in the usual way to

$$|\phi|_{2S} \;\leq\; 2^{-N}(L+1)^2(M+1)^2(7S^2)^L c^{2M} C^{2(7S)^{3/2}M}. \tag{10.9}$$

Here we need a lower bound for N instead of the upper bound (10.6). No problem: we put a square in the disc, or just the non-negative quadrant will do, to see that

$$N \;\geq\; \left(\left[\frac{S}{\sqrt{2}}\right]+1\right)^2 \;>\; \left(\frac{S}{\sqrt{2}}\right)^2 \;=\; \frac{1}{2}S^2. \tag{10.10}$$

Now instead of choosing L, M as in the previous chapter which then determine S, we will choose L, M in terms of S, which we regard as an independent parameter sufficiently large in terms of c, C. This is simply because (10.7) is slightly awkward to solve for S. Regarding (10.9) it is plausible that the last term is the worst, and so in view of the 2^{-N} and (10.10) we tentatively choose M about S^{μ} with fixed $\mu < 1/2$. Then (10.7) is fine when L is about S^{λ} with $\lambda + \mu > 2$. For example

$$L = [S^{9/5}], \quad M = [S^{2/5}]. \tag{10.11}$$

Now (10.9) for S sufficiently large implies $|\phi|_{2S} < 1$. The Fundamental Theorem of Transcendence applies on $\mathbb{G}$ as well (see the next chapter for further generalizations). We conclude indeed that $\phi(\gamma') = 0$ for all γ' in $\mathbb{G}$ with $|\gamma'| \leq 2S$.

We now argue as in the previous chapter, getting next $|\phi|_{4S} < 1$ then $\phi(\gamma'') = 0$ for all γ'' in $\mathbb{G}$ with $|\gamma''| \leq 4S$; and so on. As in the inductive section of the previous chapter, the reader can check that the size of the large S required does not get any bigger (and we will again verify the details of this sort of thing in the main part of the proof of Theorem 10.1). We end up with $\phi = 0$ on all of $\mathbb{G}$, which is implausible but not yet a contradiction in view of, for example, the sigma function. But more significantly we have $|\phi| < 1$ on $\mathbb{C}$. By Liouville's Theorem ϕ must be constant. Now the polynomial $\sum_{i=0}^{L}\sum_{j=0}^{M} p_{ij}z^iY^j$ must involve Y because a non-zero polynomial of degree at most $L = [S^{9/5}]$ in z cannot have $N > \frac{1}{2}S^2$ different zeroes (if $S \geq 32$). Thus $f(z)$ is algebraic over $\mathbb{C}(z)$. However as already observed the only entire algebraic functions are polynomials, and this completes the proof. □

The reader will be able to check that the exponent $3/2$ in $cC^{|z|^{3/2}}$ of Lemma 10.2 can be improved to any $\theta < 2$ (see Exercise 10.1). But to reach $\theta = 2$ and moreover to get the right C requires a much finer analysis.

First we indicate why the above lemma is useful; the growth there will be achieved using the following observation.

Lemma 10.3 *Let f be an entire function and suppose that there are positive integers p, q such that the functions*

$$f(z), f(z+1), \ldots, f(z+p)$$

as well as

$$f(z), f(z+i), \ldots, f(z+qi)$$

are linearly dependent over $\mathbb{C}(z)$. Then there are constants c, C with $|f|_R \leq cC^{R\log R}$ for all $R \geq 1$.

Proof We write out the linear dependence relations with non-zero polynomial coefficients P, and we can assume that those multiplying $f(z), f(z+p), f(z+qi)$ are among these. We can find constants $c_0 > 0, a, d$ such that $c_0 \leq |P|_R \leq aR^d$ for all R sufficiently large.

Consider the inequality

$$|f|_{n/2} \leq b^n(n!)^d \tag{10.12}$$

for some suitable $b \geq (p+q)a/c_0$. We show that there is a positive integer n_0, depending only on p and q, such that if (10.12) holds for some $n \geq n_0$ then it holds with $n+1$ in place of n.

Suppose $n/2 < |w| \leq (n+1)/2$ for example with $w = x + iy$ in the non-negative quadrant with $x, y \geq 0$. When $x \geq y$ it is not difficult to see with a little picture-drawing that one step westwards gives $|w-1| < n/2$; and the same is true of $w-2, \ldots, w-p$ (for suitable n_0). Now putting $z = w - p$ in the first of the above dependence relations gives

$$|f(w)| \leq c_0^{-1} pa(n/2)^d b^n (n!)^d \leq b^{n+1}((n+1)!)^d.$$

When $x \leq y$ in the same quadrant a similar argument works with steps southward and $w - i$ using the second dependence relation. In the other quadrants we navigate similarly, with $w + i$ and $w + 1$.

Now we choose $b \geq (p+q)a/c_0$ such that (10.12) holds for $n = n_0$. Then by induction it holds for all $n \geq n_0$. This plainly suffices to complete the proof. □

The conclusion of this lemma cannot be much improved (see Exercise 10.2).

Next we need finer versions of (10.6) and (10.10), and also for the Blaschke factors a weighted version. So first we present a formulation for general weights. For γ in $\mathbb{G}$ we write $U_\gamma = \gamma + [0,1)^2$ for the "half-open lattice square" in $\mathbb{R}^2 = \mathbb{C}$ of unit side whose south-west corner is γ. Write $\mathbb{D}_R$ for the set defined by $|z| \leq R$ and $\mathbb{G}_R = \mathbb{G} \cap \mathbb{D}_R$, with $\mathbb{G}'_R$ for the subset of $\mathbb{G}_R$ such that the whole U_γ lies in the interior of $\mathbb{D}_R$.

Proposition 10.4 *For $R > 0$ let F be a continuously differentiable real-valued function defined on $\mathbb{D}_{R+\sqrt{2}}$. Then*

$$\left| \sum_{\mathbb{G}_R} F(\gamma) - \int_{\mathbb{D}_R} F dx dy \right| \leq 36(R+\sqrt{2})M + \sqrt{2} \sum_{\mathbb{G}'_R} M_\gamma,$$

where

$$M = \sup_{\mathbb{D}_{R+\sqrt{2}}} |F|, \qquad M_\gamma = \sup_{U_\gamma} \sqrt{F_x^2 + F_y^2}$$

for the partial derivatives $F_x = \partial F/\partial x$ and $F_y = \partial F/\partial y$.

Proof This also needs pictures. We have a disjoint union $\mathbb{D}_R = U_R \cup P_R$, where U_R is the disjoint union of all the lattice squares U_γ for γ in $\mathbb{G}'_R$ and P_R is the disjoint union of all the pieces $U_\gamma \cap \mathbb{D}_R$ for γ in $\mathbb{G}$ such that U_γ meets the boundary of $\mathbb{D}_R$. Accordingly the integral above splits as $I_U + I_P$.

On each U_γ in U_R we approximate $F(z)$ by $F(\gamma)$. In fact with $z = x + iy$ and $\gamma = u + iv$, we have $F(z) - F(\gamma) = g(1) - g(0)$ with $g(t) = F((1-t)\gamma + tz)$. So a Mean Value Theorem gives τ with $0 < \tau < 1$ such that

$$F(z) - F(\gamma) = g'(\tau) = (x-u)F_x(w) + (y-v)F_y(w)$$

for $w = (1-\tau)\gamma + \tau z$ also in U_γ. By Cauchy–Schwarz we get $|F(z) - F(\gamma)| \le \sqrt{2}M_\gamma$. It follows that

$$|I_U - \sum_{\mathbb{G}'_R} F(\gamma)| \ \le \ \sqrt{2}\sum_{\mathbb{G}'_R} M_\gamma.$$

We treat I_P and the remaining sum Σ of $F(\gamma)$ when U_γ meets the boundary of $\mathbb{D}_R$ as follows. If $R \ge \sqrt{2}$ the U_γ lie in the annulus with radii $R \pm \sqrt{2}$ and so their number is at most the area $4\pi\sqrt{2}R$. If $R < \sqrt{2}$ a similar argument gives at most 8π. This gives the bound $(4\pi\sqrt{2}R + 8\pi)M$ for both $|I_P|$ and $|\Sigma|$, and the Proposition follows. □

We give two special cases. The first is highly classical and is the required refinement of (10.6) and (10.10).

Corollary 10.5 *We have* $|\sum_{\mathbb{G}_R} 1 - \pi R^2| \ \le \ 36(R + \sqrt{2})$.

Proof Just take $F = 1$. □

The second will be needed for Blaschke.

Corollary 10.6 *Suppose* $|w| \ge R + 5$. *Then we have*

$$\left|\sum_{\mathbb{G}_R} \log|w-\gamma| - \pi R^2 \log|w|\right| \ \le \ 36(R+\sqrt{2})\log|w| + 84|w|.$$

Proof Take

$$F(z) = \log|w-z| = \log|w| + \log\left|1 - \frac{z}{w}\right|,$$

for which the integral is the real part of

$$\int_{\mathbb{D}_R} \log|w| dxdy - \sum_{n=1}^{\infty} \frac{n}{w^n}\int_{\mathbb{D}_R} z^n dxdy \ = \ \pi R^2 \log|w|.$$

Also $\sqrt{F_x^2 + F_y^2} = 1/|w-z|$. Let γ_w be the closest lattice point to w. Then for z in U_γ with γ in $\mathbb{G}'_R$, we have

$$|w-z| \ \ge \ |\delta| - \frac{1}{2}\sqrt{2} - \sqrt{2} \ \ge \ \frac{1}{2}|\delta|$$

with $\delta = \gamma_w - \gamma$ because $|\delta| \geq 5 - \frac{1}{2}\sqrt{2} > 3\sqrt{2}$. Also

$$|\delta| \;\leq\; |w| + \frac{1}{2}\sqrt{2} + R \;\leq\; W.$$

for $W = 2|w|$. It follows that

$$\sum_{\mathbb{G}'_R} M_\gamma \;\leq 2 \sum_{0 \neq \delta \in \mathbb{G}_W} \frac{1}{|\delta|}.$$

Taking δ in the sum with strictly negative real and imaginary parts, we get at most

$$\sum_\delta \int_{U_\delta} \frac{1}{|z|} \mathrm{d}x\mathrm{d}y \;\leq\; \int_0^{\pi/2} \int_0^W \mathrm{d}r\mathrm{d}\theta \;=\; \frac{\pi}{2} W.$$

The same bound holds for the other three open quadrants, as we see by replacing δ there by $i\delta, -\delta, -i\delta$. For the remaining terms we use just $|\delta| \geq 1$ to get at most $4[W] \leq 4W$; thus the $\sum_{\mathbb{G}'_R} M_\gamma \leq 2(2\pi + 4)W$. Finally for $|z| \leq R + \sqrt{2}$ we have $\log |w - z| \leq \log(2|w|)$ and so the Proposition gives an error term of at most

$$36(R + \sqrt{2}) \log(2|w|) + 4\sqrt{2}(2\pi + 4)|w| \;\leq\; 36(R + \sqrt{2}) \log |w| + C|w|$$

with

$$C = 36 \log 2 + 4\sqrt{2}(2\pi + 4) < 84.$$

□

We can now start on the proof of Theorem 10.1. By analogy with (9.5) we assume that

$$|f(z)| \leq ce^{|z|^2/\mathcal{X}} \tag{10.13}$$

to see what absolute constant $\mathcal{X} > 0$ comes out. As hinted near the end of the previous chapter we will replace the $f(z)^j$ in (10.4) by $f(z + j)$. It turns out that the precise counting in Corollary 10.5 is not yet needed, so we are happy with the upper bound (10.6). We can even stick with the choices (10.11) of basic parameters. However the coefficients in the auxiliary function (strictly speaking it is not a polynomial) will get too big unless we use the full force of the Siegel Lemma with small exponent. We denote by $c_1, c_2, \ldots$ quantities which depend only on c and $\mathcal{X}$ (and $\mathcal{Y}$ below).

Lemma 10.7 *For any sufficiently large integer S and*

$$L = [S^{9/5}], \quad M = [S^{2/5}]$$

there are rational integers p_{ij} with

$$0 < \max |p_{ij}| \leq \exp(c_1 S^{9/5})$$

such that

$$\phi(z) = \sum_{i=0}^{L}\sum_{j=0}^{M} p_{ij} z^i f(z+j) \tag{10.14}$$

vanishes on $\mathbb{G}_S$.

Proof We get quantities like the right-hand side of (10.8), but now we take into account the Siegel exponent in Proposition 8.3, which is $1/(\sigma - 1)$ where σ is the number $(L+1)(M+1)$ of unknowns divided by the number of equations; the latter is by (10.6) at most $2(2S+1)^2$ and so $\sigma \geq (1/18)S^{1/5}$ for $S \geq 1$. We get using (10.13)

$$|p_{ij}| \;\leq\; \left((L+1)(M+1)S^L c e^{(S+M)^2/\mathcal{X}}\right)^{1/(\sigma-1)}.$$

At this stage of the argument the $\mathcal{X}$ plays no role, and so the estimate of the present lemma drops out (even for $S > 18^5$). ☐

For the next lemma we use a quantity $\mathcal{Y} > 1$ of the same nature as $\mathcal{X}$ in (10.13), also to be fixed soon (and now we need $S \geq 5/(\mathcal{Y}-1)$ too).

Lemma 10.8 *Suppose for some integer $S_1 \geq S$ that ϕ vanishes on $\mathbb{G}_{S_1}$. Then*

$$\log|\phi|_{S_1+5} \;\leq\; \left(\frac{\mathcal{Y}^2}{\mathcal{X}} - \pi \log \mathcal{Y}\right) S_1^2 + c_2 S_1^{9/5} \log S_1.$$

Proof We take some $\tilde{S}_1 \geq S_1$ and divide ϕ by the product

$$B(z) = \prod_{\gamma \in \mathbb{G}_{S_1}} \frac{\tilde{S}_1(z-\gamma)}{\tilde{S}_1^2 - \overline{\gamma} z}$$

of the individual Blaschke factors β_γ. The fact $|\beta_\gamma|_{\tilde{S}_1} = 1$ is no longer useful, but for any $\tilde{S}_1 \geq S_1 + 5$ we deduce

$$|\phi|_{S_1+5} \;\leq\; \mu^{-1}|B|_{S_1+5}|\phi|_{\tilde{S}_1}, \tag{10.15}$$

where $\mu = \inf_{|z|=\tilde{S}_1} |B(z)| > 0$. Here

$$|\phi|_{\tilde{S}_1} \;\leq\; (L+1)(M+1)\exp(c_1 S^{9/5})\tilde{S}_1^L c e^{(\tilde{S}_1+M)^2/\mathcal{X}} \tag{10.16}$$

and we will choose $\tilde{S}_1 = \mathcal{Y}S_1$. We proceed to estimate μ and $|B|_{S_1+5}$ in (10.15). We note that the denominator

$$\prod_{\gamma\in\mathbb{G}_{S_1}} (\tilde{S}_1^2 - \overline{\gamma} z) = \prod_{\gamma\in\mathbb{G}_{S_1}} |z| \left|\frac{\tilde{S}_1^2}{z} - \overline{\gamma}\right| = \prod_{\gamma\in\mathbb{G}_{S_1}} |z| \left|\frac{\tilde{S}_1^2}{z} - \gamma\right|$$

since $\mathbb{G}_{S_1}$ is symmetric about the real axis.

To begin with suppose $|z| = S_1 + 5$. By Corollaries 10.5 and 10.6 we get a first approximation

$$(\log \tilde{S}_1 - \log S_1)\pi S_1^2 + \pi S_1^2 \log S_1 - \pi S_1^2(2\log \tilde{S}_1 - \log S_1)$$

to $\log |B|_{S_1+5}$, and the error is of order at most $\tilde{S}_1 \log \tilde{S}_1$ or equivalently $S_1 \log S_1$. Similarly for $|z| = \tilde{S}_1$; the $\log S_1$ in the approximation above changes into $\log \tilde{S}_1$ giving a zero total, and so we see that $\log \mu$ is of order at most $S_1 \log S_1$. Thus we get $-\pi S_1^2 \log \mathcal{Y}$ for $\log(\mu^{-1}|B|_{S_1+5})$ with an error of order at most $S_1 \log S_1$. And now the required result follows using (10.16). □

We conclude the proof of Theorem 10.1 as follows. We fix $\mathcal{Y} > 1$ and then $\mathcal{X}$ with

$$\mathcal{X} > \frac{\mathcal{Y}^2}{\pi \log \mathcal{Y}} \tag{10.17}$$

With $\phi = 0$ on $\mathbb{G}_S$ as in Lemma 10.7, we deduce from Lemma 10.8 with $S_1 = S$ that $|\phi|_{S+5} < 1$ (for S sufficiently large). It follows that $\phi = 0$ on $\mathbb{G}_{S+5}$. Now with $S_1 = S + 5$ we get $|\phi|_{S+10} < 1$ and $\phi = 0$ on $\mathbb{G}_{S+10}$. And so on. We find that $\phi = 0$ on all of $\mathbb{G}$; but again in view of the sigma function this is not an immediate contradiction. However we also find that $|\phi| < 1$ on all of $\mathbb{C}$. So ϕ is identically constant and so identically zero, and from (10.14) we see that $f(z), f(z+1), \ldots, f(z+M)$ are linearly dependent over $\mathbb{C}(z)$.

Now the same argument all over again (yet more déja vu) with $f(z/i)$ instead of $f(z)$, which still takes $\mathbb{G}$ to $\mathbb{G}$ and satisfies a growth condition exactly as in (10.13), shows that also $f(z), f(z+i), \ldots, f(z+Mi)$ are linearly dependent over $\mathbb{C}(z)$. We deduce from Lemma 10.3 that f grows much more slowly than (10.13) and in view of Lemma 10.2 this is more than enough to conclude that f is a polynomial.

So what is our critical value of $\mathcal{X}$ in (10.13)? By (10.17) it is just the minimum of $\mathcal{Y}^2/(\pi \log \mathcal{Y})$ for $\mathcal{Y} > 1$, which occurs at $\mathcal{Y} = \sqrt{e}$ and is $2e/\pi$. This gives the required $\exp(\pi/2e)$ of Theorem 10.1, and the proof is complete.

It is not too hard to construct extremal examples. One orders the elements of $\mathbb{G}$ as $\gamma_0, \gamma_1, \gamma_2, \ldots$ and uses the Newton series

$$f(z) = a_0 + a_1\frac{z-\gamma_0}{\gamma_1-\gamma_0} + a_2\frac{(z-\gamma_0)(z-\gamma_1)}{(\gamma_2-\gamma_0)(\gamma_2-\gamma_1)} + \cdots = \sum_{k=0}^{\infty} a_k \prod_{l=0}^{k-1} \frac{z-\gamma_l}{\gamma_k-\gamma_l} \tag{10.18}$$

(compare (9.2) above and Exercise 9.6) with

$$a_k = f(\gamma_k) + \lambda_{k1}f(\gamma_{k-1}) + \lambda_{k2}f(\gamma_{k-2}) + \cdots + \lambda_{kk}f(\gamma_0)$$

for certain λ_{kl} independent of f (compare (9.3) above). One chooses the $f(\gamma_k)$ in $\mathbb{G}$ inductively to make for example a_k in the square U_{1+i}. In particular $\sqrt{2} \le |a_k| < 2\sqrt{2}$, so the resulting f is not a polynomial and its growth is not too big. The latter can be made precise using estimates as in Corollary 10.6 but with $|w| < R+5$ instead of $|w| \ge R+5$. We omit the details (see Exercises 10.5, 10.6) as auxiliary polynomials are not involved. There seems to be no natural extremal example like 2^z of the previous chapter.

Indeed there is a basic difference between the situation with $\mathbb{Z}$ in the previous chapter and with $\mathbb{G}$ here. Already we saw that there is $C > 2$ such that there are at most countably many f mapping $\{0, 1, 2, \ldots\}$ into $\mathbb{Z}$ with $|f|_R \le cC^R$. But for any $C > \exp(\pi/2e)$ the procedure above, for example choosing a_k in U_{1+i} or U_{2+2i} at random, gives uncountably many f mapping $\mathbb{G}$ into $\mathbb{G}$ with $|f|_R \le cC^{R^2}$.

See Exercise 10.12 for the general form of Gramain's Theorem; and also his original paper Gramain (1981) for more references including Masser (1980a).

Exercises

10.1 Let $\mathbb{G}_0$ be the set of γ in $\mathbb{G}$ with $\Re\gamma \ge 0$ and $\Im\gamma \ge 0$. Let f be an entire function mapping $\mathbb{G}_0$ into $\mathbb{G}$. If there are constants c, C, ρ with $\rho < 2$ and $|f(z)| \le cC^{|z|^\rho}$, adapt the proof of Lemma 10.2 to show that f is a polynomial.

10.2 Let f be an exponential polynomial of the form $\sum_{k=1}^{n} P_k(z)e^{\lambda_k z}$ for polynomials P_k and constants λ_k.

(a) Show that there are positive integers p, q such that the functions $f(z)$, $f(z+1), \ldots, f(z+p)$ as well as $f(z), f(z+i), \ldots, f(z+qi)$ are linearly dependent over $\mathbb{C}(z)$.

(b) Show that $|f|_R \le cC^R$.

10.3 Show that the polynomials in $\mathbb{C}[z]$ of degree at most 2 mapping $\mathbb{G}$ into $\mathbb{G}$ are precisely the $a_0 + a_1 z + a_2 \frac{z(z-1)}{1+i}$ $(a_0, a_1, a_2 \in \mathbb{G})$.

* * * * * * * * * * * * * * * * * * *

10.4 With $\sigma(z)$ as in (10.1) show that

$$\sigma\left(n+\frac{1}{2}\right) = (-1)^n \sigma\left(\frac{1}{2}\right) e^{-\pi/8} e^{(\pi/2)(n+\frac{1}{2})^2}$$

for any n in $\mathbb{Z}$ [*Hint*: Chapter 20].

10.5 Show that there is a constant C such that

$$\left| \sum_{\mathbb{G}_{R,*}} \log|w-\gamma| - \pi R^2 \log R + \frac{1}{2}\pi R^2 - \frac{1}{2}\pi |w|^2 \right| \leq CR\log(R+2)$$

whenever $|w| < R+5$, where the star in the summation means that any term with $|w-\gamma| < 1$ is omitted.

10.6 If all $|a_k| < 2\sqrt{2}$ in (10.18) show that $|f|_R \leq cC^{R^2}$ for any $C > \exp(\pi/2e)$.

10.7 Let $\mathcal{P}_L$ be the set of all polynomials in $\mathbb{C}[z]$ of degree at most L mapping $\mathbb{G}$ into $\mathbb{G}$. Show that $\mathcal{P}_L$ is a free $\mathbb{G}$-module of rank $L+1$ and that

$$\mathcal{P}_3 = \mathbb{G} + \mathbb{G}z + \mathbb{G}\frac{z(z-1)}{1+i} + \mathbb{G}\frac{z(z-1)(z-i)}{1+i}.$$

10.8 Is there an entire function f mapping $\mathbb{G}$ into $\mathbb{G}$, not a polynomial, with $|f(z)| \leq cC_0^{|z|^2}$ for $C_0 = \exp(\pi/2e)$? I don't know.

10.9 With $\mathbb{G}_0$ as in Exercise 10.1, is there $C_0 > 0$ such that any entire f mapping $\mathbb{G}_0$ into $\mathbb{G}$ with $|f(z)| \leq cC^{|z|^2}$ for $C < C_0$ is a polynomial? (Probably.)

10.10 For $k \geq 2$ let A_k be the area of the smallest closed disc, with centre anywhere in the plane, which contains at least k different Gaussian integers. For example $A_2 = \pi/4$, $A_3 = A_4 = \pi/2$. Show that

$$\lim_{n\to\infty}\left(\sum_{k=2}^{n}\frac{1}{A_k} - \log n\right)$$

exists.

10.11 With $\sigma(z)$ as in (10.1) show that

$$\sigma\left(\frac{1}{2}\right) = \frac{2^{5/4}\pi^{1/2}e^{\pi/8}}{\Gamma(1/4)^2}$$

in Exercise 10.4 [*Hint*: Chapter 20].

10.12 Let K be an imaginary quadratic field with ring of integers $\mathcal{O}$ and discriminant $\Delta < 0$, and let f be an entire function mapping $\mathcal{O}$ into $\mathcal{O}$. If there are constants c, C with $|f(z)| \leq cC^{|z|^2}$ and $C < \exp(\pi/(e\sqrt{|\Delta|}))$ show that f is a polynomial (Gramain, 1981).

11

Transcendence I – Mahler

In Chapters 2 and 3 we proved the irrationality of certain numbers ξ. The results of Chapters 4 and 5 can be considered as modest extensions. For example the finiteness of the number of (x, y) in $\mathbb{Z}^2$ with $x(x^2-2y^2) = y$ is equivalent to the fact that for all large η in $\mathbb{Z}$, the number ξ with $\xi(\xi^2 - 2\eta^2) = \eta$ is of degree 3 over $\mathbb{Q}$; or the irreducibility of $X^4 - 2\eta^3 X - \eta$ gives a number of degree 4 over $\mathbb{Q}$.

In the present chapter we wish to go further and prove that certain ξ are transcendental over $\mathbb{Q}$.

In Chapters 2 and 3 the irrationality of ξ was proved by first assuming ξ in $\mathbb{Q}$, then by some method constructing a sequence of small rational numbers $\xi_n \neq 0$ $(n = 0, 1, 2, \ldots)$, multiplying by suitable denominators to get into $\mathbb{Z}$, and using the Fundamental Theorem of Transcendence to get a contradiction. As it stands this will usually not suffice for stronger results; for example, if we just want to prove that ξ is not in the fixed field $\mathbb{Q}(\sqrt{2}) = \mathbb{Q}[\sqrt{2}] = \mathbb{Q}+\mathbb{Q}\sqrt{2}$, then we may be able to get into $\mathbb{Z} + \mathbb{Z}\sqrt{2} = \mathbb{Z}[\sqrt{2}]$ using denominators; however a $\omega \neq 0$ in $\mathbb{Z}[\sqrt{2}]$ certainly need not satisfy $|\omega| \geq 1$, as the example $\omega = \sqrt{2} - 1$ or even

$$\begin{aligned} 0 &< (\sqrt{2} - 1)^{50} = 6882627592338442563 - 4866752642924153522\sqrt{2} \\ &< 10^{-19} \end{aligned} \tag{11.1}$$

shows (the situation of Exercise 2.9(b) was too special). However there is a simple way of moving from $\mathbb{Z}[\sqrt{2}]$ to $\mathbb{Z}$ using the norm, defined generally on $\mathbb{Q}(\sqrt{2})$ by

$$\mathrm{Nm}(x + y\sqrt{2}) = x^2 - 2y^2$$

for x, y in $\mathbb{Q}$. It is clear that if $\omega \neq 0$ then $\mathrm{Nm}\,\omega \neq 0$; and if $\omega = x + y\sqrt{2}$ is small then possibly so is $\mathrm{Nm}\,\omega = \omega\omega'$ for $\omega' = x - y\sqrt{2}$. This means we have to take conjugates into account.

Now the transcendence of ξ is equivalent to saying that ξ does not lie in K whenever K is a finite extension of $\mathbb{Q}$, and here too there is a norm Nm from K to $\mathbb{Q}$. Its description in terms of conjugates uses field homomorphisms σ from K to a larger field, which for convenience we can take as $\mathbb{C}$. It is known that if $[K:\mathbb{Q}] = D$ then there are exactly D different σ. For example with $\mathbb{Q}(\sqrt{2})$ as above, one of these satisfies $\sigma(\sqrt{2}) = \sqrt{2}$ and the other one satisfies $\sigma(\sqrt{2}) = -\sqrt{2}$. Or with $\mathbb{Q}(\sqrt[3]{2})$ again one of these satisfies $\sigma(\sqrt[3]{2}) = \sqrt[3]{2}$ and the other two satisfy

$$\sigma(\sqrt[3]{2}) = \zeta\sqrt[3]{2} \tag{11.2}$$

where $\zeta = (-1 \pm \sqrt{-3})/2$ are the two primitive cube roots of unity.

For ξ in K one defines the norm

$$\mathrm{Nm}\,\xi = \prod_{\sigma} \sigma(\xi),$$

which is known to lie in $\mathbb{Q}$ and is clearly non-zero when ξ is.

And for ξ in K we define

$$||\xi|| = \max_{\sigma} |\sigma(\xi)| \tag{11.3}$$

(called variously size, house, maison or Haus). As the notation suggests this depends only on ξ and not K (however that is not so for the norm). For example ξ could also lie in a finite extension $\tilde{K}$ of $\mathbb{Q}$ containing K. Then it is known that every σ on K extends to at least one $\tilde{\sigma}$ on $\tilde{K}$ (and in fact exactly $[\tilde{K}:K]$ different $\tilde{\sigma}$), and clearly every $\tilde{\sigma}$ restricts on K to a σ, so that the two definitions of $||\xi||$ are the same. This argument extends to any finite extension containing ξ simply by considering the composite field of two such extensions.

For example $\xi = (\sqrt{2}-1)^{50}$ in (11.1) has conjugate

$$\xi' = (-\sqrt{2}-1)^{50} = 688262759233844256 3 + 486675264292415352 2\sqrt{2} > 10^{19}$$

and so $||\xi|| = \xi'$. Or Ramanujan gave

$$\xi = \frac{63}{25}\left(\frac{17+15\sqrt{5}}{7+15\sqrt{5}}\right) = 3.14159265380568820189839000\ldots \tag{11.4}$$

in $\mathbb{Q}(\sqrt{5})$ as a good approximation to $\pi = 3.1415926535\ldots$. Now the conjugate is

$$\xi' = \frac{63}{25}\left(\frac{17-15\sqrt{5}}{7-15\sqrt{5}}\right) = 1.57052630530211848955142412\ldots \tag{11.5}$$

(actually not a bad approximation to $\pi/2 = 1.5707\ldots$) and so now $||\xi|| = \xi$.

In general it is easy to see that

$$||\xi+\eta|| \leq ||\xi||+||\eta||, \quad ||\xi\eta|| \leq ||\xi||.||\eta|| \tag{11.6}$$

with more precise versions $||a\xi|| = |a|||\xi||$ for a in $\mathbb{Q}$ and $||\xi^m|| = ||\xi||^m$ for m in $\mathbb{N}$; the latter even gives $||\xi^{1/m}|| = ||\xi||^{1/m}$ for any determination of $\xi^{1/m}$.

To get a version of the Fundamental Theorem for algebraic numbers, we use the ring $\mathbb{Z}_K$ of all ω in K which are integral over $\mathbb{Z}$; that is, for which there are $a_1, \ldots, a_m$ in $\mathbb{Z}$ with

$$\omega^m + a_1\omega^{m-1} + \cdots + a_m = 0. \tag{11.7}$$

It is known that $\mathbb{Z}_K$ is a finitely generated abelian group of rank D; that is, there are $\omega_1, \ldots, \omega_D$, linearly independent over $\mathbb{Q}$, such that

$$\mathbb{Z}_K = \mathbb{Z}\omega_1 + \cdots + \mathbb{Z}\omega_D. \tag{11.8}$$

For any ω in $\mathbb{Z}_K$ the norm $\mathrm{Nm}\,\omega$ lies in $\mathbb{Z}$; for example it lies in $\mathbb{Q}$ and also in $\mathbb{Z}_{\mathcal{K}}$ for any extension $\mathcal{K}$ of $\mathbb{Q}$ containing all the conjugates of ξ, so also in $\mathbb{Q} \cap \mathbb{Z}_{\mathcal{K}} = \mathbb{Z}_{\mathbb{Q}}$ which is well-known to be $\mathbb{Z}$.

Note that any ξ in K satisfies $\sum_{i=0}^{m} b_i\xi^{m-i} = 0$ for some $b_0 \neq 0, b_1, \ldots, b_m$ in $\mathbb{Z}$; and as (11.7) follows with $a_i = b_0^{i-1}b_i$ for $\omega = b_0\xi$, we see that b_0 is a denominator for ξ in the sense that $b_0\xi$ lies in $\mathbb{Z}_K$ (as in Exercise 5.2(a) too).

Proposition 11.1 *Suppose K in $\mathbb{C}$ has $[K : \mathbb{Q}] = D$ and $\omega \neq 0$ is in $\mathbb{Z}_K$. Then $||\omega|| \geq 1$ and*

$$|\omega| \geq ||\omega||^{-(D-1)}. \tag{11.9}$$

Proof As K lies in $\mathbb{C}$ one of the complex embeddings σ is the identity map σ_0, and so $\mathrm{Nm}\,\omega = \omega\prod_{\sigma\neq\sigma_0}\sigma(\omega)$. Now $N = \mathrm{Nm}\,\omega \neq 0$ lies in $\mathbb{Z}$, and so everything follows from the Fundamental Theorem $|N| \geq 1$ (which is itself just the case $D = 1$). □

This sort of estimate is often known as a "Liouville inequality". Here it can be shown that the exponent $D - 1$ is best possible (see Exercise 11.6).

We will illustrate the application of (11.9) to transcendence in a moment, but for a five-finger exercise now consider the equation

$$x^5 + 15x + 12 = 0.$$

My Maple can compute its Galois group as metacyclic of order 20, well-known to be solvable; but it cannot compute a solution ξ in radicals. The standard procedure gives after some effort $5\xi = \eta_1 + \eta_2$, where

$$\eta_1 = -\sqrt[5]{1875 + 525\sqrt{10}} - \sqrt[5]{1875 - 525\sqrt{10}}$$

$$\eta_2 = -\sqrt[5]{-5625 + 1800\sqrt{10}} + \sqrt[5]{5625 + 1800\sqrt{10}}$$

with all fifth roots real, but it would be nice to check this directly. Maple can indeed simplify $\eta = \xi^5 + 15\xi + 12$ algebraically as 0 but it can also help purely numerically with

$$|\eta| < 10^{-700}, \tag{11.10}$$

which will contradict (11.9) as follows.

It is clear that each fifth root ζ in 5ξ has degree at most 10 over $\mathbb{Q}$, so for $K = \mathbb{Q}(\xi)$ we have $D \le 10000$ (of course if really $\xi^5 + 15\xi + 12$ vanishes then $D \le 5$). Also each ζ is integral over $\mathbb{Z}$, so 5ξ is in $\mathbb{Z}_K$ (if $\xi^5 + 15\xi + 12$ vanishes then ξ itself is integral). Now clearly $\omega = 3125\eta$ is in $\mathbb{Z}_K$. And the last ζ has

$$||\zeta|| = ||5625 + 1800\sqrt{10}||^{1/5} = (5625 + 1800\sqrt{10})^{1/5}.$$

Similarly for the other three, so that

$$||\xi|| \le \frac{2}{5}\left((1875 + 525\sqrt{10})^{1/5} + (5625 + 1800\sqrt{10})^{1/5}\right) = 4.637\ldots$$

and we find

$$||\omega|| \le 3125(||\xi||^5 + 15||\xi|| + 12) < 7000000.$$

Thus $|\omega| \le 7000000^{-10000}$ implies $\omega = 0$. This is not good enough for (11.10) but the fault lies clearly with the estimate for D, which we improve by noting that the fifth roots are related; for example in η_1

$$(1875 + 525\sqrt{10})(1875 - 525\sqrt{10}) = 759375 = 15^5$$

(similarly in η_2) and this leads to $D \le 100$ (and even $D \le 50$). Now (11.10) implies

$$|\omega| < (3125)10^{-700} < 7000000^{-100}$$

and so $\omega = 0$ as required.

Now we give a simple transcendence proof based on the truncation method of Chapter 2. The number $e = \sum_{k=0}^{\infty} 1/k!$ will have to wait a bit, but it was Liouville who in 1844 exploited factorials in a more powerful way to construct the first explicit transcendental number

$$\sum_{k=1}^{\infty} \frac{1}{10^{k!}} = 0.110001000000000000000001000000000000000000000000\ldots.$$

Here we generalize this to

Theorem 11.2 *Suppose α is algebraic in $\mathbb{C}$ with $0 < |\alpha| < 1$. Then $\sum_{k=1}^{\infty} \alpha^{k!}$ is transcendental.*

Proof If not, then both α and $\beta = \sum_{k=1}^{\infty} \alpha^{k!}$ are algebraic, and so $K = \mathbb{Q}(\alpha, \beta)$ is a finite extension of $\mathbb{Q}$. We consider the truncation

$$\beta_n = \beta - \sum_{k=1}^{n} \alpha^{k!} = \sum_{k=n+1}^{\infty} \alpha^{k!}.$$

Clearly $|\beta_n| \leq 2|\alpha|^{(n+1)!}$ for sufficiently large n. We can get into $\mathbb{Z}_K$ by choosing denominators a, b in $\mathbb{N}$ of α, β; then $\omega_n = a^{n!}b\beta_n$ is in $\mathbb{Z}_K$ with

$$|\omega_n| < 2a^{n!}b|\alpha|^{(n+1)!}. \tag{11.11}$$

Also

$$||\omega_n|| = a^{n!}b||\beta_n|| \leq a^{n!}b(||\beta|| + \sum_{k=1}^{n} ||\alpha||^{k!}) \leq a^{n!}b(||\beta|| + ||\alpha|| + 1)^{n!}.$$

So if $\omega_n \neq 0$ then (11.9) gives $|\omega_n| \geq \Omega^{-n!}$ for some Ω independent of n, and this contradicts (11.11) if n is sufficiently large.

In other words, $\omega_n = 0$ for all sufficiently large n, and so also $\beta_n = 0$; but as $\beta_{n+1} - \beta_n = -\alpha^{(n+1)!} \neq 0$ this contradiction is final. □

No-one knows how to prove the transcendence of the numbers

$$\sum_{k=0}^{\infty} \frac{\alpha^k}{2^{k(k-1)/2}}$$

considered in Chapter 2, even for say $\alpha = 1/3$. But we can handle the values of Mahler's $f(z) = \sum_{k=0}^{\infty} z^{2^k}$.

Recall here that genuine auxiliary polynomials were needed; for example

$$\phi(z) = P(z, f(z)) = 2z^6 + \cdots$$

as in (8.6) with

$$P(X, Y) = (3 - 2X)Y^2 + (1 - 4X)Y - X$$

as in (8.5) was used to prove the irrationality of $f(2/3)$. It will not suffice for the transcendence of $f(2/3)$ or even to show that it is not quadratic over $\mathbb{Q}$; however it does suffice to show that $\beta = f(1/2)$ (trivially irrational) is not quadratic, a fact also visible from the non-periodicity of the surprising continued fraction

$$\beta = [0; 1, 4, 2, 4, 4, 6, 4, 2, 4, 6, 2, 4, 6, 4, 4, 2, 4, 6, 2, 4, 4, 6, 4, 2, 6, 4, 2, 4, 6, 4, \ldots]$$

(in the standard notation) discovered independently by Kmošek and Shallit (see Exercise 3.14).

To see this assume that β is quadratic with denominator b, and consider for $\alpha = 1/2$ the familiar quantity

$$\beta_n = \phi(\alpha^{2^{n+1}}) = P(\alpha^{2^{n+1}}, \beta - \alpha - \cdots - \alpha^{2^n}).$$

At once $|\beta_n| \leq 3\alpha^{6.2^{n+1}}$ for large n and $\omega_n = 2^{2^{n+1}}(2^{2^n}b)^2\beta_n$ is in $\mathbb{Z}_K$ for $K = \mathbb{Q}(\beta)$ with

$$|\omega_n| \leq 2^{2^{n+1}}3(2^{2^n}b)^2\alpha^{6.2^{n+1}}.$$

Also $||\beta_n|| \leq \overline{P}(1, ||\beta|| + 1)$ for the majorizing polynomial

$$\overline{P}(X, Y) = (3 + 2X)Y^2 + (1 + 4X)Y + X$$

(compare Exercise 4.39). We obtain with $D = 2$ in (11.9)

$$|\omega_n|||\omega_n|| \leq 33b^4(||\beta|| + 1)^2 \left(\frac{1}{16}\right)^{2^n},$$

which implies $\omega_n = 0$ for all sufficiently large n. We already saw several ways of disproving this (see also below).

Now it is only a short jump to

Theorem 11.3 *Suppose α is algebraic in $\mathbb{C}$ with $0 < |\alpha| < 1$. Then $\sum_{k=0}^{\infty}\alpha^{2^k}$ is transcendental.*

Proof We choose P as in Lemma 3.3. Assume then that $\beta = \sum_{k=0}^{\infty}\alpha^{2^k}$ is algebraic, and work in $K = \mathbb{Q}(\alpha, \beta)$. Then for

$$\beta_n = \phi(\alpha^{2^{n+1}}) = P(\alpha^{2^{n+1}}, \beta - \alpha - \cdots - \alpha^{2^n})$$

and denominators a, b as above we get $\omega_n = a^{2^{n+1}L}(a^{2^n}b)^M\beta_n$ in $\mathbb{Z}_K$, with

$$|\omega_n| \leq C_1 a^{2^{n+1}L}(a^{2^n}b)^M|\alpha|^{2^{n+1}(LM+L+M)}$$

by Lemma 3.4, where C_1 is independent of n. Also

$$\begin{aligned} ||\beta_n|| &\leq \overline{P}(||\alpha||^{2^{n+1}}, ||\beta|| + ||\alpha|| + \cdots + ||\alpha||^{2^n}) \\ &\leq C_2(||\alpha|| + 1)^{2^{n+1}L}(||\beta|| + ||\alpha|| + 1)^{2^n M} \end{aligned}$$

for C_2 independent of n. Thus for $D = [K : \mathbb{Q}]$ in (11.9), we get for $|\omega_n|||\omega_n||^{D-1}$ the upper bound

$$C_1C_2^{D-1}(a^{2^{n+1}L}(a^{2^n}b)^M)^D\left((||\alpha|| + 1)^{2^{n+1}L}(||\beta|| + ||\alpha|| + 1)^{2^n M}\right)^{D-1}|\alpha|^N$$

with $N = 2^{n+1}(LM + L + M)$. This bound is $C_3\Theta^{2^n}$ with C_3 independent of n and

$$\Theta = a^{2LD}a^{MD}\left((||\alpha|| + 1)^{2L}(||\beta|| + ||\alpha|| + 1)^M\right)^{D-1}|\alpha|^{2(LM+L+M)}.$$

As L, M tend to infinity this Θ tends to zero; thus we just have to choose L, M (for example with $L = M$) such that $\Theta < 1$. Then we conclude $\beta_n = 0$ for all sufficiently large n. Again we know how to handle this, and now Proposition 8.1 is especially convenient although not really necessary, as we could even find the asymptotics for β_n using (2.26). □

Nearly all the comments about irrationality at the end of Chapter 3 extend to transcendence. For example we follow up on the Mandelbrot set $\mathbb{M}$ briefly mentioned there. This is of course the set of c in $\mathbb{C}$ for which the iterates of $z^2 + c$ starting with $z = 0$ do not go to infinity.

There is a conformal map $w = \Phi(z)$ from the complement of $\mathbb{M}$ to the set of w with $|w| > 1$. One has $\Phi(c) = f_c(c)$ for a suitable $f_c(z)$ given by

$$z + \frac{\frac{c}{2}}{z} + \frac{\frac{c}{4} - \frac{c^2}{8}}{z^3} + \frac{-\frac{3c^2}{8} + \frac{c^3}{16}}{z^5} + \frac{\frac{c}{8} - \frac{3c^2}{32} + \frac{15c^3}{32} - \frac{5c^4}{128}}{z^7} + \cdots$$

with functional equation $f_c(z^2 + c) = f_c(z)^2$. Becker and Bergweiler (see also Nishioka (1996) for references) showed that $\Phi(\alpha)$ is transcendental for every algebraic α in $\mathbb{C} \setminus \mathbb{M}$.

For some time it was thought that Mahler's Method might be able to prove that the digits in the decimal expansion of any irrational algebraic number like

$$\sqrt{2} = 1.41421356237309504880168872420969807856967187537694807 31...$$

cannot be generated by a specific class of Turing Machines called automata. But in the end this was proved by Adamczewski and Bugeaud using instead Schlickewei's extension of Schmidt's Subspace Theorem.

Just as for irrationality, one can ask for transcendence measures. But we don't give any in this book. See however Exercise 11.15.

Exercises

11.1 Let $\varepsilon_1, \varepsilon_2, \ldots, \varepsilon_n$ be in $\{+1, -1\}$ with $\alpha = \varepsilon_1\sqrt{1} + \varepsilon_2\sqrt{2} + \cdots + \varepsilon_n\sqrt{n} \neq 0$. Show that $|\alpha| \geq n^{-2^n}$.

11.2 Let $1 \leq f_1 < f_2 < \cdots$ be in $\mathbb{N}$ with $f_{k+1}/f_k \to \infty$, and let α in $\mathbb{C}$ be algebraic with $0 < |\alpha| < 1$. Show that $\sum_{k=1}^{\infty} \alpha^{f_k}$ is transcendental.

11.3 Using $P(X, Y) = (3 - 2X)Y^2 + (1 - 4X)Y - X$ and $f(z) = \sum_{k=0}^{\infty} z^{2^k}$, show that $f(a/b)$ is not quadratic irrational for all a, b in $\mathbb{N}$ with $a < b^{1/3}$.

11.4 We want to show, using P in $\mathbb{Z}[X, Y]$ of degree at most L in X and of degree at most L in Y, that $f(1/2)$ is not cubic irrational. How big must L be?

* * * * * * * * * * * * * * * * * * * *

11.5 Show that

$$\sqrt{11+2\sqrt{29}}+\sqrt{16-2\sqrt{29}+2\sqrt{55-10\sqrt{29}}} \;=\; \sqrt{5}+\sqrt{22+2\sqrt{5}}$$

(Shanks, 1974).

11.6 For an integer $D \geq 2$ and an even integer $t \geq 4$ let $P(X) = X^D - tX + 2$.

(a) Show that $P(X)$ is irreducible over $\mathbb{Q}$ [*Hint*: Eisenstein].

(b) Show that there is real ω with $P(\omega) = 0$ and $0 < \omega < 4/t$ [*Hint*: calculate two values of P].

(c) Show that $||\omega||^{D-1} \leq 2t$.

(d) Deduce that Proposition 11.1 cannot be essentially improved for any fixed $D \geq 2$.

(e) Show that $\omega \geq 1/(2t)$ and $||\omega||^{D-1} \geq t/4$ [*Hint*: Proposition 11.1].

11.7 Let $g(z) = \sum_{k=0}^{\infty} z^{2^k}/(1+z^{2^{k+1}})$. Show that z and $g(z)$ are algebraically independent over $\mathbb{C}$; that $g(\alpha)$ is transcendental for all algebraic α in $\mathbb{C}$ with $0 < |\alpha| < 1$; and that $\sum_{m=0}^{\infty} 1/V_{2^m}$ is transcendental for the V_n in Exercise 3.7.

11.8 Let $\varepsilon_1, \varepsilon_2, \ldots, \varepsilon_{1948}$ be in $\{+1, -1\}$. Show that $\varepsilon_1\sqrt{1} + \varepsilon_2\sqrt{2} + \cdots + \varepsilon_{1948}\sqrt{1948} \neq 0$ [*Hint*: 1933].

11.9 (a) If $\xi^n = \eta^n = 1$ for some positive integer n, and $\alpha = \xi + \eta \neq 0$ show that $|\alpha| \geq 2^{-n}$.

(b) Improve this to $|\alpha| \geq \pi/n$.

11.10 If $\xi^n = \eta^n = \zeta^n = 1$ for some positive integer n, and $\alpha = \xi+\eta+\zeta \neq 0$ show that $|\alpha| \geq c/n$ for some absolute constant $c > 0$. This sort of thing can be done also for four roots of unity (Myerson, 1986), but not yet for five, where only exponential bounds are known.

11.11 Does Exercise 11.2 still hold with "f_{k+1}/f_k unbounded" instead of "$f_{k+1}/f_k \to \infty$"?

11.12 Let $\alpha, \alpha_1, \ldots, \alpha_m$ be algebraic with $\alpha^m + \alpha_1\alpha^{m-1} + \cdots + \alpha_m = 0$. Show that $\| \alpha \| \leq \max\{1, \| \alpha_1 \| + \cdots + \| \alpha_m \|\}$.

11.13 Let $\mathcal{D} \geq 1, A \geq 1, S \geq 1$ be in $\mathbb{R}$, and let $\overline{\mathbb{Q}}(\mathcal{D}, A, S)$ be the set of all algebraic α with degree at most $\mathcal{D}$, a denominator at most A, and $\| \alpha \|$ at most S. Show that $\overline{\mathbb{Q}}(\mathcal{D}, A, S)$ is empty or finite.

11.14 Let τ be transcendental. Show that there is a function $F : [1,\infty)^3 \to (0,\infty)$ such that $|\tau - \alpha| \geq F(\mathcal{D}, A, S) > 0$ for all α in $\overline{\mathbb{Q}}(\mathcal{D}, A, S)$.

11.15 Find such a $F(\mathcal{D}, A, S)$ in Exercise 11.14 for $\tau = \sum_{k=1}^{\infty} 10^{-k!}$ [*Hint*: $|\tau - \alpha| \geq |\alpha_n - \alpha| - |\tau - \alpha_n|$].

11.16 Let $\overline{\mathbb{Z}}(\mathcal{D}, S) = \overline{\mathbb{Z}} \cap \overline{\mathbb{Q}}(\mathcal{D}, A, S) = \overline{\mathbb{Q}}(\mathcal{D}, 1, S)$. Show that $\overline{\mathbb{Z}}(\mathcal{D}, 1) = \{0, \text{all roots of unity}\}$ (Kronecker, 1857).

11.17 (a) Given D show there exists $\omega(D)$ such that any root of unity in any number field K with $[K : \mathbb{Q}] = D$ has order at most $\omega(D)$.

(b) Write $\overline{\omega}(\mathcal{D}) = \max_{D \leq \mathcal{D}} \omega(D)$. Suppose α has degree d and there is a positive integer n with α^n in $\mathbb{Q}$. Show that there is such an n with $n \leq \overline{\omega}(d^2)!$ for the factorial.

(c) Suppose β has degree d and there is a positive integer n with β^n in a quadratic field. Show that there is such an n with $n \leq \overline{\omega}(d^2)!$.

11.18 Show that $\xi = \sum_{k=1}^{\infty} 2^{-k!}$ and $\eta = \sum_{k=1}^{\infty} 3^{-k!}$ are algebraically independent over $\mathbb{Q}$ [*Hint*: let $P(X, Y)$ be in $\mathbb{Z}[X, Y]$ with $P(\xi, \eta) = 0$; with $\xi_n = \sum_{k=1}^{n} 2^{-k!}$ and $\eta_n = \sum_{k=1}^{n} 3^{-k!}$ show that $P(\xi_n, \eta_n) = 0$ for all $n \geq n_0$, and then $\frac{\partial P}{\partial X}(\xi, \eta) = 0$].

11.19 Is there $C \geq 1$ such that $|\alpha| \geq C^{-n}$ in Exercise 11.1? No-one knows.

11.20 Does $\sum_{k=1}^{\infty} z^{k!}$ have infinitely many zeroes on $\mathbb{D}$? I don't know.

11.21 (Theorem of the Six Exponentials.) Let ξ, η be complex numbers linearly independent over $\mathbb{Q}$, and let $\sigma_1, \sigma_2, \sigma_3$ be complex numbers also linearly independent over $\mathbb{Q}$. Using $P(e^{\xi z}, e^{\eta z})$ on $\mathbb{Z}\sigma_1 + \mathbb{Z}\sigma_2 + \mathbb{Z}\sigma_3$ show that at least one among the six numbers $e^{\xi\sigma_1}, e^{\eta\sigma_1}, e^{\xi\sigma_2}, e^{\eta\sigma_2}, e^{\xi\sigma_3}, e^{\eta\sigma_3}$ is transcendental.

11.22 Let c be irrational. Show that at least one of $2^c, 3^c, 5^c$ is transcendental.

11.23 In Exercise 11.22 must at least one of $2^c, 3^c$ be transcendental? No-one knows.

11.24 Show that $||\eta|| \leq \max\{1, ||\xi||^2 + ||\xi||\}$ for any algebraic numbers with $\eta(\eta^2 - \xi^2) = \xi$.

11.25 Show that there are at most finitely many algebraic integers ξ, η with $[\mathbb{Q}(\xi, \eta) : \mathbb{Q}] \leq 2$ and $\eta(\eta^2 - \xi^2) = \xi$ [*Hint*: Exercise 5.1 with $e = 1$ and $M = 2$, Proposition 11.1 and Exercise 11.13 – why can we assume $|\xi| = ||\xi||$?].

11.26 Let f be in $\mathbb{Z}[X, Y]$ of degree at most 5 such that $F(X, Y) = X^6 + Y^6 + f(X, Y)$ is irreducible over $\mathbb{Q}$. Show that there are at most finitely many algebraic integers ξ, η with $[\mathbb{Q}(\xi, \eta) : \mathbb{Q}] \leq 2$ and $F(\xi, \eta) = 0$ (Nitsch Master Thesis 2006).

11.27 Let f be in $\mathbb{Z}[X,Y]$ of degree at most 29 such that $F(X,Y) = X^{30} + Y^{30} + f(X,Y)$ is irreducible over $\mathbb{Q}$. Show that there are at most finitely many algebraic integers ξ, η with $[\mathbb{Q}(\xi,\eta) : \mathbb{Q}] \leq 3$ and $F(\xi,\eta) = 0$ (Nitsch Master Thesis 2006).

11.28 Let K be a subfield of $\mathbb{C}$ with $[K : \mathbb{Q}] = e$, and let λ be in the quotient field $K((1/X))$ of $K[[1/X]]$ (see Exercise 4.33), convergent for all large $|X|$, with $[K(X,\lambda) : K(X)] = d$. Show that there are at most finitely many x in $\mathbb{Z}$ such that $\lambda(x)$ converges with $[\mathbb{Q}(\lambda(x)) : \mathbb{Q}] \leq f$ provided $d > ef$. The case $e = d-1, f = 1$ (to be found in Schucan's Master Thesis 2004) implies Theorem 4.7 and the case $e = 1, f = d-1$ implies Theorem 5.1.

11.29 Let m be in $\mathbb{N}$ and let f be monic in $\mathbb{Z}[X]$, not of the form g^m with g in $\mathbb{Z}[X]$, of degree divisible by m. Show that there are at most finitely many integers x, y with $y^m = f(x)$ (compare the opening salvo of Chapter 4 together with several exercises there).

12
Irrationality measures II – Thue

The irrationality of algebraic numbers like $\sqrt{2}$ or $\sqrt[3]{2}$ is rather trivial, and so we can jump straight to the problem of irrationality measures in the sense of Chapter 8. One argument is fast, using the notation of the previous chapter. Namely if $\alpha = \sqrt{2}$ is near r/s (with r in $\mathbb{Z}$ and s in $\mathbb{N}$), then $\omega = r - s\alpha$ is small and as soon as $|\omega| \leq 1$ we get $|r| \leq (\alpha+1)s$ so $||\omega|| \leq |r| + s||\alpha|| \leq (2\alpha+1)s$. Thus by (11.9) with $d = 2$ we get

$$\left|\sqrt{2} - \frac{r}{s}\right| = \frac{1}{s}|\omega| \geq \frac{c}{s^2} \tag{12.1}$$

with $c = 1/(2\alpha + 1)$ (even if $|\omega| > 1$). We saw in (8.4) near the beginning of Chapter 8 that approximation to within order at most $1/s^2$ is possible for any real irrational, and consequently (12.1) is essentially best possible.

A similar argument with $\alpha = \sqrt[3]{2}$ and (11.2) leads to

$$\left|\sqrt[3]{2} - \frac{r}{s}\right| \geq \frac{c}{s^3} \tag{12.2}$$

with $c = 1/(2\alpha + 1)^2$, no longer clearly best possible. It gets worse with general algebraic numbers: if α has degree d then the same procedure leads easily to Liouville's

$$\left|\alpha - \frac{r}{s}\right| \geq \frac{c}{s^d}. \tag{12.3}$$

Of course behind (12.2) lies the equation $|N| \geq 1$ for the norm $N = r^3 - 2s^3$ of $r - s\alpha$. This leads to an interesting possibility of reversing the argument: if we could somehow improve (12.2) to

$$\left|\sqrt[3]{2} - \frac{r}{s}\right| \geq \frac{c}{s^\kappa} \tag{12.4}$$

for some $c > 0$ and some $\kappa < 3$, then it would follow that

$$|r^3 - 2s^3| \geq \frac{cs}{s^\kappa}|r - \zeta\alpha s||(r - \zeta^2\alpha s| \geq \frac{3}{4}c\alpha^2 s^{3-\kappa}$$

because $|z| \geq |\Im z|$ for the imaginary part. Thus for example the diophantine equation

$$x^3 - 2y^3 = m \tag{12.5}$$

for fixed m in $\mathbb{Z}$ would have at most finitely many solutions (x, y) in $\mathbb{Z}^2$. This cannot be deduced from the main result of Chapter 4 or the techniques used there.

Thue was the first to establish things like (12.4). We state his general result as follows, which substantially improves (12.3) when $d \geq 3$.

Theorem 12.1 *Suppose α is algebraic in $\mathbb{C}$ with degree $d \geq 3$. Then for any*

$$\kappa > \frac{d}{2} + 1 \tag{12.6}$$

there is $c > 0$ such that

$$\left|\alpha - \frac{r}{s}\right| \geq \frac{c}{s^\kappa} \tag{12.7}$$

for all integers r and $s \geq 1$.

We note that if α is not real with imaginary part $\eta \neq 0$ then trivially $|\alpha - r/s| \geq |\eta|$. So in the proof we may (although in fact we don't) assume that α is real.

How do we prove things like (12.4) or (12.7)?

The irrationality of e is a fairly immediate consequence of its good rational approximations $\sum_{k=0}^{n} 1/k!$, and these lead also to an irrationality measure (see Exercise 2.5).

Another approach comes from continued fractions. It is known that if $|\theta - r/s| < 1/(2s^2)$ then r/s is a convergent. Here

$$e = [2; 1, 2, 1, 1, 4, 1, 1, 6, 1, 1, 8, 1, 1, 10, 1, 1, 12, 1, 1, 14, 1, 1, 16, 1, 1, 18, ...]$$

is rather regular, involving arithmetic progressions, and this leads to $|e - r/s| \geq c/s^\kappa$ for any $\kappa > 2$. Similarly with

$$e^2 = [7; 2, 1, 1, 3, 18, 5, 1, 1, 6, 30, 8, 1, 1, 9, 42, 11, 1, 1, 12, 54, 14, 1, 1, 15, 66, ...].$$

However I don't know if

$$e^3 = [20; 11, 1, 2, 4, 3, 1, 5, 1, 2, 16, 1, 1, 16, 2, 13, 14, 4, 6, 2, 1, 1, 2, 2, 2, 3, 5, ...]$$

is similarly regular (doesn't look it), even though

$$2e = [5; 2, 3, 2, 3, 1, 2, 1, 3, 4, 3, 1, 4, 1, 3, 6, 3, 1, 6, 1, 3, 8, 3, 1, 8, 1, 3, 10, ...],$$
$$3e = [8; 6, 2, 5, 2, 5, 1, 2, 5, 1, 2, 1, 5, 4, 5, 5, 1, 4, 5, 1, 4, 1, 5, 6, 5, 1, 6, 5, 1, 6, ...]$$

are (Theorem of Hurwitz).

But none of this will work for algebraic numbers.

Generally if real θ has approximations p_n/q_n $(n = 0, 1, 2, \ldots)$ (with p_n in $\mathbb{Z}$ and q_n in $\mathbb{N}$), we may argue (as in Exercise 11.15) that

$$\left|\theta - \frac{r}{s}\right| \geq \left|\frac{p_n}{q_n} - \frac{r}{s}\right| - \left|\theta - \frac{p_n}{q_n}\right| \tag{12.8}$$

and the first term on the right is $N_n/(q_n s)$ for N_n in $\mathbb{Z}$. Thus if we can find n with

$$N_n \neq 0 \tag{12.9}$$

we deduce

$$\left|\theta - \frac{r}{s}\right| \geq \frac{1}{q_n}\left(\frac{1}{s} - |q_n\theta - p_n|\right);$$

and as long as

$$|q_n\theta - p_n| \to 0 \tag{12.10}$$

we can hope for a positive lower bound.

For example with

$$\theta = e, \quad q_n = n!, \quad p_n = n!\sum_{k=0}^{n}\frac{1}{k!},$$

indeed

$$|q_n\theta - p_n| = n!\sum_{k=n+1}^{\infty}\frac{1}{k!} \to 0,$$

and all reduces to the usual snag (12.9), here totally innocuous.

But how do we find p_n/q_n for $\theta = \sqrt[3]{2}$ (or even $\sqrt{2}$ for that matter)?

The classical method of Newton provides such approximations. Generally to solve $f(X) = 0$, one picks a first approximation x_0 and then defines inductively

$$x_n = x_{n-1} - \frac{f(x_{n-1})}{f'(x_{n-1})} \quad (n = 1, 2, \ldots). \tag{12.11}$$

There is a well-known simple geometric interpretation using tangents.

With $f(X) = X^3 - 2$ this is $x_n = R(x_{n-1})$ for

$$R(X) = \frac{2X^3 + 2}{3X^2}. \tag{12.12}$$

As

$$\sqrt[3]{2} = 1.2599210498948731647672106072782283505\,7\ldots$$

we could tentatively try $x_0 = 5/4$ (it is well-known that the wrong choice here can lead to chaos), and we get

$$x_0 = 1.25000000000000000000000000000000000000\ldots,$$

$$x_1 = 1.26000000000000000000000000000000000000\ldots,$$

$$x_2 = 1.25992105484168976232468295960359452422\ldots,$$

$$x_3 = 1.25992104989487318418985167181551599157\ldots,$$

$$x_4 = 1.25992104989487316476721060727822864998\ldots,$$

$$x_5 = 1.25992104989487316476721060727822835057\ldots,$$

$$x_6 = 1.25992104989487316476721060727822835057\ldots,$$

already stable to this accuracy. Of course these are all rational, for example $x_1 = 63/50$ and $x_2 = 375047/297675$.

The convergence to $\alpha = \sqrt[3]{2}$ is indeed rapid. In fact the main feature of Newton's Method is that

$$\left|\alpha - \frac{p_n}{q_n}\right| \le c\left|\alpha - \frac{p_{n-1}}{q_{n-1}}\right|^2$$

the so-called "quadratic convergence"; in this example we have the identity

$$R(X) - \alpha \;=\; \frac{2X+\alpha}{3X^2}(X-\alpha)^2. \tag{12.13}$$

Putting $X = p_{n-1}/q_{n-1}$ shows

$$\frac{p_n}{q_n} - \alpha = c_n\left(\frac{p_{n-1}}{q_{n-1}} - \alpha\right)^2, \tag{12.14}$$

with presumably $c_n \to (2\alpha+\alpha)/(3\alpha^2) = 1/\alpha$.

Thus $|\alpha - p_n/q_n| \to 0$ very quickly; for example $|\alpha - p_6/q_6| < 10^{-134}$. However this is not (12.10); for example $|q_6\alpha - p_6| > 10^{466}$. And indeed

$$\frac{p_n}{q_n} = \frac{2p_{n-1}^3 + 2q_{n-1}^3}{3p_{n-1}^2 q_{n-1}},$$

making it unlikely that q_n is of order much less than $p_{n-1}^2 q_{n-1}$, which is itself of order at least q_{n-1}^3. Thus while the p_n/q_n converge quadratically, the q_n

appear to "grow cubically". And now (12.10) looks rather improbable. In fact our wonderful approximations do not suffice to prove even the irrationality of $\sqrt[3]{2}$!

In the classical literature there are refinements of Newton's (12.11). One is Halley's

$$x_n = x_{n-1} - \frac{2f(x_{n-1})f'(x_{n-1})}{2f'(x_{n-1})^2 - f(x_{n-1})f''(x_{n-1})} \quad (n = 1, 2, \ldots). \tag{12.15}$$

This can be obtained geometrically as in Newton's Method, but instead of a tangent line $x = ay + b$ one uses a hyperbola $x = (ay + b)/(cy + 1)$ with triple contact. It leads to "cubic convergence".

With $f(X) = X^3 - 2$ this is $x_n = R(x_{n-1})$ for

$$R(X) = \frac{X^4 + 4X}{2X^3 + 2}. \tag{12.16}$$

With $x_0 = 5/4$ we get

$$x_0 = 1.25000000000000000000000000000000000\ldots,$$

$$x_1 = 1.25992063492063492063492063492063492063\ldots,$$

$$x_2 = 1.25992104989487316473719924558857105814\ldots,$$

$$x_3 = 1.25992104989487316476721060727822835057\ldots,$$

$$x_4 = 1.25992104989487316476721060727822835057\ldots,$$

indeed converging faster than before.

Now we have

$$R(X) - \alpha \;=\; \frac{X + \alpha}{2X^3 + 2}(X - \alpha)^3 \tag{12.17}$$

supporting cubic convergence, as well as

$$\frac{p_n}{q_n} = \frac{p_{n-1}^4 + 4p_{n-1}q_{n-1}^3}{2p_{n-1}^3 q_{n-1} + 2q_{n-1}^4}$$

suggesting that the q_n might "grow quartically". And now (12.10) still looks unlikely; but an optimist might think the situation a little better than before. However we still cannot deduce the irrationality of $\sqrt[3]{2}$.

Actually Newton's (12.11) and Halley's (12.15) are just the first two in a sequence involving general $f(X)$; but no single member of this sequence is good enough for our purposes. The key to further progress lies in (12.13) and (12.17) for the special $f(X) = X^3 - 2$. We try to find a rational function R

directly such that $R(X) - \alpha$ has a multiple factor $(X - \alpha)^n$. This apparently changes the role of the parameter n, but in fact it will end up much as before.

To get a factor $(X - \alpha)^n$ for $R(X) = P(X)/Q(X)$ probably means that

$$\mathrm{ord}_{X=\alpha}(P(X) - \alpha Q(X)) \geq n$$

or

$$P^{(j)}(\alpha) = \alpha Q^{(j)}(\alpha) \quad (j = 0, 1, \ldots, n-1) \tag{12.18}$$

for the derivatives. These are n linear equations over the field $\mathbb{Q}(\alpha)$ of degree 3 but as we saw in previous chapters they become $3n$ equations over $\mathbb{Q}$. So they are solvable non-trivially with P, Q of degree at most L provided only

$$2(L+1) > 3n. \tag{12.19}$$

The examples $n = 2, L = 3$ and $n = 3, L = 4$ correspond to (12.12) and (12.16). If we simply fix n, L and iterate, then we get nowhere. But if we find $P = P_n$ and $Q = Q_n$ with (12.18), say in $\mathbb{Z}[X]$, and define the integers

$$p_n = 4^L P_n\left(\frac{5}{4}\right), \quad q_n = 4^L Q_n\left(\frac{5}{4}\right), \tag{12.20}$$

then we are in a different game, the big one started by Thue.

The factor $(X - \alpha)^n$ of $P_n(X) - \alpha Q_n(X)$ suggests that $P_n(5/4) - \alpha Q_n(5/4)$ may be of order at most $(\alpha - 5/4)^n$ and so exponentially small in terms of n. By (12.19) we may hope to be able to keep the factors 4^L growing at most exponentially fast in n. But these considerations ignore the coefficients of P_n, Q_n themselves, which now have to be found with Siegel's Lemma. And in (12.18) we may well see factorials. Thanks to binomial coefficients these hold no Terror for us, and as in (9.1) we simply divide out by $j!$. This restricts the quantity U in Proposition 8.3 to at most exponential growth in n. We still have to take into account the Siegel exponent $M/(N-M)$, and once we used $N = 2M$ to tame this, and Gramain even made N/M large.

Here we can take N only a bit larger than M. The reason lies in the eternal snag (12.9), that is

$$p_n s \neq q_n r. \tag{12.21}$$

With (12.20) we have no chance of proving this. We cannot use our old method of varying n, as the connexion between p_n, q_n and p_{n-1}, q_{n-1} has been lost along with the iterative set-up.

Instead, we note for example that the derivative $P_n'(X) - \alpha Q_n'(X)$ has a factor $(X-\alpha)^{n-1}$ and so the quantities

$$P_n'(5/4), \quad Q_n'(5/4) \tag{12.22}$$

may furnish in (12.20) approximations which are nearly as good (for Halley's (12.16) the top and bottom even differentiate identically to those in Newton's (12.12) apart from a numerical factor). If the same snag occurs then we can try second derivatives and $(X-\alpha)^{n-2}$. But we cannot go too far for fear of differentiating our key factor $(X-\alpha)^n$ out of existence. The argument ruling this out can be expressed as a zero estimate for complex polynomials in the style of Proposition 8.1 (but with a much simpler proof).

It seems best if we now present this analytic zero estimate, and then afterwards carry on with the arithmetic.

Proposition 12.2 *For $K \geq 1$ let $\theta_1, \ldots, \theta_K$ and $x_1, \ldots, x_K$ be complex numbers with $x_1, \ldots, x_K$ different, and let $P(X), Q(X)$ in $\mathbb{C}[X]$ be polynomials of degree at most $L \geq 1, M \geq 1$. If each $P(X) - \theta_k Q(X) \neq 0$ then the quantities*

$$w_k = \mathrm{ord}_{X=x_k}(P(X) - \theta_k Q(X))$$

satisfy

$$w_1 + \cdots + w_K \leq K + L + M - 1.$$

Proof If $\phi = P - \theta Q$ has a zero of order w at x, then $\phi' = P' - \theta Q'$ has a zero of order at least $w-1$ there. Thus so does the Wronskian

$$W = \phi Q' - \phi' Q = PQ' - P'Q,$$

which is a polynomial of degree at most $L+M-1$. If $W \neq 0$ then with $x = x_1, \ldots, x_K$ and orders $w_1, \ldots, w_K$ we find that $\sum_{k=1}^{K}(w_k - 1) \leq L+M-1$, which is what we want. If $W = 0$ then on general grounds this implies the linear dependence of P, Q over $\mathbb{C}$; but that can be seen here by noting that if $Q \neq 0$ then the derivative of P/Q vanishes. In that case $P = \theta Q$ for some complex θ, and by hypothesis $\theta \neq \theta_k$. Now the $w_k = \mathrm{ord}_{X=x_k} Q(X)$ satisfy $\sum_{k=1}^{K} w_k \leq M$, which is even better. And if $Q = 0$ we get at once $\sum_{k=1}^{K} w_k \leq L$. □

In fact we need this only for $M = L$ in the present chapter, but for $M \neq L$ in Chapter 17.

We are now set up to prove things like (12.4). But the above discussion should make it clear that success depends strongly on the choice of the starting approximation, taken above as $5/4$. In fact this is not good enough to make Thue's Method work (unless it is supplemented by other considerations like hypergeometric identities, which avoid Siegel's Lemma). Furthermore no-one has been able to find an approximation to $\sqrt[3]{2}$ that is good enough! Even worse, there is almost certainly no such approximation!!

This gloomy state of affairs can be rescued at the considerable expense of effectivity. Up to now, all the material of this book has been effective in the sense that all constants can be worked out in principle, even though the task is often difficult and sometimes pointless. But the constant c in Thue's main result Theorem 12.1 is not known to be effective, and this situation represents a fundamental problem of diophantine analysis untouched for over a century. In order to make the situation clear we shall stay effective for some time, and to labour the issue we shall keep the constants explicit.

From now on let α be a real algebraic number of degree d. The quadratic case $d = 2$ can be handled as we did for $\sqrt{2}$ and so we shall assume $d \geq 3$. We shall calculate the constants in terms of the quantity $||\alpha||$ of the previous chapter and so it is handy to assume, right up to the final (at the end short) proof of Theorem 12.1, that α is an algebraic integer.

We start with an easy observation useful for the splitting of the equations (12.18).

Lemma 12.3 *For each $i \geq 0$ there are integers $a(i, e)$ with absolute values at most $(1 + ||\alpha||)^{di}$ such that $\alpha^i = \sum_{e=0}^{d-1} a(i, e)\alpha^e$.*

Proof This is trivial for $i = 0, 1, \ldots, d - 1$; but after that we have to use the minimal polynomial $A = a_0X^d + \cdots + a_d$ of α over $\mathbb{Z}$. This divides the polynomial corresponding to some monic equation (11.7) and so (for example by Gauss's Lemma) we can assume that $a_0 = 1$. We deduce

$$a(i, e) = a(i - 1, e - 1) - a(i - 1, d - 1)a_{d-e} \quad (e = 1, \ldots, d - 1)$$

holding also for $e = 0$ with a term missing. So for $A(i) = \max_{e=0,1,\ldots,d-1} |a(i, e)|$ we get $A(i) \leq aA(i - 1)$ for $a = 1 + |a_1| + \cdots + |a_d|$. Thus $A(i) \leq a^i$. Finally the zeroes of A are precisely the conjugates of α and it follows easily that $a \leq (1 + ||\alpha||)^d$. □

Next we construct our auxiliary polynomial as in (12.18). The condition (12.19) now reads $2(L + 1) > dn$, but as already explained we need to tighten this a bit. Thus we introduce a δ with $0 < \delta \leq 1$.

Lemma 12.4 *For any $L \geq 1, n \geq 1$ in $\mathbb{Z}$ with*

$$2(L + 1) > (d + 2\delta)n \tag{12.23}$$

there are P, Q in $\mathbb{Z}[X]$, of degree at most L and with coefficients of absolute values at most $(1 + ||\alpha||)^{2d^2L/\delta}$, such that

$$\phi(X) = P(X) - \alpha Q(X) \tag{12.24}$$

is not identically zero and

$$\mathrm{ord}_{X=\alpha}\,\phi(X) \geq n. \tag{12.25}$$

Proof We write

$$P(z) = \sum_{i=0}^{L} u_i z^i, \quad Q(z) = \sum_{i=0}^{L} v_i z^i.$$

The condition (12.25) written as in (12.18) and divided by $j!$ is

$$\sum_{i=0}^{L} \binom{i}{j} (u_i \alpha^{i-j} - v_i \alpha^{i-j+1}) \;=\; 0 \quad (j = 0, 1, \ldots, n-1).$$

After using Lemma 12.3 to split, we get

$$\sum_{i=0}^{L} \binom{i}{j} (u_i a(i-j, e) - v_i a(i-j+1, e)) = 0$$
$$(j = 0, 1, \ldots, n-1; e = 0, 1, \ldots, d-1).$$

These are $M = dn$ equations in $N = 2(L+1)$ unknowns, and by (12.23) the Siegel exponent

$$\varsigma = \frac{M}{N-M} < \frac{d}{2\delta}. \tag{12.26}$$

(that is "varsigma", not "zeta"). The coefficients in the equations have absolute values at most $U = 2^L(1 + ||\alpha||)^{d(L+1)}$ and so

$$NU \;\leq\; 2^{d(L+1)}(1 + ||\alpha||)^{d(L+1)} \;\leq\; (1 + ||\alpha||)^{4dL}$$

(recall $||\alpha|| \geq 1$ from Proposition 11.1). Now Proposition 8.3 gives the upper bound $(NU)^{\varsigma}$ and the lemma follows, because ϕ cannot be identically zero otherwise the irrationality of α would imply $P = Q = 0$. □

We will soon choose L minimal in (12.23) so of order at most n, justifying the discussion above that the coefficients in ϕ grow at most exponentially in n. But if we had used just the trivial estimate $\varsigma \leq M = dn$, then we might have seen an exponential dependence in n^2. Bombieri and Cohen have shown that in other analogous situations this may be unavoidable.

We are now in business to get our rational approximations as in (12.20). But due to the discussion around (12.22) we have to differentiate further; and at this stage it should be a knee-jerk reaction to divide by factorials. On plots Raymond Chandler advised "When in doubt, have a man come through the door with a gun in his hand." but the mathematician has the option of not only differentiating but also in extreme circumstances hyperdifferentiating.

Thus for integers p_0 and $q_0 \geq 1$ (soon such that p_0/q_0 is an approximation to α) we define the integers

$$p^{(m)} = \frac{1}{m!}q_0^L P^{(m)}\left(\frac{p_0}{q_0}\right), \quad q^{(m)} = \frac{1}{m!}q_0^L Q^{(m)}\left(\frac{p_0}{q_0}\right) \quad (m = 0, 1, 2, \ldots) \tag{12.27}$$

in terms of the higher derivatives.

For convenience we assume that we don't have a stupid approximation; that is

$$\left|\alpha - \frac{p_0}{q_0}\right| \leq 1; \tag{12.28}$$

of course we will soon demand something much better than this.

Lemma 12.5 *We have* $|q^{(m)}| \leq q_0^L(1 + ||\alpha||)^{4d^2L/\delta}$.

Proof We have

$$q^{(m)} = q_0^L \sum_{i=0}^{L} \binom{i}{m} v_i \left(\frac{p_0}{q_0}\right)^{i-m},$$

and using the estimates for v_i together with for example $|p_0/q_0| \leq 1 + ||\alpha||$, we find the required bound. □

Now we start to see our approximations in the shape (12.10).

Lemma 12.6 *For* $0 \leq m \leq n$ *we have*

$$|q^{(m)}\alpha - p^{(m)}| \leq q_0^L(1 + ||\alpha||)^{4d^2L/\delta} \left|\alpha - \frac{p_0}{q_0}\right|^{n-m}.$$

Proof We could do this by straightforward estimation but as we have some analytic machinery we may as well use it. The function $\psi(z) = \phi^{(m)}(\alpha - z)/m!$ has a zero of order at least $n - m$ at $z = 0$ and we deduce by the maximum principle in the usual way that

$$|q^{(m)}\alpha - p^{(m)}| = q_0^L \left|\psi\left(\alpha - \frac{p_0}{q_0}\right)\right| \leq q_0^L \left|\alpha - \frac{p_0}{q_0}\right|^{n-m} |\psi|_1$$

in the notation of Chapter 9. Here

$$\psi(z) = \sum_{i=0}^{L} \binom{i}{m} (z + \alpha)^{i-m}(u_i - \alpha v_i)$$

and we find the required bound much as in the proof of the previous lemma. □

We pause here to note that the analytic machinery can be pushed further to avoid the Herculean effort of differentiating. In fact we use Cauchy to integrate instead. Namely for any function f analytic on an open set containing a disc say for simplicity that given by $|z| \leq R$, we have

$$\frac{f^{(m)}(w)}{m!} = \frac{1}{2\pi i} \int_{|z-w|=R-r} \frac{f(z)dz}{(z-w)^{m+1}}$$

provided $|w| < r < R$. There follows at once

$$\frac{\lfloor f^{(m)} \rfloor_r}{m!} \leq \frac{\lfloor f \rfloor_R}{(R-r)^m}. \tag{12.29}$$

The next remark overcomes the snag (12.21) as long as L is chosen minimally in (12.23); that is

$$L = \left[\frac{(d+2\delta)n}{2}\right]. \tag{12.30}$$

Lemma 12.7 *For any complex θ there is m with*

$$0 \leq m \leq 2\delta n + d, \quad p^{(m)} \neq \theta q^{(m)}.$$

Proof If by bad luck $P = \theta Q$ then $Q \neq 0$ and θ must be rational. But then (12.25) would imply $\mathrm{ord}_{X=\alpha} Q(X) \geq n$. This holds at each of the d distinct conjugates of α, and so we deduce that $dn \leq L \leq (d+2\delta)n/2$, a contradiction. Thus $P \neq \theta Q$, and if our lemma is false then $\mathrm{ord}_{X=p_0/q_0}(P(X) - \theta Q(X)) > 2\delta n + d$. As above, already (12.25) holds with α replaced by each of its d conjugates. As these are different from p_0/q_0 we have the situation of Proposition 12.2 with $K = d+1, M = L$ and

$$w_1 + \cdots + w_K > 2\delta n + d + nd \geq K + 2L - 1;$$

and this is our decisive contradiction. □

Now we can give an explicit irrationality measure for α provided it has a sufficiently good approximation p_0/q_0.

Theorem 12.8 *Suppose α is an algebraic integer in $\mathbb{C}$ with degree $d \geq 3$, and let $0 < \delta \leq 1/3$. Suppose that p_0 and $q_0 \geq 1$ in $\mathbb{Z}$ are such that*

$$E = (1 + ||\alpha||)^{-4d^3/\delta} q_0^{-(d/2+\delta)} \left|\alpha - \frac{p_0}{q_0}\right|^{-(1-3\delta)} > 1. \tag{12.31}$$

Then with

$$A = (1 + ||\alpha||)^{4d^3/\delta} q_0^{d/2+\delta}, \tag{12.32}$$

we have for any r and $s \geq 1$ in $\mathbb{Z}$

$$\left|\alpha - \frac{r}{s}\right| \geq A^{-2d/\delta}\frac{1}{(2s)^{\lambda}} \tag{12.33}$$

with

$$\lambda = \frac{\log EA}{\log E}. \tag{12.34}$$

Proof For a positive integer n soon to be chosen in terms of α, δ and more significantly r, s, we define L by (12.30) and then ϕ as in Lemma 12.4. We define $p^{(m)}, q^{(m)}$ by (12.27) and then select m as in Lemma 12.7 with $\theta = r/s$. To bring things more in line with the Newton-Halley discussion, we denote the resulting $p^{(m)}, q^{(m)}$ simply by p_n, q_n.

Following (12.8) but without knowing that $q_n \neq 0$, we have

$$\left|q_n\left(\alpha - \frac{r}{s}\right)\right| \geq \left|p_n - q_n\frac{r}{s}\right| - |q_n\alpha - p_n| \geq \frac{1}{s} - |q_n\alpha - p_n|.$$

Certainly (12.31) implies (12.28). And as long as

$$n \geq \frac{d}{\delta}, \tag{12.35}$$

we have $m \leq 3\delta n \leq n$, so Lemma 12.6 gives (after using $L \leq dn$ in the exponent of $1 + ||\alpha||$ but not elsewhere) the bound $|q_n\alpha - p_n| \leq E^{-n}$. Thus we should choose n with (12.35) and $E^n \geq 2s$, for example

$$n = \left[\frac{d}{\delta} + \frac{\log(2s)}{\log E}\right] + 1. \tag{12.36}$$

We find after all $q_n \neq 0$ and $|\alpha - r/s| \geq 1/(2s|q_n|)$. By Lemma 12.5 we have $|q_n| \leq A^n$ and using (12.36) leads to our result (12.33). □

This result is completely effective. But what about the exponent λ? Let us return to $\alpha = \sqrt[3]{2}$ and try to achieve the modest-looking $\lambda = 2.955$. A short calculation using (12.34) shows that the approximation p_0/q_0 must satisfy some bound

$$\left|\alpha - \frac{p_0}{q_0}\right| \leq C^{-1}q_0^{-\mu} \tag{12.37}$$

with

$$\mu = \frac{\lambda}{(\lambda - 1)(1 - 3\delta)}\left(\frac{3}{2} + \delta\right). \tag{12.38}$$

As δ tends to 0 then μ tends to $2.267\ldots$, which is not too bad but then

$$C = (1 + \alpha)^{\frac{216\mu}{\delta(3+2\delta)}}$$

blows up. We had better not have $\mu \geq 3$ because then (12.2) almost certainly rules the approximation out. So we try the compromise $\delta = .03$ giving $\mu = 2.541\ldots$; and then

$$C > 10^{2117}!$$

It is hopeless!! And the continued fraction

$$\sqrt[3]{2} = [1; 3, 1, 5, 1, 1, 4, 1, 1, 8, 1, 14, 1, 10, 2, 1, 4, 12, 2, 3, 2, 1, 3, \ldots] \quad (12.39)$$

is not exactly encouraging.

But it cannot be theoretically ruled out; and life is full of surprises. In 1965 Brillhart discovered that the real number

$$\beta = \frac{1}{3}\sqrt[3]{135 + 3\sqrt{489}} + \frac{8}{\sqrt[3]{135 + 3\sqrt{489}}} = [3; a_1, a_2, \ldots]$$

with $\beta^3 - 8\beta - 10 = 0$ has for $n = 17, 33, 59, 81, 103, 121, 139, 161$ the partial quotients

$$a_n = 22986, 1501790, 35657, 49405, 53460, 16467250, 48120, 325927$$

(here $489 = 3 \times 163$ and indeed (10.2) is involved, for example $\beta + 2$ is within 10^{-16} of $e^{\pi\sqrt{163}/24}$ and even $e^{7\pi\sqrt{163}}$ is within $\frac{1}{800}$ of an integer).

Anyway, if there is such a p_0/q_0 with (12.37), then we will indeed get a result

$$\left|\alpha - \frac{r}{s}\right| > cs^{-2.955}$$

with

$$c = 10^{-254949} q_0^{-306}. \quad (12.40)$$

As this depends on q_0 it is not effective until we have actually found the probably-non-existent approximation.

Nevertheless Baker succeeded in proving

$$\left|\sqrt[3]{2} - \frac{r}{s}\right| \geq \frac{10^{-6}}{s^{2.955}} \quad (12.41)$$

using our very first approximation $p_0/q_0 = 5/4$ (in the disguise $5^3 - 2.4^3 = -3$), even though this spectacularly fails (12.37). He followed part of Thue's proof but used hypergeometric identities to construct ϕ explicitly without using Siegel's Lemma. That avoided the trouble caused with the blowing up of the Siegel exponent in (12.26) with N near M. Dividing by factorials is implicit in this approach but Baker found it necessary to take into account also powers of 3.

That approach failed with $\sqrt[3]{5}$ but Baker and Stewart later proved

$$\left|\sqrt[3]{5} - \frac{r}{s}\right| \geq \frac{10^{-12900}}{s^{2.9999999999998}} \tag{12.42}$$

using the completely different method of linear forms in logarithms. The reader can check that even this sort of thing almost certainly cannot be done with Theorem 12.8. In fact that gives at best only exponents bigger than 4, worse than (12.3), because for $\lambda = 4$ we get $\mu \to 2$ in (12.38).

And what about our Theorem 12.1? If α is an algebraic integer then it follows from Theorem 12.8 with a logical twist. Take any

$$\kappa > \frac{d}{2} + 1$$

and then try $\lambda = \kappa$ in (12.33). If we cannot, then $E^{\kappa-1} < A$ for all p_0/q_0 – and we take the liberty of calling them r/s now – then we find (compare (12.37) but reversed)

$$\left|\alpha - \frac{r}{s}\right| > C^{-1}s^{-\mu}$$

with now

$$\mu = \mu(\delta) = \frac{\kappa}{(\kappa-1)(1-3\delta)}\left(\frac{d}{2} + \delta\right).$$

Thus in either case we obtain an exponent $\max\{\kappa, \mu\}$ for any δ with $0 < \delta \leq 1/3$. As δ tends to 0 this tends to

$$\max\left\{\kappa, \frac{\kappa d}{2(\kappa-1)}\right\} = \kappa,$$

and so for sufficiently small δ we get Theorem 12.1, at least for algebraic integers. In general we just fix a denominator a in $\mathbb{N}$ for α and now the inequality (12.7) with $a\alpha$ instead of α and ar instead of r implies (12.7) with c/a instead of c.

It should be clear from the discussion around (12.40) that the resulting c is not effective.

Finally here is the application back to equations like (12.5). We imitate the style of our earlier result Theorem 4.7. There the F_0 was reducible and the degree of f was allowed to go up to $d-1$. Further the result was effective in that all the solutions could be found in principle and even in practice. Here not, even for the equation

$$x^4 - 2y^4 + xy + x = 1948. \tag{12.43}$$

Theorem 12.9 *Let F in $\mathbb{Q}[X,Y]$ be a polynomial of the shape*

$$F(X,Y) = F_0(X,Y) + f(X,Y)$$

for $F_0(X,Y)$ homogeneous of degree $d \geq 3$ and irreducible over $\mathbb{Q}$, and a polynomial f of total degree strictly less than $d/2-1$. Then there are at most finitely many (x,y) in $\mathbb{Z}^2$ with $F(x,y)=0$.

Proof Let (x,y) in $\mathbb{Z}^2$ be any solution of $F(x,y) = 0$. We may assume that $x \neq 0$, $|x| \geq |y|$ and even $x \geq 1$. After dividing by a leading coefficient we can write

$$F_0(X,Y) = (Y-\alpha_1 X)\cdots(Y-\alpha_d X) \tag{12.44}$$

for distinct complex $\alpha_1,\ldots,\alpha_d$, of course algebraic over $\mathbb{Q}$ of degree exactly d. We may also assume that the minimum of the $|\alpha_i - y/x|$ occurs at $i=1$. As their product is $x^{-d}|f(x,y)| \leq c_1 x^{-(d-e)}$ for the degree e of f and some c_1 independent of x,y we deduce $|\alpha_1 - y/x| \leq (c_1 x^{-(d-e)})^{1/d}$. As the exponent of x is at least -1 this is not yet enough to apply Theorem 12.1; but now for $i>1$ we have

$$\left|\alpha_i - \frac{y}{x}\right| \geq |\alpha_i - \alpha_1| - \left|\alpha_1 - \frac{y}{x}\right| \geq |\alpha_i - \alpha_1| - \left|\alpha_i - \frac{y}{x}\right|$$

so $|\alpha_i - y/x| \geq \frac{1}{2}|\alpha_i - \alpha_1|$, and we get the improvement

$$\left|\alpha_1 - \frac{y}{x}\right| \leq \frac{c_2}{x^{d-e}} \tag{12.45}$$

for some c_2 also independent of x,y. As the exponent here is $d-e > d/2+1$, we contradict Theorem 12.1 if x is large. □

Thus indeed we can handle (12.5), and indeed any $ax^3 + by^3 = m$ with $aX^3 + bY^3$ irreducible. The irreducibility can be easily relaxed (see for example Exercise 12.33), and then this is enough to solve equations of an apparently quite different sort involving so-called "S-units", for example

$$2^p 7^q + (-1)^u 3^v 5^w = 1 \tag{12.46}$$

in which the unknowns are the integers p,q,u,v,w, here naturally non-negative (see Exercise 12.36). These equations are also associated with the nametag Mordell–Lang; see for example Hindry and Silverman (2000) or Bombieri and Gubler (2006).

The inequality (12.45) explains why (12.43) is ineffective; we have $d = 4, e = 2$ and then we would need an effective

$$\left|\sqrt[4]{2} - \frac{r}{s}\right| > \frac{c}{s^\kappa}$$

with $\kappa < 2$.

Our Theorem 12.1 is by no means the last word on the subject (actually in Thue (1909) the first word). The bound $d/2+1$ in (12.6) was improved several times, most notably by Siegel (1921) to

$$\min_{M \in \mathbb{N}} \left\{ \frac{d}{M+1} + M \right\} < 2\sqrt{d}$$

using an auxiliary polynomial

$$\phi(X,Y) = P_0(X) + P_1(X)Y + \cdots + P_M(X)Y^M.$$

Dyson (1947) (and independently round the same time Gelfond – see Gelfond (1960) for much more) improved this using suitable conditions

$$\left(\frac{\partial}{\partial X}\right)^j \left(\frac{\partial}{\partial Y}\right)^k \phi(\alpha,\alpha) = 0.$$

Finally Roth (1955) achieved the bound 2, essentially best possible in view of (8.4), and so indeed the last word, at least in this respect. Thus for any $\epsilon > 0$ there is $c > 0$ with

$$\left|\alpha - \frac{r}{s}\right| \geq \frac{c}{s^{2+\epsilon}}. \tag{12.47}$$

But here the number of variables $X, Y, Z, \ldots$ must increase without limit. The account of this result in Cassels (1965) is excellent.

That allows the degree e of f in Theorem 12.9 to go up to $e < d - 2$. However by appealing to Siegel's Theorem (Siegel, 1929) on integral points (see also Zannier (2014b) for a modern point of view) Schinzel (1969) could relax to the maximum with $e < d$.

Many people think that (12.47) can be improved further, with s^ϵ replaced by a power $(\log s)^\kappa$ of a logarithm, even with any $\kappa > 1$. But we still know no improvement at all.

It seems that Thue's work was essentially independent of anything earlier (if indeed there was anything). I cannot find any evidence in his papers that he knew the work of Runge, for example. After experiments with explicit P, Q in Lemma 12.4 (essentally with $\delta = 0$) obtained with hypergeometric functions (see Exercises 12.41,12.42), he conceived the notion of inexplicit P, Q using the Box Principle implicit in Proposition 8.3. He had the idea of dividing by factorials. He recognized the need for a zero estimate; however his version of Proposition 12.2 took into account that P, Q are not only in $\mathbb{Z}[X]$ but also that their coefficients are not too large. It could be considered an "impure" zero estimate; such things culminated in Roth's Lemma. Finally he discovered the "logical twist" explained above, which incidentally is the first appearance in mathematics of ineffectivity in a natural context (the next came from Siegel with lower bounds for L-functions (Siegel, 1935) and the existence of their real zeroes). Compare Schinzel's Question "But how can it be, how can it be in *number theory*, that one could prove the finiteness of a set of natural numbers, without being able to give a bound for its cardinality?" quoted in Schmidt (1991) (p. 57, with Answers); see also Exercise 12.6.

It was Dyson who "purified" the zero estimates to polynomials in $\mathbb{C}[X, Y]$, and after the work of Viola (1985) these were generalized to many variables $X, Y, Z, \ldots$ by Esnault and Viehweg (1984) (see the account in Bombieri (1982) for example).

However the ineffectivity situation is not hopeless. One can give an effective upper bound for the number of r/s occurring in any

$$\left|\alpha - \frac{r}{s}\right| < \frac{c}{s^{2+\epsilon}}$$

(for simple examples see Exercises 12.8, 12.27, 12.28). Or one can construct special α that do have a sufficiently good starting approximation. For example, when $a \geq 3$ the unique real $\alpha > 1$ with $\alpha^d - a\alpha^{d-1} + 1 = 0$ satisfies $|\alpha - a| < |a|^{-(d-1)}$, and using this Bombieri (1982) proved that for each d there is an effective positive integer a and an effective $c > 0$ such that

$$\left|\alpha - \frac{r}{s}\right| \geq \frac{c}{s^{39.2575}}$$

even though α has degree d.

Also for special α one can use hypergeometric functions, as in (12.41); an example is that for any $\epsilon > 0$ there is an effective positive integer a (please note the quantifiers) and an effective $c > 0$ such that

$$\left|\sqrt[3]{1 + \frac{1}{a}} - \frac{r}{s}\right| \geq \frac{c}{s^{2+\epsilon}} \tag{12.48}$$

for all r and $s \geq 1$ (where also an obvious starting approximation is available). For general α one can use linear forms in logarithms, as in (12.42), to prove that for every α of degree d there are effective $\gamma > 0$ and $c > 0$ (quantifiers again) such that

$$\left|\alpha - \frac{r}{s}\right| \geq \frac{c}{s^{d-\gamma}}. \tag{12.49}$$

This result, due to Feldman (1971), improves Liouville's (12.3).

These effective approximation results naturally transfer to diophantine equations as in Theorem 12.9, so that we get effective bounds for the number of (x,y) (see for example Exercise 12.39) or the $|x|,|y|$ themselves (see for example Exercises 12.43, 12.44). In particular (12.49) enables Theorem 12.9 to be made effective for any F_0 and any constant f; that is, for general Thue equations of the form $F_0(x,y) = m$. We cannot resist quoting

$$\max\{|x|,|y|\} \; < \; \exp\{d^{65536d^4}|F_0|^{256d^5} + (\log|m|)^{2d+2}\}$$

for $m \neq 0$, where $|F_0|$ is the maximum modulus of the coefficients of F_0. See Baker's game-changing (Baker, 1968b) paper. With a bit of algebraic number theory we could now prove that the Mordell equation $y^2 = x^3 + k$ $(k \neq 0)$ has at most finitely many integer solutions (see Exercise 12.35). We are equally powerless to omit

$$\max\{|x|,|y|\} < \exp\{(10000000000|k|)^{10000}\}$$

also proved in Baker (1968b) (in fact with reduction theory).

In spite of all this, one still cannot find all the solutions of the equation (12.43). Their finiteness is guaranteed by Siegel's Theorem mentioned above, which involves naturally the genus g of the associated algebraic curve. It can be made effective for $g = 1$; but not yet for $g = 2$. And (12.43) has $g = 3$.

Things by no means ended with Roth, and a sequence of developments by Schmidt led to his famous Subspace Theorem. An extremely simply-stated and highly non-trivial case is that for any $\kappa > 3/2$ there is $c > 0$ such that

$$\max\left\{\left|\sqrt{2} - \frac{r}{s}\right|, \left|\sqrt{3} - \frac{\tilde{r}}{s}\right|\right\} \geq \frac{c}{s^\kappa}$$

for any integers $r, \tilde{r}$ and $s \geq 1$. Compare with the upper bound (8.4) for $\kappa = 2$ (and see Exercise 12.49). There are applications to diophantine equations in many variables $x, y, z, \ldots$. We have no room to go further into this; see for example the account in Bombieri and Gubler (2006), which is still the only book (as opposed to lecture notes) version. It even includes the important extensions to general valuations such as p-adic or $\mathfrak{p}$-adic (see (14.16) later) pioneered by Schlickewei. Naturally these can be applied back to diophantine equations: a simple example with $p = 3, 5$ is the finiteness of $x, y, u \geq 0, v \geq 0$ in $\mathbb{Z}$ with $x^3 - 2y^3 = 3^u 5^v$ (compare (12.46) and Exercise 14.41), otherwise x/y would be too close to $\sqrt[3]{2}$ in the 3-adic or 5-adic (or ∞-adic) sense. This can be formulated in terms of the ring $\mathbb{Z}[\frac{1}{3}, \frac{1}{5}]$ of S-integers with $S = \{3, 5\}$; and one can even handle finitely generated rings like $\mathbb{Z}[r_1, \ldots, r_m]$ (or $\mathbb{F}_p[r_1, \ldots, r_m]$ as well).

In Chapter 17 we will see ideas very similar to Thue's applied to a completely different problem.

Exercises

12.1 Let $\alpha = 1.32471795\ldots = \beta/6 + 2/\beta$ with $\beta = \sqrt[3]{108 + 12\sqrt{69}} > 0$ be the smallest Pisot-Vijayaraghavan number, with $\alpha^3 = \alpha + 1$ and $[\mathbb{Q}(\alpha) : \mathbb{Q}] = 3$. Find $c > 0$ such that $|\alpha - r/s| \geq c/s^3$ for all r, s in $\mathbb{Z}$ and $s \geq 1$.

12.2 Let α be in $\mathbb{C}$ with $[\mathbb{Q}(\alpha) : \mathbb{Q}] = d \geq 2$. Show that there is $c > 0$ such that $|\alpha - r/s| \geq c/s^d$ for all r, s in $\mathbb{Z}$ and $s \geq 1$.

12.3 Let θ be in $\mathbb{R}$. Show that the following are equivalent.

(a) The number θ is irrational.

(b) There are p_n, q_n in $\mathbb{Z}$ with $q_n \geq 1$ $(n = 1, 2, \ldots)$, such that $0 < |q_n\theta - p_n| \to 0$.

12.4 Let $\alpha = \sqrt[3]{2}$, let x_0 be in $\mathbb{R}$, and let $x_n = R(x_{n-1})$ $(n = 1, 2, \ldots)$ with $R(X) = (2X^3 + 2)/(3X^2)$. If $x_n \neq 0, \alpha$ $(n = 1, 2, \ldots)$ and $x_n \to \alpha$, show that there is $\xi > 0$ with $|x_n - \alpha| \geq \xi^{2^n}$ for all n large enough.

12.5 In Exercise 12.4 let x_0 be in $\mathbb{Q}$, and let $x_n = p_n/q_n$ with $q_n \geq 1$ and p_n $(n = 1, 2, \ldots)$ coprime in $\mathbb{Z}$. Show that there is $\Xi > 1$ such that $q_n \geq \Xi^{3^n}$ for all n large enough [*Hint*: $3q(2(p^3 + q^3)) - 2p(3p^2q) = 6q^4$, $3p^2(2(p^3 + q^3)) - 2q^2(3p^2q) = 6p^5$].

12.6 Let $\mathcal{S}$ be in $\mathbb{N}$ such that

(a) $s, s' \in \mathcal{S}$ implies $s' \leq Cs^{\Theta}$

(b) $s, s' \in \mathcal{S}$, $s' > s$ implies $s' > cs^{\theta}$

with $c > 0, \theta > 1$. Let $n = 1 + [\log\Theta/\log\theta]$, $\phi = (\theta^n - 1)/(\theta - 1)$ and $\epsilon = \theta^n - \Theta > 0$. Show that $s < (Cc^{-\phi})^{1/\epsilon}$ for all s in $\mathcal{S}$ with at most n exceptions [*Hint*: Let $s_1 < s_2 < \cdots < s_n$ be maximal in $\mathcal{S}$, and let $s < s_1$ in $\mathcal{S}$]. Compare Schinzel's Question, as well as Schinzel's Answer and Schmidt's Extended Answer.

12.7 For $\kappa > 2$ let r, s, r', s' be in $\mathbb{Z}$ with $1 \leq s < s'$, $r/s \neq r'/s'$ and

$$\left|\alpha - \frac{r}{s}\right| \leq \frac{1}{s^{\kappa}}, \quad \left|\alpha - \frac{r'}{s'}\right| \leq \frac{1}{s'^{\kappa}}.$$

Show that $s' > \frac{1}{2}s^{\kappa-1}$ [*Hint*: $|r/s - r'/s'|$ and the triangle inequality].

12.8 Find S and n with the following properties. If

$$\left|\alpha - \frac{r}{s}\right| \leq \frac{1}{s^{\kappa}} \tag{**}$$

with $\alpha = 2^{1/3}$ and $\kappa = 2.51$ then $s \leq S$ with at most n exceptions [*Hint*: Let $\lambda = 2.505$ and $\delta = 1/1103$ with $(\frac{3}{2} + \delta)/(1 - 3\delta) = \lambda - 1$. If there are no p_0, q_0 in Theorem 12.8 with $\log EA/\log E \leq \lambda$ find straightaway $s < S_1$

without exceptions. Otherwise let q_0 be minimal in Theorem 12.8, let $\mathcal{S}$ be the set of $s \geq q_0$ in (**), and find $s < S_2$ with exceptions. And what if $s < q_0$?].

* * * * * * * * * * * * * * * * * * *

12.9 Why is $\alpha^{200} = 2658793989922287946990250.000000000001012\ldots$ in Exercise 12.1?

12.10 Let $\theta = \sum_{k=0}^{\infty} 2^{-3^k}$. Show that there is $c > 0$ such that $|\theta - r/s| \geq c/s^3$ for all r, s in $\mathbb{Z}$ with $s \geq 1$.

12.11 In Exercise 12.10 show that there are infinitely many $s \geq 1$ in $\mathbb{Z}$ such that $|\theta - r/s| < 2/s^3$ for some r in $\mathbb{Z}$.

12.12 Show that θ in Exercise 12.10 is transcendental.

12.13 In Exercise 12.4 show that there is x_0 in $\mathbb{R}$ with $|x_{1948} - \alpha| > 1$.

12.14 In Exercise 12.4 must $x_n \to \alpha$ for all x_0 in $\mathbb{R}$ with at most countably many exceptions?

12.15 In Exercise 12.4 must $x_n \to \alpha$ for all x_0 in $\mathbb{C}$ with at most countably many exceptions?

12.16 Let $P(X, Y), Q(X, Y)$ be homogeneous and coprime in $\mathbb{Z}[X, Y]$. Show that there is $C = C(P, Q)$ such that the highest common factor $\text{hcf}(P(p, q), Q(p, q)) \leq C$ for all p, q coprime in $\mathbb{Z}$.

12.17 Let $\alpha^3 - 2\alpha^2 + 4\alpha - 4 = 0$. Show that $\alpha^{52} = a + b\alpha$ with a, b in $\mathbb{Z}$ (Berstel).

12.18 Let α be algebraic of degree d, integral over $\mathbb{Z}$, and let $a(i, e)$ be in $\mathbb{Z}$ with $\alpha^i = \sum_{e=0}^{d-1} a(i, e)\alpha^e$ $(i = 0, 1, \ldots)$. Show that

$$\lim_{i \to \infty} \frac{1}{i} \log \max_{e=0,\ldots,d-1} |a(i, e)| = \log ||\alpha||.$$

12.19 Using $x = ay^2 + by + c$ find an alternative Halley for $f(X) = 0$ with cubic convergence.

12.20 Find a formula with quartic convergence.

12.21 Let P, Q, R be in $\mathbb{C}[X]$ with Wronskian

$$\begin{vmatrix} P & Q & R \\ P' & Q' & R' \\ P'' & Q'' & R'' \end{vmatrix} = 0. \qquad (*)$$

Show that P, Q, R are linearly dependent over $\mathbb{C}$.

12.22 Let P, Q, R be in $\mathbb{F}_p[X]$ with $(*)$. Show that P, Q, R are linearly dependent over $\mathbb{F}_p(X^p)$.

12.23 Let A, B, C be in $\mathbb{C}[X]$, not all constant and with no common factor, such that

$$A + B + C = 0.$$

Show that

$$\max\{\deg A, \deg B, \deg C\} \leq N - 1,$$

where N is the number (without multiplicities) of x in $\mathbb{C}$ with $A(x)B(x)C(x) = 0$ [*Hint*: show that

$$\frac{A}{C} = -\frac{B'/B - C'/C}{B'/B - A'/A}, \quad \frac{B}{C} = -\frac{A'/A - C'/C}{A'/A - B'/B}$$

and note that

$$\frac{P'}{P} = \sum_{x \in \mathbb{C}} \frac{\mathrm{ord}_{X=x} P}{X - x}$$

for the derivatives].

12.24 Let $\alpha = 2^{1/3}$, let n be in $\mathbb{N}$, and let L, M be in $\mathbb{N}$ with $L + M = 3n - 1$.

(a) Show that there are P, Q in $\mathbb{Z}[X]$, not both zero, with $\deg P \leq L$, $\deg Q \leq M$ and $\mathrm{ord}_{X=\alpha}(P(X) - \alpha Q(X)) \geq n$.

(b) Let x, θ be in $\mathbb{C}$ with $x^3 \neq 2$. Show that

$$P(x) = \theta Q(x), \;\; P'(x) = \theta Q'(x), \;\; P''(x) = \theta Q''(x)$$

are not all possible.

12.25 For $K \geq 1$ let $\chi_1, \ldots, \chi_K, \theta_1, \ldots, \theta_K$ and $x_1, \ldots, x_K$ be complex numbers with $x_1, \ldots, x_K$ different, and let $P(X), Q(X)$ in $\mathbb{C}[X]$ be polynomials of degree at most $L \geq 1, M \geq 1$. If each $\chi_k P(X) - \theta_k Q(X) \neq 0$, show that the quantities

$$w_k = \mathrm{ord}_{X=x_k}(\chi_k P(X) - \theta_k Q(X))$$

satisfy

$$w_1 + \cdots + w_K \leq K + L + M - 1.$$

12.26 For $K \geq 1$ let $\pi_1, \ldots, \pi_K, \kappa_1, \ldots, \kappa_K, \rho_1, \ldots, \rho_K$ and $x_1, \ldots, x_K$ be complex numbers with $x_1, \ldots, x_K$ different, and let $P(X), Q(X), R(X)$ in $\mathbb{C}[X]$

be polynomials of degree at most $L \geq 2, M \geq 2, N \geq 2$. If each $\pi_k P(X) + \kappa_k Q(X) + \rho_k R(X) \neq 0$, show that the quantities

$$w_k = \mathrm{ord}_{X=x_k}(\pi_k P(X) + \kappa_k Q(X) + \rho_k R(X))$$

satisfy

$$w_1 + \cdots + w_K \leq 2K + L + M + N - 3.$$

12.27 Let α be an algebraic integer of degree $d \geq 3$, and let $\kappa > \frac{d}{2} + 1$. Show that there are effective S and n such that $s \leq S$ in $(**)$ of Exercise 12.8 with at most n exceptions.

12.28 Let α be an algebraic integer of degree $d \geq 5$. Show that there are effective $\kappa < d$ and effective S such that $s \leq S$ in $(**)$ of Exercise 12.8 with at most 3 exceptions.

12.29 Show that there are at most finitely many (x, y) in $\mathbb{Z}^2$ with $x \neq 0$ and $x^3 - 2y^3 = \pm[\log|x|]$.

12.30 Show that there are infinitely many (x, y) in $\mathbb{Z}^2$ with $x \neq 0$ and $x^2 - 2y^2 = \pm[\log|x|]$.

12.31 Show that there are at most finitely many (x, y) in $\mathbb{Z}^2$ with

$$(x^3 - 2y^3)^3 = x + y + 1948.$$

12.32 Show that there are at most finitely many (x, y) in $\mathbb{Z}^2$ with

$$(x^5 - 2y^5)^7 (x^7 - 2y^7)^5 = x^{45} + y^{45} + 1948.$$

12.33 Let F_0 in $\mathbb{Q}[X, Y]$ be homogeneous and not of the form aG^e for G of degree at most 2. Show that for any $m \neq 0$ there are at most finitely many (x, y) in $\mathbb{Z}^2$ with $F_0(x, y) = m$ [*Hint*: Exercise 5.9].

12.34 Show that there are at most finitely many (x, y) in $\mathbb{Z}^2$ with

$$y^3 = x(x-1)\cdots(x-1948).$$

12.35 (a) If the equation

$$v^2 = u^3 + 13 \tag{$*$}$$

has a solution (u, v) in $\mathbb{Z}^2$, show that at least one of the equations

$$x^3 + 9x^2y + 39xy^2 + 39y^3 = 16, \; 3x^3 + 33x^2y + 117xy^2 + 143y^3 = 16 \tag{$**$}$$

has a solution (x, y) in $\mathbb{Z}^2$ [*Hint*: $\mathbb{Q}(\sqrt{13})$ and the units].

(b) Show that neither of the equations (∗∗) has a solution in $(\mathbb{Z}/32\mathbb{Z})^2$ (even though (∗) has solutions in $(\mathbb{Z}/m\mathbb{Z})^2$ for every m in $\mathbb{N}$ and by the way (∗∗) has solutions in $(\mathbb{Z}/m\mathbb{Z})^2$ for every $m < 32$ in $\mathbb{N}$).

12.36 (a) Show that there are at most finitely many non-negative integers p, q, u, v, w with $2^p 7^q + (-1)^u 3^v 5^w = 1$ [*Hint*: consider the exponents modulo 3 and use Exercise 12.33 with cubic F_0].

(b) Show that there are at most finitely many units x, y in $\mathbb{Z}[\frac{1}{2}, \frac{1}{3}, \frac{1}{5}, \frac{1}{7}] = \mathbb{Z}[\frac{1}{210}]$ ("$\{2, 3, 5, 7\}$-units") with $x + y = 1$.

12.37 If u, v, w are non-negative integers with $3^u + 5^v - 7^w = 1$ show that $u = v = w = 0$ or $u = v = w = 1$ (Brenner and Foster, 1982; Leitner, 2011) [*Hint*: congruences].

12.38 If y, u, v are non-negative integers with $y^2 = 1 + 2^u + 3^v$ show that $y = 2$ or $y = 6$ (Leitner, 2011).

12.39 Let m be in $\mathbb{Z}$. Show that there is effective $C = C(m)$ such that $x^3 - 2y^3 = m$ has at most C solutions (x, y) in $\mathbb{Z}^2$.

12.40 If $d \geq 25$ is odd and $|a| \geq 2^{62d}$, show that there are at most eleven (x, y) in $\mathbb{Z}^2$ with

$$(x - ay)(x^2 + y^2)^{\frac{d-1}{2}} - y^d = x + y$$

(Dill Master Thesis 2015). There can be five.

12.41 Let $d, m \geq 1$ and

$$U(z) = \sum_{k=0}^{m} (-1)^{m-k} \binom{m + \frac{1}{d}}{k} \binom{-m-1}{m-k} z^k,$$

$$V(z) = -\sum_{k=0}^{m} \binom{m + \frac{1}{d}}{k} \binom{-m-1}{m-k} z^k (1+z)^{m-k}.$$

Show that $\mathrm{ord}_{z=0}(U(z) - (1+z)^{1/d} V(z)) \geq 2m + 1$.

12.42 With $\alpha \neq 0, n = 2m + 1, z = (X/\alpha)^d - 1$ and $P(X) = XV(z), Q(X) = U(z)$, show that $\mathrm{ord}_{X=\alpha}(P(X)) - \alpha Q(X)) \geq n$ (compare Lemma 12.4 for α^d in $\mathbb{Q}$ without Siegel for L about $dn/2$ so "$\delta = 0$").

12.43 Use the estimate (12.41) to find explicit C, κ such that $\max\{|x|, |y|\} \leq C|m|^\kappa$ for all integers x, y, m satisfying $x^3 - 2y^3 = m$.

12.44 Use the estimate (12.42) to find explicit C, κ such that $\max\{|x|, |y|\} \leq C|m|^\kappa$ for all integers x, y, m satisfying $x^3 - 5y^3 = m$.

12.45 Let $f(z) = e^{z^2}$. Show that $\lim_{r\to\infty} |f'|_r/|f|_r = \infty$.

12.46 Is there an entire function $f \neq 0$ with $\lim_{r\to\infty} |f'|_r/|f|_r^2 = \infty$? I don't know.

12.47 Does Newton's formula (12.11) converge for $f(x) = e^x - x$ with $x_0 = 2\pi ik + \log(2\pi ik)$ for large enough k?

12.48 Let P be non-zero in $\mathbb{Z}[X, Y]$ of degree at most L in each variable. Let α, η be in $\mathbb{C}$ with $|\alpha| = 2/3$ and $|\eta - \theta| = \epsilon \leq 1/2$ for $\theta = f(\alpha)$, where $f(z)$ is Mahler's $\sum_{k=0}^{\infty} z^{2^k}$. Define

$$f_m = P(\alpha^{2^{m+1}}, \eta - \alpha - \cdots - \alpha^{2^m}) \quad (m = 0, 1, 2, \ldots)$$

and suppose for some $n \geq 1$ and K that

$$f_n = f_{n+1} = \cdots = f_{n+K} = 0$$

as in (8.18).

(a) Deduce that

$$|\phi(\alpha^{2^{m+1}})| \leq 3^L \mathcal{L}(P)\epsilon \quad (m = n, n+1, \ldots, n+K)$$

for $\phi(z) = P(z, f(z))$ and the length.

(b) For the (Lagrange) polynomial

$$Q(X) = \sum_{m=n}^{n+K} \prod_{l=n, l\neq m}^{n+K} \left(\frac{X - \alpha^{2^{l+1}}}{\alpha^{2^{m+1}} - \alpha^{2^{l+1}}} \right) \phi(\alpha^{2^{m+1}})$$

deduce that $|Q|_1 \leq 3^L \mathcal{L}(P) C\epsilon$ for some $C = C(n, K)$.

(c) Prove that $|\phi - Q|_{1/27} \leq 2^{-K-1} |\phi - Q|_{2/3}$ [*Hint:* $|\alpha^{2^{m+1}}| \leq 16/81$].

(d) Deduce that $|\phi|_{1/27} \leq \delta$ for

$$\delta = 6^L \mathcal{L}(P) C\epsilon + 4^L 2^{-K-1} \mathcal{L}(P).$$

(e) Deduce further that

$$|\phi^{(t)}(0)| \leq t! 27^t \delta \quad (t = 0, 1, \ldots, 12L^2).$$

(f) Conclude using Proposition 8.1 that

$$|\eta - \theta| \geq \frac{1}{2} \left((12L^2)! 27^{12L^2} 6^L \mathcal{L}(P) C \right)^{-1},$$

provided K is chosen with

$$2^K < 2(12L^2)! 27^{12L^2} 4^L \mathcal{L}(P) \leq 2^{K+1}.$$

This explains the "couple of things on trust" in Chapter 8.

12.49 Let $\theta, \tilde{\theta}$ be real. Show that for any real $S > 1$, there is an integer s with $0 < s < S$ and

$$\epsilon(s) = \min_{r\in\mathbb{Z}} \min_{\tilde{r}\in\mathbb{Z}} \max\left\{ \left|\theta - \frac{r}{s}\right|, \left|\tilde{\theta} - \frac{\tilde{r}}{s}\right| \right\} \leq \frac{1}{sS^{1/2}}$$

and so infinitely many s with $\epsilon(s) < 1/s^{3/2}$ [*Hint*: Minkowski's First Theorem].

13

Transcendence II – Hermite–Lindemann

By now we have stockpiled so many weapons of mass destruction that the transcendence of e, first proved by Hermite in an amazing work no less astonishing for the fact of his age 51, is a relatively defenceless target. We can even prove the following

Theorem 13.1 *Suppose $\alpha \neq 0$ is algebraic in $\mathbb{C}$. Then e^α is transcendental.*

This includes the transcendence of π (up to now not proved even to be irrational in these pages), because if it were algebraic then so would $2\pi i$ be; but $e^{2\pi i} = 1$. Similarly $e^{\log 2} = 2$ so we get the transcendence of $\log 2$. More generally we deduce the transcendence of any non-zero determination of $\log \beta$ for any non-zero algebraic β.

The above theorem has the same shape as Theorems 11.2 and 11.3 of Chapter 11. By writing $f(z) = e^z$ we can even make a connexion with Chapters 9 and 10: if $f(\alpha)$ lies in a finite field extension K of $\mathbb{Q}$ then so does $f(k\alpha) = f(\alpha)^k$ $(k = 0, 1, 2, \ldots)$. In particular the proof of Lemma 10.2 suggests that we may be able to use an auxiliary polynomial in z and e^z, perhaps with the points $z = 0, \alpha, \ldots, S\alpha$ as in (9.7) even though we are no longer working over $\mathbb{Z}$ or even $\mathbb{Q}$. In fact we have a functional equation $f(2z) = f(z)^2$ as for the functions in Chapters 2 and 3, or come to that a whole array $f(nz) = f(z)^n$ of such things; and this does make such an approach possible.

We also have a differential equation $f'(z) = f(z)$ and so we can hope to use derivatives as in Chapter 6 (see (6.4) there) and Chapter 7 (see (7.5) there); not to mention Chapter 12. We choose this approach because it works for more general rather classical functions, as in Chapter 20.

We are working over a finite extension of $\mathbb{Q}$ and so we can always split the equations as in (4.25) and (4.35) (and so on) to land up over $\mathbb{Q}$ or $\mathbb{Z}$; but this is becoming increasingly awkward and so we do it in general with the following extension of Siegel's Proposition 8.3. We will later obtain a more streamlined version.

Proposition 13.2 *Suppose* $[K : \mathbb{Q}] = D$ *and* $\mathbb{Z}_K = \mathbb{Z}\omega_1 + \cdots + \mathbb{Z}\omega_D$. *For positive integers* M, N *with* $DM < N$ *and real* $W \geq 1$ *let* w_{mn} $(m = 1, \ldots, M;\ n = 1, \ldots, N)$ *be elements of* $\mathbb{Z}_K$ *with* $||w_{mn}|| \leq W$. *Then there are rational integers* $x_1, \ldots, x_N$, *not all zero, with*

$$\sum_{n=1}^{N} w_{mn} x_n = 0 \quad (m = 1, \ldots, M)$$

and

$$|x_n| \leq (D! ||\omega_1|| \cdots ||\omega_D|| NW)^{\frac{DM}{N-DM}} \qquad (n = 1, \ldots, N).$$

Proof Any w in $\mathbb{Z}_K$ can be written as $u_1\omega_1 + \cdots + u_D\omega_D$ with rational integers $u_1, \ldots, u_D$. Applying any one of the complex embeddings $\sigma_1, \ldots, \sigma_D$ of K, we get equations $\sum_{j=1}^{D} \sigma_i(\omega_j) u_j = \sigma_i(w)$ $(i = 1, \ldots, D)$ for the u_j, whose matrix Ω is well-known to be non-singular; in fact the entries of $\Omega^t\Omega$ (for the transpose Ω^t) are the traces $\mathrm{Tr}(\omega_i\omega_k) = \sum_{j=1}^{D} \sigma_j(\omega_i\omega_k)$ which lie in $\mathbb{Z}$ and its determinant is non-zero (in fact the discriminant Δ_K of K). In particular $|\det \Omega| \geq 1$. Now solving the equations with Cramér's Rule gives easily

$$|u_j| \leq ||\omega_j||.|u_j| \leq D! ||\omega_1|| \cdots ||\omega_D||.||w|| \ (j = 1, \ldots, D).$$

We use this remark to split the original M equations over $\mathbb{Z}_K$ into DM equations over $\mathbb{Z}$, to which we then apply Proposition 8.3. □

We need the following analogue of Lemma 3.2.

Lemma 13.3 *The functions* z, e^z *are algebraically independent over* $\mathbb{C}$.

Proof There are many ways. For example we could imitate the proof of Lemma 3.2 by using the functional equations above or even, most simply, just the periodicity (also a very useful concept later) $f(z + 2\pi i) = f(z)$; we leave all this to the reader. Here we show that if $\lambda_1, \ldots, \lambda_n$ are different complex numbers then $e^{\lambda_1 z}, \ldots, e^{\lambda_n z}$ are linearly independent over $\mathbb{C}(z)$. This is stronger because we can write a polynomial in z, e^z as a linear combination of e^{jz} $(j = 0, 1, \ldots, M)$ with coefficients in $\mathbb{C}(z)$. The simple argument is by induction on n, the case $n = 1$ being trivial. So suppose $R_1 e^{\lambda_1 z} + \cdots + R_n e^{\lambda_n z} = 0$ for $R_1, \ldots, R_n$, by induction assumed to be non-zero, in $\mathbb{C}(z)$. Dividing by the last term allows us to take $R_n = 1$ and also $\lambda_n = 0$. Then differentiating gives

$$(R_1' + \lambda_1 R_1)e^{\lambda_1 z} + \cdots + (R_{n-1}' + \lambda_{n-1} R_{n-1})e^{\lambda_{n-1} z} = 0.$$

Here the coefficients are non-zero because for example by factorizing R_1, we see that R_1'/R_1 cannot be a constant $-\lambda_1 \neq 0 = -\lambda_n$ (see also Exercise 12.23). This proves the result. □

It is not difficult to see with similar arguments (see Exercise 13.19) that further for any w we have

$$\mathrm{ord}_{z=w}\left(\sum_{i=1}^{n} R_i e^{\lambda_i z}\right) \;\le\; -1 + \sum_{i=1}^{n}(1+\deg R_i) \tag{13.1}$$

for non-zero polynomials $R_1, \ldots, R_n$ in $\mathbb{C}[z]$. This is best possible because the right-hand side counts the number of coefficients in this exponential polynomial.

Our proof of Theorem 13.1 kicks off rightaway with the opposite assumption: we suppose that $e^\alpha = \beta$ is algebraic, and we work in the finite extension $K = \mathbb{Q}(\alpha, \beta)$ of $\mathbb{Q}$. So we need a contradiction. Write $D = [K : \mathbb{Q}]$.

The first step, involving the auxiliary polynomial, imitates Lemmas 10.2 of Chapter 10 and 12.4 of Chapter 12 (I told you not to mention that!), but we change the parameter n there to T. Everything in this chapter is effective so we no longer feel the compulsion to calculate everything explicitly and we will use $c_1, c_2, \ldots$ for quantities depending only on α, β.

Lemma 13.4 *For any $L \geq 2, S \geq 1, T \geq 1$ in $\mathbb{Z}$ with*

$$(L+1)^2 \geq 2DST \tag{13.2}$$

there is P in $\mathbb{Z}[X, Y]$, of degree at most L in each variable, and with coefficients of absolute values at most $L^{3T} c_1^{LS}$, such that

$$\phi(z) = P(z, e^z) \tag{13.3}$$

is not identically zero and

$$\mathrm{ord}_{z=s\alpha}\, \phi(z) \geq T \quad (s = 1, \ldots, S). \tag{13.4}$$

Proof With $P(X, Y) = \sum_{i=0}^{L} \sum_{j=0}^{L} p_{ij} X^i Y^j$ we have

$$\phi(z) = \sum_{i=0}^{L} \sum_{j=0}^{L} p_{ij} z^i e^{jz}$$

and we use Leibniz's Rule on $\phi^{(t)}(s\alpha)$ to obtain the equations

$$\sum_{i=0}^{L} \sum_{j=0}^{L} E_{ij}(s,t) p_{ij} = 0 \quad (s = 1, \ldots, S;\; t = 0, 1, \ldots, T-1)$$

with

$$E_{ij}(s,t) = \sum_{u=0}^{t} \binom{t}{u} i(i-1)\cdots(i-u+1)(s\alpha)^{i-u}j^{t-u}\beta^{js} \tag{13.5}$$

(this time no need to divide by factorials). If a, b in $\mathbb{N}$ are denominators (soon we will be able to rid ourselves of such mild irritations) for α, β respectively, then the $w = a^L b^{LS} E_{ij}(s,t)$ are in $\mathbb{Z}_K$ with

$$||w|| \le 2^T L^T S^L a^L \max\{1, ||\alpha||\}^L L^T b^{LS} \max\{1, ||\beta||\}^{LS} \le L^{3T} c_2^{LS} = W.$$

By (13.2) the Siegel exponent in Proposition 13.2 is at most 1 (no need for the refinements of Chapters 10 or 12) and so we need only multiply W by an extra $c_3(L+1)^2$, as usual quite harmless (up to now), to get a bound for the $|p_{ij}|$. As these are not all zero we conclude the proof by appealing to Lemma 13.3. □

We could now increase the range of s in (13.4) as in Chapter 9 and 10, but it is technically easier also to increase the order of vanishing; this will actually be unavoidable in Chapter 19. We do it in the inductive style of Chapters 9 and 10, in particular Lemma 10.8. The next result is a preliminary step towards doubling both ranges.

Lemma 13.5 *Suppose for some integers $S_1 \ge S$ and $T_1 \ge T$ that*

$$\mathrm{ord}_{z=s\alpha}\, \phi(z) \ge T_1 \quad (s = 1, \ldots, S_1).$$

Then the derivatives up to order $2T_1 - 1$ satisfy

$$|\phi^{(t)}(s\alpha)| \le 2^{-S_1T_1} c_4^{T_1} T_1^{2T_1} L^{3T} c_5^{LS_1} \quad (s = 1, \ldots, 2S_1;\ t = 0, 1, \ldots, 2T_1 - 1).$$

Proof As in (9.11) of Chapter 9 we divide out by zeroes but now taking into account multiplicities as in earlier chapters; thus

$$\tilde{\phi}(z) = \frac{\phi(z)}{(z-\alpha)^{T_1} \cdots (z - S_1\alpha)^{T_1}} \tag{13.6}$$

is entire, and we have

$$|\tilde{\phi}|_{4S_1|\alpha|} \le |\tilde{\phi}|_{11S_1|\alpha|}$$

in our usual maximum modulus notation. The right-hand side here is at most $(10S_1|\alpha|)^{-S_1T_1}|\phi|_{11S_1|\alpha|}$ and the left-hand side is at least $(5S_1|\alpha|)^{-S_1T_1}|\phi|_{4S_1|\alpha|}$. We deduce

$$|\phi|_{4S_1|\alpha|} \le 2^{-S_1T_1}|\phi|_{11S_1|\alpha|}.$$

Now the supremum on the right-hand side can be estimated as we did for the application of the Siegel Lemma (but without derivatives), and we get say $L^{3T}c_5^{LS_1}$.

Finally we use (12.29) with $m = t, r = 2S_1|\alpha|, R = 4S_1|\alpha|$ to get

$$|\phi^{(t)}|_{2S_1|\alpha|} \leq t!\frac{|\phi|_{4S_1|\alpha|}}{(2S_1|\alpha|)^t} \leq T_1^{2T_1}c_4^{T_1}|\phi|_{4S_1|\alpha|} \quad (t = 0, 1, \ldots, 2T_1 - 1);$$

and putting these together finishes the proof. □

We will soon show that all these small values $\phi^{(t)}(s\alpha)$ can be made to vanish. The key remark here is

Lemma 13.6 *For any integers $S_1 \geq S, T_1 \geq T$ and each of the quantities*

$$\xi = \phi^{(t)}(s\alpha) \quad (s = 1, \ldots, 2S_1;\ t = 0, 1, \ldots, 2T_1 - 1)$$

we have either $\xi = 0$ or

$$|\xi| \geq L^{-9DT_1}c_6^{-LS_1}.$$

Proof We see as in the proof of Lemma 13.4 that $\omega = a^L b^{LS_1}\xi$ lies in $\mathbb{Z}_K$, with

$$||\omega|| \leq a^L b^{LS_1}(L+1)^2 L^{3T} c_1^{LS_1} L^{6T_1} c_7^{LS_1} \leq L^{9T_1} c_8^{LS_1}.$$

The desired result now follows from this and Proposition 11.1. □

Now Theorem 13.1 follows quickly after choosing the parameters in the following way.

We see from the previous two lemmas that this time the plucky little $2^{-S_1T_1}$ has to beat powers of $T_1^{T_1}, L^{T_1}$ and e^{LS_1}, subject to $S_1 \geq S, T_1 \geq T$ and $(L+1)^2 \geq 2DST$ in (13.2). We could for example try at first to express L, S, T as powers of an additional integer parameter n. If we take

$$L = [n^{3/4}], \quad S = [n^{1/4}], \quad T = n,$$

then for all n large enough in terms of α, β we do indeed deduce all $\xi = 0$ in Lemma 13.6 for $S_1 = S, T_1 = T$. Thus from the hypothesis of Lemma 13.5 we infer

$$\mathrm{ord}_{z=s\alpha}\,\phi(z) \geq 2T \quad (s = 1, \ldots, 2S).$$

Then for $S_1 = 2S, T_1 = 2T$, then $4S, 4T$ and so on by induction; here it is convenient to deal with the irritating implicit $\log T$ actually by increasing it via say $T^T \leq e^{T^{9/8}}$ for large enough n. At the end of this procedure, we now get zeroes of infinite order, contradicting the fact that ϕ is not identically zero (and so here at the end just a single $s\alpha$ suffices, say $s = 1$). This establishes at last the Hermite–Lindemann Theorem, proved in 1882 by Lindemann with a generalization of Hermite's method.

It is of course not the shortest proof of the transcendence of e, which can be carried out on a single page (see Baker (1990) for example) without our well over 150 pages of preparation. But it smooths the path for the proof of the transcendence of $2^{\sqrt{2}}$ (and much more) in Chapter 19. In fact the Gelfond–Schneider Theorem on α^β can now be proved without much extra effort, and to illustrate this we give some details in the exercises below.

We may also remark that (13.1) implies that the above orders are strictly less than $(L+1)^2 \leq 2T^{3/2}$, so there is no need to go to infinity. This would also make it easier to give transcendence measures. But such a step is advisable in Chapter 19, as the corresponding zero estimates are much harder to prove.

Finally in order to introduce some sort of classification into the "zoo" of various results to be exhibited in Chapter 20, we let drop the remark that Theorem 13.1 corresponds to the commutative group variety $\mathbb{G}_{\mathrm{a}} \times \mathbb{G}_{\mathrm{m}}$, where $\mathbb{G}_{\mathrm{a}}$ denotes $\mathbb{C}$ under addition and $\mathbb{G}_{\mathrm{m}}$ denotes $\mathbb{C}^*$ under multiplication. So it is a commutative group surface, with exponential map (w, e^z) (see Exercise 13.21). But neither Hermite nor Lindemann knew this.

Exercises

13.1 Let α, β, γ in $\mathbb{C}$ be algebraic with $D = [\mathbb{Q}(\alpha, \beta, \gamma) : \mathbb{Q}]$ and $\gamma = \alpha^\beta = e^{\beta \log \alpha}$ for some branch of the logarithm. Show that there is c with the following properties. Let $L \geq 2, S \geq 1, T \geq 1$ be in $\mathbb{Z}$ with $(L+1)^2 \geq 2DST$. Then there is $0 \neq P$ in $\mathbb{Z}[X, Y]$, of degree at most L in X and of degree at most L in Y, and coefficients of absolute values at most $c^T L^T c^{LS}$, such that for $\phi(z) = P(e^z, e^{\beta z})$ we have $\phi^{(t)}(s \log \alpha) = 0$ $(s = 1, \ldots, S;\ t = 0, 1, \ldots T-1)$.

13.2 In Exercise 13.1 let β be irrational. Show that $\phi \neq 0$.

13.3 Let $\alpha \neq 0, \log \alpha \neq 0$ in Exercises 13.1,13.2; and let $T_1 \geq T$ be in $\mathbb{Z}$ with $\mathrm{ord}_{z=s\log\alpha}\, \phi(z) \geq T_1$ $(s = 1, \ldots, S)$. Show that

$$|\phi^{(t)}(s \log \alpha)| \leq 2^{-ST_1} c^{T_1} L^{T_1} c^{LS} T_1^{2T_1} \quad (s = 1, \ldots, S;\ t = 0, 1, \ldots 2T_1 - 1)$$

(perhaps with a different c).

13.4 With T_1 in Exercise 13.3 and $\xi = \phi^{(t)}(s \log \alpha)$ $(s = 1, \ldots, S;\ t = 0, 1, \ldots, 2T_1 - 1)$, show that $\xi = 0$ or $|\xi| \geq c^{-T_1} L^{-DT_1} c^{-LS}$.

13.5 Conclude that $2^{\sqrt{2}}$ is transcendental.

* * * * * * * * * * * * * * * * * * *

13.6 Let α, β in $\mathbb{C}$ be algebraic with β irrational and $\alpha \neq 0, \log \alpha \neq 0$. Show that α^β is transcendental.

13.7 Show that e^π is transcendental.

13.8 Show that

$$e^{\pi\sqrt{163}} = 262537412640768743.99999999999925007259719818568\ldots$$

and

$$\sqrt[3]{e^{\pi\sqrt{163}} - 744} = 640319.99999999999999999999999390317352319 47\ldots$$

are transcendental.

13.9 Let $\alpha \neq 0, \gamma \neq 0$ in $\mathbb{C}$ be algebraic with $\log\alpha \neq 0$. Show that $\frac{\log\gamma}{\log\alpha}$ is either rational or transcendental.

13.10 Show that $\frac{\log 3}{\log 2}$ is transcendental.

13.11 Is

$$\frac{\log(30122754096401 + 21300003689580\sqrt{2})}{\log(1607521 + 1136689\sqrt{2})} = 2.1176470588235294\ldots$$

transcendental? (Solution later).

13.12 Is

$$\frac{\log(30122754096402 + 21300003689580\sqrt{2})}{\log(1607521 + 1136689\sqrt{2})} = 2.1176470588235305\ldots$$

transcendental? (Solution later).

13.13 Let $\alpha = 3.318\ldots$ with $\alpha^3 - 8\alpha - 10 = 0$. Why is

$$|\alpha + 2 - e^{\pi\sqrt{163}/24}| < 10^{-16}?$$

13.14 Why is

$$e^{7\pi\sqrt{163}} = 859\ldots(116\,\textit{places})\ldots951.998\ldots$$

slightly near an integer?

13.15 For D in $\mathbb{N}$ show that there are C, κ depending only on D with the following property. For any algebraic integers $\alpha \neq 0, \gamma \neq 0$ in $\mathbb{C}$ with $D = [\mathbb{Q}(\alpha,\gamma) : \mathbb{Q}]$, $\log\alpha \neq 0$ and $\frac{\log\gamma}{\log\alpha}$ algebraic, there are r, s in $\mathbb{Z}$ with $\alpha^r = \gamma^s$ and

$$0 < \max\{|r|, |s|\} \leq C\max\{||\alpha||, ||\gamma||\}^\kappa$$

(see also Exercises 19.22 and 19.28).

13.16 Let γ be Euler's constant. Show that at least one of $\gamma + \pi, \gamma\pi$ is transcendental.

13.17 (a) Show that for any $P \neq 0$ in $\mathbb{C}[X, Y]$, of degree at most L in X and of degree at most M in Y, and any a, we have

$$\mathrm{ord}_{z=a} P(z, e^z) \leq (L + 1)(M + 1) - 1.$$

(b) Show that for any L and M and any a there is $P \neq 0$ in $\mathbb{C}[X, Y]$, of degree at most L in X and of degree at most M in Y, such that

$$\mathrm{ord}_{z=a} P(z, e^z) = (L + 1)(M + 1) - 1.$$

13.18 (a) Show that for any $P \neq 0$ in $\mathbb{C}[X, Y]$, of degree at most 1 in X and of degree at most 1 in Y, and any $a \neq b$, we have

$$\mathrm{ord}_{z=a} P(z, e^z) + \mathrm{ord}_{z=b} P(z, e^z) \leq 4. \qquad (*)$$

(b) Find $w \neq 0$ such that there exists $P \neq 0$ in $\mathbb{C}[X, Y]$, of degree at most 1 in X and of degree at most 1 in Y, with

$$\mathrm{ord}_{z=0} P(z, e^z) = 2, \quad \mathrm{ord}_{z=w} P(z, e^z) = 2.$$

(c) Thus $(*)$ is best possible. What is the best possible estimate for general degrees L, M? I don't know.

13.19 Show that for any non-zero $R_1, \ldots, R_n$ in $\mathbb{C}[z]$ of degrees at most $L_1, \ldots,$ L_n and any distinct $\lambda_1, \ldots, \lambda_n$ in $\mathbb{C}$, we have

$$\mathrm{ord}_{z=a} \left(\sum_{i=1}^{n} R_i(z) e^{\lambda_i z} \right) \leq -1 + \sum_{i=1}^{n} (L_i + 1)$$

for any a in $\mathbb{C}$. We already remarked that this is best possible.

13.20 Show that for any non-zero $R_1, \ldots, R_n$ in $\mathbb{R}[z]$ of degrees at most $L_1, \ldots,$ L_n and any distinct $\lambda_1, \ldots, \lambda_n$ in $\mathbb{R}$, the function $\sum_{i=1}^{n} R_i(z) e^{\lambda_i z}$ has at most $-1 + \sum_{i=1}^{n} (L_i + 1)$ real zeroes (with multiplicity) [*Hint*: Rolle]. This too is clearly best possible.

13.21 Show that $\exp(w, z) = (w, e^z)$ defines a homomorphism from the group $\mathbb{C} \times \mathbb{C}$ to the group $G = \mathbb{C} \times \mathbb{C}^*$ whose kernel is $\{0\} \times 2\pi i\mathbb{Z}$. (This G is the group variety $\mathbb{G}_{\mathrm{a}} \times \mathbb{G}_{\mathrm{m}}$ with its exponential map.)

14

Heights

As the exercises in the previous chapter show, we are now in a good position to prove the Gelfond–Schneider Theorem on α^β (and much more). But the reader (not to mention the writer) deserves a rest from all this intricate estimating. So before going further with transcendence we will introduce another way of measuring algebraic numbers α. The norm $||\alpha||$ defined in (11.3) of Chapter 11 is quite useful, but essentially only for algebraic integers, and we would like to take denominators into account, while at the same time preserving handy properties like (11.6). We will define a height $H(\alpha)$ which also has the important property, sometimes associated with the name of Northcott, that for any d and any $\mathcal{H}$ there are at most finitely many algebraic numbers α of degree d with $H(\alpha) \leq \mathcal{H}$. This would be false with $||\alpha|| \leq \mathcal{H}$, even for $d = 1$, as the example $||1/n|| = 1/n \leq 1$ $(n = 1, 2, \ldots)$ shows (but it becomes true for algebraic integers – see Exercise 11.13 for $A = 1$).

In fact it is rather easy to define $H(\alpha)$. As α is algebraic over $\mathbb{Q}$, it satisfies an equation $A(\alpha) = 0$ with non-zero A in $\mathbb{Q}[X]$. We can assume that A is irreducible over $\mathbb{Q}$; then its degree $d = [\mathbb{Q}(\alpha) : \mathbb{Q}]$. By multiplying up by a least common denominator we can further assume that

$$A(X) = a_0X^d + a_1X^{d-1} + \cdots + a_d \tag{14.1}$$

with $a_0, a_1, \ldots, a_d$ coprime in $\mathbb{Z}$ (in fact it is then irreducible over $\mathbb{Z}$). We may further suppose that $a_0 \geq 1$, and then this fixes A uniquely. Over $\mathbb{C}$ there is a factorization

$$A(X) = a_0(X - \alpha_1) \cdots (X - \alpha_d) \tag{14.2}$$

and finally we define

$$H(\alpha)^d \;=\; a_0 \max\{1, |\alpha_1|\} \cdots \max\{1, |\alpha_d|\} \;\geq\; 1. \tag{14.3}$$

This involves only the zeroes outside the unit circle.

In fact the α_i are exactly the $\sigma(\alpha)$ for the complex embeddings σ of $\mathbb{Q}(\alpha)$, so this is certainly related to (11.3). Further we know from Chapter 11 that a_0 is a denominator for α.

Sometimes one uses different norms to get for example $\sqrt{1+|\alpha_i|^2}$ above (l^2 instead of l^∞). This is unimportant when considering certain upper bounds for the height, but the problems about lower bounds are especially interesting for (14.3) itself, as we will see in Chapter 16.

Here are some examples.

First inside $\mathbb{Q}$. Then $\alpha = r/s$ with coprime r and $s \geq 1$ in $\mathbb{Z}$. Now $A = sX - r = s(X - \alpha)$ and so

$$H(\alpha) = s\max\{1, |\alpha|\} = \max\{s, |r|\}.$$

Thus $H(1) = 1, H(0) = H(0/1) = 1$, and

$$H(1947/1948) = H(-1947/1948) = H(1948/1947) = 1948.$$

Or $\alpha = (1+i)/2$ inside the unit circle. Now $(\alpha - 1/2)^2 = -1/4$ so

$$A = 2X^2 - 2X + 1 = 2(X-\alpha)(X-\overline{\alpha})$$

for the complex conjugate also inside the unit circle, so $H(\alpha)^2 = 2$ and $H(\alpha) = \sqrt{2} = H(\overline{\alpha})$.

Or $\alpha = 1 + \sqrt{2}$ outside the unit circle, with $(\alpha - 1)^2 = 2$ so

$$A = X^2 - 2X - 1 = (X - \alpha)(X - \alpha')$$

for the conjugate α' now inside the unit circle, so $H(\alpha)^2 = \alpha$ and $H(\alpha) = \sqrt{1+\sqrt{2}} = H(\alpha')$.

Or $\alpha = 1 + \sqrt{-6}$ with $H(\alpha) = \sqrt{7}$.

Or $H(\sqrt[3]{2}) = \sqrt[3]{2}$, $H(1+\sqrt[3]{2}) = \sqrt[3]{3}$ and $H(-1+\sqrt[3]{2}) = \sqrt[3]{1+\sqrt[3]{2}+\sqrt[3]{4}}$.

Or α as a primitive fifth root of unity, say $\zeta_5 = e^{2\pi i/5}$. Here A is the corresponding cyclotomic

$$\frac{X^5-1}{X-1} = X^4 + X^3 + X^2 + X + 1 = (X-\zeta_5)(X-\zeta_5^2)(X-\zeta_5^3)(X-\zeta_5^4),$$

so $H(\zeta_5) = 1$.

Or $\alpha = 1 + \zeta_5$. Now only the conjugates $1+\zeta_5, 1+\zeta_5^4$ lie outside the unit circle, and so

$$H(\alpha)^4 = |(1+\zeta_5)(1+\zeta_5^4)| = \frac{3+\sqrt{5}}{2} = \left(\frac{1+\sqrt{5}}{2}\right)^2$$

that is

$$H(1+\zeta_5) = \sqrt{\frac{1+\sqrt{5}}{2}}. \tag{14.4}$$

Or $H(\sqrt{2}+\sqrt{3}) = \sqrt{\sqrt{2}+\sqrt{3}}$.

Or $H(\sqrt{5}+\sqrt{7-\sqrt{11}})$; but this needs some work.

With an arbitrary root of unity ζ the corresponding cyclotomic polynomial may not be easy to write explicitly, but in $\mathbb{Z}[X]$ it divides some $X^n - 1$ whose zeroes are all on the unit circle and so

$$H(\zeta) = 1 \tag{14.5}$$

in general.

However $H(1+\zeta)$ also needs work, and it is a fact that

$$\lim_{n\to\infty} H(1+e^{2\pi i/n}) = \exp\left(\frac{1}{2\pi}\int_0^{2\pi} \log\max\{1, |1+e^{i\theta}|\}d\theta\right) = 1.38135\ldots. \tag{14.6}$$

Or $\alpha = 2^{1/d}$. It is known, for example by Eisenstein's Criterion, that this has degree d, with $A = X^d - 2$ whose zeroes $e^{2\pi i m/d}\alpha$ $(m = 0, 1, \ldots, d-1)$ all have absolute value α. It follows that

$$H(2^{1/d}) = 2^{1/d}. \tag{14.7}$$

Our final example involves Ramanujan's

$$\alpha = \frac{63}{25}\left(\frac{17+15\sqrt{5}}{7+15\sqrt{5}}\right). \tag{14.8}$$

With (11.4) and the conjugate α' in (11.5) we find

$$(X-\alpha)(X-\alpha') = X^2 - \frac{31689}{6725} + \frac{829521}{168125} \tag{14.9}$$

and $A = 168125X^2 - 792225X + 829521$, so

$$H(\alpha) = \sqrt{168125\alpha\alpha'} = \sqrt{829521} = 63\sqrt{209}.$$

We can estimate the height from above using the following result on lengths, which also gives an explicit version of Lemma 5.2 for several polynomials in one variable.

Proposition 14.1 *(i) For any $P_1, \ldots, P_m$ in $\mathbb{C}[X]$ with $P = P_1 \cdots P_m$ of degree at most $L \geq 0$ we have*

$$\mathcal{L}(P_1)\cdots\mathcal{L}(P_m) \leq 8^L \mathcal{L}(P_1\cdots P_m).$$

(ii) If further $P \neq 0$ is in $\mathbb{Z}[X]$ with $P(\alpha) = 0$ for some α of degree d, then $H(\alpha)^d \leq 4^L \mathcal{L}(P)$.

Proof The key to (*i*) is the Mahler measure $\mathcal{M}(A)$ of a general A in $\mathbb{C}[X]$, which for A formally as in (14.1) and (14.2) with $a_0 \neq 0$ is defined by the left-hand side of (14.3) with $|a_0|$ in place of a_0.

Clearly

$$|a_i| \leq \binom{d}{i} \mathcal{M}(A) \quad (i = 0, 1, \ldots, d)$$

and so $\mathcal{L}(A) \leq 2^d \mathcal{M}(A)$.

To get an inequality in the opposite sense, write $A = A_0 A_\infty$ for A_0 monic with zeroes inside the unit circle and A_∞ with zeroes α outside. Then, in the by now standard notation,

$$\mathcal{M}(A) = \mathcal{M}(A_\infty) = |a_0| \prod_\alpha |\alpha| \leq 2^d |A_\infty|_{1/2}.$$

Here

$$|A_\infty|_{1/2} \leq |A_\infty|_2 \leq |A_0 A_\infty|_2 \leq 2^d \mathcal{L}(A).$$

In summary

$$\mathcal{L}(A) \leq 2^d \mathcal{M}(A), \quad \mathcal{M}(A) \leq 4^d \mathcal{L}(A).$$

Now the Mahler measure is clearly multiplicative in the sense that $\mathcal{M}(AB) = \mathcal{M}(A)\mathcal{M}(B)$. It follows that in (*i*) above

$$\mathcal{L}(P_1) \cdots \mathcal{L}(P_m) \leq 2^L \mathcal{M}(P_1 \cdots P_m) \leq 8^L \mathcal{L}(P_1 \cdots P_m).$$

As for (*ii*), we take A literally as in (14.1) and then there is Q in $\mathbb{Z}[X]$ with $AQ = P$. We conclude

$$H(\alpha)^d = \mathcal{M}(A) \leq \mathcal{M}(A)\mathcal{M}(Q) = \mathcal{M}(P) \leq 4^L \mathcal{L}(P).$$

□

Now we prove the Northcott property in the following form.

Proposition 14.2 *Given $\mathcal{D}$ and $\mathcal{H}$ there are at most finitely many algebraic α of degree at most $\mathcal{D}$ with $H(\alpha) \leq \mathcal{H}$.*

Proof As above we get

$$|a_i| \leq \binom{d}{i} H(\alpha)^d \leq \binom{d}{i} \mathcal{H}^d \qquad (i = 0, 1, \ldots, d) \tag{14.10}$$

for the coefficients of A. So there are at most $\prod_{i=0}^d (1 + 2\binom{d}{i}\mathcal{H}^d)$ possibilities for A; and then each A has at most d different zeroes. Finally $d \leq \mathcal{D}$. □

The resulting upper bound

$$\sum_{d=1}^{\mathcal{D}} d \prod_{i=0}^{d} \left(1 + 2\binom{d}{i}\mathcal{H}^d\right) \leq 36^{\mathcal{D}^2}\mathcal{H}^{\mathcal{D}(\mathcal{D}+1)} \tag{14.11}$$

for the number of α in Proposition 14.2 (when $\mathcal{H} \geq 1$) is not bad, at least in its dependence on $\mathcal{H}$.

We will soon show that

$$H(\alpha+\beta) \leq 2H(\alpha)H(\beta), \quad H(\alpha\beta) \leq H(\alpha)H(\beta) \tag{14.12}$$

(and much more) which is almost as good as (11.6), and furthermore which takes care of denominators. But even the identity $H(\alpha^m) = H(\alpha)^m$ is not so easy to prove from scratch, already for $m = 2$. The trouble lies precisely in the a_0 in (14.2). First, it may not be the smallest denominator for α. An example is $\alpha = (1+\sqrt{2})/2$ with $4\alpha^2 - 4\alpha - 1 = 0$, but 2α is an algebraic integer (see Exercise 14.5 for further precision). Also the quantity $a_0 = a_0(\alpha)$ can behave a bit irregularly as a function of α. For example let $\alpha = \sqrt{2}/2$. Then the values of $a_0(\alpha^m)$ $(m = 0, 1, 2, 3, 4, 5, 6, 7)$ are

$$1, 2, 2, 8, 4, 32, 8, 128 \tag{14.13}$$

corresponding to the following polynomials A

$$X-1,\ 2X^2-1,\ 2X-1,\ 8X^2-1,\ 4X-1,\ 32X^2-1,\ 8X-1,\ 128X^2-1. \tag{14.14}$$

Those also show that the degree $d = d(\alpha)$ is not quite regular.

To explain these things it is necessary to re-introduce number fields K. The complex embeddings σ of K give rise to functions on K defined by $|\alpha|_\sigma = |\sigma(\alpha)|$, which are valuations in the sense that

$$|\alpha+\beta|_\sigma \leq |\alpha|_\sigma + |\beta|_\sigma, \quad |\alpha\beta|_\sigma = |\alpha|_\sigma|\beta|_\sigma \tag{14.15}$$

for all α, β in K.

These are not the only valuations on K, but we have to go a bit teutonic to find others. Namely let $\mathfrak{p}$ be a prime ideal of $\mathfrak{o} = \mathbb{Z}_K$. It is known that for any $\alpha \neq 0$ in $\mathfrak{o}$ there is an ideal factorization $\alpha\mathfrak{o} = \mathfrak{p}^e\mathfrak{b}$ with $e \geq 0$ and $\mathfrak{p}$ not dividing $\mathfrak{b}$. Now we compensate with some more exotic greek: given any real $\Xi_\mathfrak{p} > 1$ the function $|\alpha|_\mathfrak{p} = \Xi_\mathfrak{p}^{-e}$ satisfies the analogue of (14.15). But in checking this, one finds the stronger

$$|\alpha+\beta|_\mathfrak{p} \leq \max\{|\alpha|_\mathfrak{p}, |\beta|_\mathfrak{p}\}, \quad |\alpha\beta|_\mathfrak{p} = |\alpha|_\mathfrak{p}|\beta|_\mathfrak{p}, \tag{14.16}$$

the point being that $\mathfrak{p}^{\min\{e,f\}}$ divides both $\mathfrak{p}^e$ and $\mathfrak{p}^f$.

The first of these is known as the ultrametric inequality. Then one can use the second to define $|\alpha|_\mathfrak{p}$ for any $\alpha \neq 0$ in K by writing it as a quotient of integers of K (for example $(a_0\alpha)/a_0$). Finally one defines $|0|_{\mathfrak{p}} = 0$ and then checks that (14.16) extends to all α, β in K.

Ostrowski proved that these are essentially all the valuations on K (see Exercises 14.8, 14.43), but we don't need to know this (or exactly what the definition of valuation is – but see later).

We will have to take all $\mathfrak{p}$ together, and for this it is necessary to choose $\Xi_{\mathfrak{p}}$ properly; also we prefer symbols that are topologically connected. The best choice is $\Xi_{\mathfrak{p}} = [\mathfrak{o} : \mathfrak{p}]$, the so-called norm $N(\mathfrak{p})$ of $\mathfrak{p}$.

We give some examples.

Even $K = \mathbb{Q}$ is worthwhile. Then there is a unique σ, and one usually writes $|\alpha|_\sigma$ as $|\alpha|_\infty$. Also the prime ideals $\mathfrak{p}$ correspond to positive prime numbers $p = N(\mathfrak{p})$, and one usually writes $|\alpha|_{\mathfrak{p}}$ as $|\alpha|_p$. For $\alpha = 1948 = 2^2 487$ we find

$$|\alpha|_2 = \frac{1}{4}, \quad |\alpha|_{487} = \frac{1}{487}$$

with $|\alpha|_p = 1$ for all other p.

Next in $K = \mathbb{Q}(\sqrt{5})$ with Ramanujan's (14.8). One finds for the numerator $\alpha_1 = 63(17 + 15\sqrt{5})$ and the denominator $\alpha_2 = 25(7 + 15\sqrt{5})$ the factorizations

$$\alpha_1\mathfrak{o} = \mathfrak{p}_2\mathfrak{p}_3^2\mathfrak{p}_7\mathfrak{p}_{11}\mathfrak{p}_{19}, \quad \alpha_2\mathfrak{o} = \mathfrak{p}_2\mathfrak{p}_5^4\mathfrak{p}_{269},$$

where the primes satisfy

$$2\mathfrak{o} = \mathfrak{p}_2, \; 3\mathfrak{o} = \mathfrak{p}_3, \; 5\mathfrak{o} = \mathfrak{p}_5^2, \; 7\mathfrak{o} = \mathfrak{p}_7,$$

$$11\mathfrak{o} = \mathfrak{p}_{11}\mathfrak{p}_{11}', \; 19\mathfrak{o} = \mathfrak{p}_{19}\mathfrak{p}_{19}', \; 269\mathfrak{o} = \mathfrak{p}_{269}\mathfrak{p}_{269}'.$$

It follows that

$$|\alpha|_{\mathfrak{p}_3} = N(\mathfrak{p}_3)^{-2} = 9^{-2} = \frac{1}{81}, \quad |\alpha|_{\mathfrak{p}_5} = N(\mathfrak{p}_5)^4 = 5^4 = 625,$$

$$|\alpha|_{\mathfrak{p}_7} = N(\mathfrak{p}_7)^{-1} = 49^{-1} = \frac{1}{49}, \quad |\alpha|_{\mathfrak{p}_{11}} = N(\mathfrak{p}_{11})^{-1} = 11^{-1} = \frac{1}{11},$$

$$|\alpha|_{\mathfrak{p}_{19}} = N(\mathfrak{p}_{19})^{-1} = 19^{-1} = \frac{1}{19}, \quad |\alpha|_{\mathfrak{p}_{269}} = N(\mathfrak{p}_{269}) = 269.$$

with $|\alpha|_{\mathfrak{p}} = 1$ for all other $\mathfrak{p}$.

The key to the behaviour of $a_0(\alpha)$ and the properties of $H(\alpha)$ lies in the following two results for general K. The first is the famous Product Formula.

Proposition 14.3 *Given $\alpha \neq 0$ in K there are at most finitely many $\mathfrak{p}$ with $|\alpha|_\mathfrak{p} \neq 1$, and we have*

$$\prod_\mathfrak{p} |\alpha|_\mathfrak{p} \prod_\sigma |\alpha|_\sigma = 1. \tag{14.17}$$

Proof Assume first that α is an integer in K, with a prime factorization $\alpha\mathfrak{o} = \prod \mathfrak{p}^e$. Then $|\alpha|_\mathfrak{p} \neq 1$ implies that $\mathfrak{p}$ occurs in this finite product. Also $\prod_\mathfrak{p} |\alpha|_\mathfrak{p}^{-1} = \prod N(\mathfrak{p})^e = N(\prod \mathfrak{p}^e) = N(\alpha\mathfrak{o})$, which is well-known to be the absolute value $|\operatorname{Nm}(\alpha)| = \prod_\sigma |\sigma(\alpha)| = \prod_\sigma |\alpha|_\sigma$ of the norm. This explains the choice of $\Xi_\mathfrak{p}$, and proves (14.17) for an integer; it follows in general by looking at a quotient. □

The Ramanujan calculations just above, together with (14.9), give indeed in (14.17)

$$\left(\frac{1}{81}\right) 625 \left(\frac{1}{49}\right)\left(\frac{1}{11}\right)\left(\frac{1}{19}\right) 269 \left(\frac{829521}{168125}\right) = 1.$$

The second key result expresses everything in (14.3) entirely in terms of valuations.

Proposition 14.4 *Given α in K with $a_0, \alpha_1, \ldots, \alpha_d$ as above, we have*

$$a_0^{[K:\mathbb{Q}(\alpha)]} = \prod_\mathfrak{p} \max\{1, |\alpha|_\mathfrak{p}\}, \quad \prod_{i=1}^d \max\{1, |\alpha_i|\}^{[K:\mathbb{Q}(\alpha)]} = \prod_\sigma \max\{1, |\alpha|_\sigma\} \tag{14.18}$$

so that

$$H(\alpha)^{[K:\mathbb{Q}]} = \prod_\mathfrak{p} \max\{1, |\alpha|_\mathfrak{p}\} \prod_\sigma \max\{1, |\alpha|_\sigma\}. \tag{14.19}$$

Proof We have already seen just after (11.3) how complex embeddings behave when one goes to a larger number field. This yields immediately the second of (14.18). It follows incidentally that

$$\prod_\sigma \max\{1, |\alpha|_\sigma\}^{1/[K:\mathbb{Q}]} = \prod_{i=1}^d \max\{1, |\alpha_i|\}^{1/d}$$

is independent of K.

Now by observing how a prime ideal $\mathfrak{p}$ in $\mathbb{Z}_K$, or rather its extension $\mathfrak{p}\mathbb{Z}_{\tilde{K}}$, splits into primes $\tilde{\mathfrak{p}}$ in a larger field $\tilde{K}$, and taking into account the crucial normalization involving $N(\mathfrak{p})$, one can prove the analogous fact that

$$\prod_\mathfrak{p} \max\{1, |\alpha|_\mathfrak{p}\}^{1/[K:\mathbb{Q}]}$$

is independent of K. For more of the details see Exercise 14.35. One can also consult section 1.5 of Bombieri and Gubler (2006), although this, like several other texts, uses more systematically the language of valuations.

Next the well-known Gauss Lemma over $\mathbb{Z}[X]$ has a local version for each prime p, best expressed in terms of p-adic valuations. In fact the latter extends to any ultrametric valuation (that is, satisfying (14.16) without $\mathfrak{p}$, as well as $|c| \geq 0$ with equality if and only if $c = 0$) on any field $\mathcal{K}$ as follows. One defines $|P|$ for any P in $\mathcal{K}[X]$ in terms of its coefficients c as $\max |c|$. Then $|PQ| = |P||Q|$. See for example section 1.6 of Bombieri and Gubler (2006).

We take $\mathcal{K} = \tilde{K}$ as above, with any $\tilde{K}$ containing K and the conjugates $\alpha_1, \ldots, \alpha_d$ in (14.2). Then for any $\tilde{\mathfrak{p}}$ we deduce

$$|A|_{\tilde{\mathfrak{p}}} = |a_0|_{\tilde{\mathfrak{p}}} \prod_{i=1}^{d} \max\{1, |\alpha_i|_{\tilde{\mathfrak{p}}}\}.$$

Because the coefficients in A are coprime in $\mathbb{Z}$, they cannot all be divisible by any $\tilde{\mathfrak{p}}$ when considered in $\mathbb{Z}_{\tilde{K}}$. It follows that $|A|_{\tilde{\mathfrak{p}}} = 1$.

Next taking the product over all $\tilde{\mathfrak{p}}$ and using Proposition 14.3 gives

$$1 = a_0^{-[\tilde{K}:\mathbb{Q}]} \prod_{i=1}^{d} \prod_{\tilde{\mathfrak{p}}} \max\{1, |\alpha_i|_{\tilde{\mathfrak{p}}}\}. \tag{14.20}$$

Now for each product over $\tilde{\mathfrak{p}}$, we descend from $\tilde{K}$ to $\mathbb{Q}(\alpha_i)$ using the independence noted just above. We get $\prod_{\mathfrak{p}_i} \max\{1, |\alpha_i|_{\mathfrak{p}_i}\}^{[\tilde{K}:\mathbb{Q}]/d}$ taken over the primes of $\mathbb{Q}(\alpha_i)$. However $\mathbb{Q}(\alpha_i)$ and $\mathbb{Q}(\alpha)$ are isomorphic, so this is just $\prod_{\mathfrak{p}_\alpha} \max\{1, |\alpha|_{\mathfrak{p}_\alpha}\}^{[\tilde{K}:\mathbb{Q}]/d}$ over the primes of $\mathbb{Q}(\alpha)$. Again by independence this is in turn $\prod_{\mathfrak{p}} \max\{1, |\alpha|_{\mathfrak{p}}\}^{[\tilde{K}:K]}$ over the primes of K. And now inserting into (14.20) gives the first of (14.18); which as a whole implies (14.19). □

We note that the above isomorphism remark extends also to the σ, and implies

$$H(\alpha) = H(\alpha') \tag{14.21}$$

for any conjugate $\alpha' = \alpha_i$ of α.

We can now explain (14.13). From the first of (14.18), the quantity $f(\alpha) = a_0(\alpha)^{[K:\mathbb{Q}(\alpha)]}$ for any fixed K containing α satisfies $f(\alpha^m) = f(\alpha)^m$ for all m in $\mathbb{N}$. Choosing naturally $K = \mathbb{Q}(\alpha)$ gives $a_0(\alpha^m) = a_0(\alpha)^{m/d_m}$, where

$$d_m = [\mathbb{Q}(\alpha) : \mathbb{Q}(\alpha^m)]. \tag{14.22}$$

For $\alpha = \sqrt{2}/2$ the latter alternate between 1 and 2 as in (14.14), and as now $a_0(\alpha) = 2$ we discover the pattern in (14.13).

Similarly it is now clear from (14.19) that

$$H(\alpha^m) = H(\alpha)^m. \tag{14.23}$$

But also

$$H(\alpha^{-1}) = H(\alpha), \tag{14.24}$$

(when $\alpha \neq 0$) because for any K we can take out a factor $\prod_{\mathfrak{p}} |\alpha|_{\mathfrak{p}} \prod_{\sigma} |\alpha|_{\sigma} = 1$ in (14.19) to get

$$H(\alpha)^{[K:\mathbb{Q}]} = \prod_{\mathfrak{p}} \max\{|\alpha^{-1}|_{\mathfrak{p}}, 1\} \prod_{\sigma} \max\{|\alpha^{-1}|_{\sigma}, 1\} = H(\alpha^{-1})^{[K:\mathbb{Q}]}. \tag{14.25}$$

Now (14.23) and (14.24) can be combined into the single

$$H(\alpha^n) = H(\alpha)^{|n|} \tag{14.26}$$

(when $\alpha \neq 0$) for any n in $\mathbb{Z}$.

Already (14.23) suffices for a classical result ascribed to Kronecker (but it can also be proved from (14.3) directly, and even without heights as in Exercise 11.16).

Theorem 14.5 *Suppose α is non-zero with $H(\alpha) = 1$. Then α is a root of unity.*

Proof By Proposition 14.2 there are at most finitely many β with degree at most $d = [\mathbb{Q}(\alpha) : \mathbb{Q}]$ and $H(\beta) \leq 1$; let M be their number. As

$$H(\alpha^m) = H(\alpha)^m = 1 \quad (m = 0, 1, \ldots, M),$$

these $M + 1$ powers α^m cannot all be different. Now $\alpha^{m'} = \alpha^{m''}$ for $m' \neq m''$ implies the conclusion. □

The following is an immediate consequence; it will be used in Chapter 16.

Lemma 14.6 *Suppose α is non-zero with minimum polynomial A. If there is an integer $m \geq 2$ with $A(\alpha^m) = 0$, then α is a root of unity.*

Proof In this situation α^m must be one of the conjugates $\alpha' = \alpha_i$. In particular

$$H(\alpha)^m = H(\alpha^m) = H(\alpha') = H(\alpha)$$

using (14.23) and (14.21). Thus $H(\alpha) = 1$. □

We next give generalized versions of (14.12).

Proposition 14.7 *Given P in $\mathbb{Z}[X_1, \ldots, X_n]$ of degree at most $L_1 \geq 0$ in $X_1, \ldots, L_n \geq 0$ in X_n and algebraic numbers $\xi_1, \ldots, \xi_n$, we have*

$$H(P(\xi_1, \ldots, \xi_n)) \leq \mathcal{L}(P) H(\xi_1)^{L_1} \cdots H(\xi_n)^{L_n} \tag{14.27}$$

for the length $\mathcal{L}(P)$.

Proof Fix any K containing $\xi_1, \ldots, \xi_n$. For any $\mathfrak{p}$ the ultrametric inequality implies

$$|P(\xi_1, \ldots, \xi_n)|_{\mathfrak{p}} \le \max\{1, |\xi_1|_{\mathfrak{p}}\}^{L_1} \cdots \max\{1, |\xi_n|_{\mathfrak{p}}\}^{L_n}.$$

Thus clearly also for $\max\{1, |P(\xi_1, \ldots, \xi_n)|_{\mathfrak{p}}\}$. But for σ we get an extra $\mathcal{L}(P)$. Multiplying up over all $\mathfrak{p}, \sigma$ and taking the $[K : \mathbb{Q}]$ root as in (14.25), we get the result. □

Now (14.12) follows by taking P as $X_1 + X_2, X_1X_2$. The fact that the heights are multiplied in the first inequality cannot be avoided; for example it would not suffice to take the sum (see Exercise 14.55).

From (14.5) we see that the height attains its minimum 1 at every root of unity ζ, for example $-e^{2\pi i/n}$. But from (14.6) then $H(1-\zeta)$ is not near 1. We cannot resist presenting a special case of a result of Zhang which shows that this phenomenon has hardly anything to do with roots of unity, even though the auxiliary polynomial $X^5 + (1-X)^5 - 1$ used in the proof is not quite in the sense of this book (being far too explicit).

Theorem 14.8 *For any algebraic $\alpha \neq 0, 1, e^{\pm\pi i/3}$ we have*

$$H(\alpha)H(1-\alpha) \ge \left(\frac{5}{3}\right)^{1/5} = 1.10756634\ldots. \tag{14.28}$$

Proof We consider

$$\beta = \alpha^5 + (1-\alpha)^5 - 1 = 5\alpha(\alpha-1)(\alpha^2 - \alpha + 1), \tag{14.29}$$

which is plucked from the same hat as long ago in (2.9). Thus $\beta \neq 0$. We are going to use the Product Formula, so we fix any K containing α. For any $\mathfrak{p}$ the ultrametric inequality on the first expression in (14.29) gives

$$|\beta|_{\mathfrak{p}} \le \max\{1, |\alpha|_{\mathfrak{p}}\}^5 \max\{1, |1-\alpha|_{\mathfrak{p}}\}^5.$$

But for non-ultrametric σ we get an extra 3, the number of terms in the first expression of (14.29). Multiplying up over all $\mathfrak{p}, \sigma$ and taking the $[K : \mathbb{Q}]$ root we do not get the result, only the triviality $H(\alpha)H(1-\alpha) \ge (1/3)^{1/5}$.

The trick is to use instead the second expression in (14.29) to get

$$|\beta|_{\mathfrak{p}} \le |5|_{\mathfrak{p}} \max\{1, |\alpha|_{\mathfrak{p}}\}^4 \le |5|_{\mathfrak{p}} \max\{1, |\alpha|_{\mathfrak{p}}\}^5 \max\{1, |1-\alpha|_{\mathfrak{p}}\}^5$$

(this is an improvement if $\mathfrak{p}$ divides 5). Now the same procedure gives

$$1 \le 3^{[K:\mathbb{Q}]} C H(\alpha)^{5[K:\mathbb{Q}]} H(1-\alpha)^{5[K:\mathbb{Q}]},$$

where $C = \prod_{\mathfrak{p}} |5|_{\mathfrak{p}}$ can be evaluated again with the Product Formula to give $\prod_\sigma |5|_\sigma^{-1} = 5^{-[K:\mathbb{Q}]}$. And now we see what we want. □

A trivial consequence (also trivial directly – see Exercise 14.10) is that the only roots of unity α, β with $\alpha + \beta = 1$ are $\alpha = e^{\pm\pi i/3}, \beta = e^{\mp\pi i/3}$. This is a very special case of a conjecture associated with the nametag Manin–Mumford; see for example Hindry and Silverman (2000) or Bombieri and Gubler (2006) (see also Exercise 14.12(c), as well as Exercise 14.63(d) for three roots of unity, and later Exercise 15.11).

Some sort of lower bound bigger than 1 in (14.28) can be obtained using only the original definition (14.3) and geometric considerations about the distribution of the conjugates (see the next chapter for equidistribution, and Exercise 15.9). Zhang (1992) was the first to obtain such a bound, even in a much more general context. Then Zagier (1993) proved that

$$H(\alpha)H(1-\alpha) \geq \sqrt{\frac{1+\sqrt{5}}{2}},$$

which is best possible in view of (14.4) and $\alpha = -\zeta_5$. Our argument is taken from Bombieri and Zannier (1995).

We are equally unable to resist proving a height upper bound of a sort useful in recent developments.

Theorem 14.9 *For algebraic $\alpha \neq 0, 1$ suppose that there are integers r, s not both zero with $\alpha^r(1-\alpha)^s = 1$. Then we have*

$$H(\alpha) \leq 8. \tag{14.30}$$

Proof Here we can argue just using formal properties of the height function without needing the definition. There are a few cases.

Suppose first that s is small compared with r, more precisely $|s| \leq 2|r|/3$. Then using (14.26) we get

$$H(\alpha)^{|r|} = H(\alpha^r) = H((1-\alpha)^{-s}) = H(1-\alpha)^{|s|} \leq H(1-\alpha)^{2|r|/3}.$$

And by Proposition 14.7 (or the first in (14.12) above) we have $H(1-\alpha) \leq 2H(\alpha)$. Cancelling $|r| \neq 0$, we get even $H(\alpha) \leq 4$.

If $|r| \leq 2|s|/3$ we get in a similar way $H(1-\alpha) \leq 4$ and then (14.30).

Thus we can assume that $2|r|/3 < |s| < 3|r|/2$, so that $r \neq 0, s \neq 0$ have about the same size.

If r, s have opposite signs then $|r+s| = ||s| - |r|| < |r|/2$, and since $\beta^{-r}(1-\alpha)^{r+s} = 1$ for $\beta = \alpha^{-1} - 1$, we get

$$H(\beta) \leq H(1-\alpha)^{1/2} \leq (2H(\alpha))^{1/2};$$

on the other hand

$$H(\alpha) = H(\alpha^{-1}) \leq 2H(\beta) \tag{14.31}$$

leading to $H(\alpha) \leq 8$.

If r, s have the same signs then $|r-s| < |r|/2$, and since now $\gamma^r(1-\alpha)^{s-r} = 1$ for $\gamma = \alpha(1-\alpha)$, we get $H(\gamma) \leq (2H(\alpha))^{1/2}$.

This time

$$H(\alpha)^2 = H(\alpha^2) = H(\alpha - \gamma) \leq 2H(\alpha)H(\gamma)$$

so

$$H(\gamma) \geq \frac{1}{2}H(\alpha) \tag{14.32}$$

leading again to $H(\alpha) \leq 8$. □

The reader may see another strategy for proving this: namely we have a perfectly explicit equation for α, so we may be able to use the definition (14.3). The obstacle is that we don't know that the equation is irreducible – indeed already $X^2(1-X)^2 - 1$ is reducible – not to mention

$$X^5 - (1-X)^7 \;=\; (X^2 - X + 1)(X^5 - 6X^4 + 15X^3 - 14X^2 + 6X - 1);$$

and Proposition 14.1(ii) is useless if the degree of α is too small.

Of course Theorem 14.9 can at once be improved a bit, but Cohen and Zannier obtained $H(\alpha) \leq 2$, which is best possible in view of the values $\alpha = 1/2$ and $\alpha = 2$. They even showed that there is $c < 2$ such that $H(\alpha) \leq c$ apart from these values. Their proof used Bilu's Equidistribution Theorem, which we will meet in the next chapter.

We note in passing that (14.32) can be substantially improved for large $H(\alpha)$. For example $\gamma - 1/4 = -\delta^2$ with $\delta = \alpha - 1/2$, so

$$H(\delta)^2 \leq 2H(\gamma)H(1/4) = 8H(\gamma)$$

and $H(\alpha) \leq 4H(\delta)$ leading to

$$H(\alpha^2 - \alpha) \geq \frac{1}{128}H(\alpha)^2. \tag{14.33}$$

Here the exponent of $H(\alpha)$ cannot be further improved, because by Proposition 14.7 we have in the opposite direction

$$H(\alpha^2 - \alpha) \leq 2H(\alpha)^2. \tag{14.34}$$

Similarly the observation $H((1-\alpha)/\alpha) \geq H(\alpha)/2$ in (14.31) cannot be much improved, because

$$H\left(\frac{1-\alpha}{\alpha}\right) \leq 2H(\alpha). \tag{14.35}$$

For later use, we record the following simple generalization of Proposition 14.7 from a polynomial to a rational function. It implies both (14.34) and (14.35).

Proposition 14.10 *Given P, Q in $\mathbb{Z}[X_1, \ldots, X_n]$ of degree at most $L_1 \geq 0$ in $X_1, \ldots, L_n \geq 0$ in X_n and algebraic numbers $\xi_1, \ldots, \xi_n$ with $Q(\xi_1, \ldots, \xi_n) \neq 0$, we have*

$$H\left(\frac{P(\xi_1, \ldots, \xi_n)}{Q(\xi_1, \ldots, \xi_n)}\right) \leq \max\{\mathcal{L}(P), \mathcal{L}(Q)\} H(\xi_1)^{L_1} \cdots H(\xi_n)^{L_n}. \qquad (14.36)$$

Proof Fix any K containing $\xi_1, \ldots, \xi_n$. For any $\mathfrak{p}$ the ultrametric inequality implies

$$\max\{|P(\xi_1, \ldots, \xi_n)|_{\mathfrak{p}}, |Q(\xi_1, \ldots, \xi_n)|_{\mathfrak{p}}\} \leq \max\{1, |\xi_1|_{\mathfrak{p}}\}^{L_1} \cdots \max\{1, |\xi_n|_{\mathfrak{p}}\}^{L_n}.$$

And for σ we get an extra $\max\{\mathcal{L}(P), \mathcal{L}(Q)\}$. Now we get the required result using the Product Formula for $Q(\xi_1, \ldots, \xi_n)$. □

For $n = 1$ this upper bound cannot essentially be improved in its dependence on $H(\xi_1)$; see Exercise 14.77.

It is sometimes convenient not to distinguish between the valuations with $\mathfrak{p}$ and those with σ, even though only the former are ultrametric. We do this with the symbol v for valuation. Thus for example (14.19) reads

$$H(\alpha)^{[K:\mathbb{Q}]} = \prod_v \max\{1, |\alpha|_v\} \qquad (14.37)$$

and if $\alpha \neq 0$ the Product Formula becomes

$$\prod_v |\alpha|_v = 1. \qquad (14.38)$$

Here a small warning: not all the valuations are different. Any complex embedding σ has a conjugate $\overline{\sigma}$ defined by $\overline{\sigma}(\alpha) = \overline{\sigma(\alpha)}$, and then

$$|\alpha|_{\overline{\sigma}} = |\overline{\sigma}(\alpha)| = |\overline{\sigma(\alpha)}| = |\sigma(\alpha)| = |\alpha|_\sigma$$

so we get the same valuation. But it can happen that σ does not map K into $\mathbb{R}$, and then $\overline{\sigma} \neq \sigma$. Some authors write v for both σ and $\overline{\sigma}$, so that an exponent 2 is sometimes needed in (14.37) and (14.38).

We will soon return to auxiliary polynomials, and to construct those we need a version of Siegel's Lemma for heights. Its proof will follow that of the prototype Proposition 8.3, in which some easy counting was done. Here it is done as follows (compare Mahler (1964) and his "ceiling").

Lemma 14.11 *Let K be a number field and suppose that for each v we are given real $R_v > 0$ and μ_v in K, such that $R_{\overline{\sigma}} = R_\sigma$ and $\mu_{\overline{\sigma}} = \mu_\sigma$ for each σ and $R_{\mathfrak{p}} \neq 1$ for at most finitely many $\mathfrak{p}$. Then there are at most*

$$(2R^{1/[K:\mathbb{Q}]} + 1)^{[K:\mathbb{Q}]} \qquad (14.39)$$

elements ξ of K with

$$|\xi - \mu_v|_v \leq R_v \tag{14.40}$$

for all v, where $R = \prod_v R_v$.

Proof Let Σ be a maximal set of σ whose different elements are not complex conjugates. For σ in Σ write $K_\sigma = \mathbb{R}$ if $\sigma(K)$ lies in $\mathbb{R}$ and otherwise $K_\sigma = \mathbb{C}$ (these correspond to certain topological completions). We use y_σ for the corresponding real or complex variables, considered as coordinates of a vector $\mathbf{y}$ in $\mathcal{K} = \prod_{\sigma\in\Sigma} K_\sigma$ (isomorphic to $\mathbb{R}^r \times \mathbb{C}^s$ in a standard notation).

For ξ in K we define a sort of open ball $B(\xi)$ by

$$|y_\sigma - \sigma(\xi)| < \frac{1}{2}R^{-1/D}R_\sigma \tag{14.41}$$

where for brevity $D = [K : \mathbb{Q}]$. We show first that the various $B(\xi)$ for ξ in (14.40) do not overlap.

So suppose some $\mathbf{y}$ lies in both $B(\xi_1)$ and $B(\xi_2)$. By eliminating y_σ from the two (14.41), we get

$$|\xi_1 - \xi_2|_\sigma = |\sigma(\xi_1) - \sigma(\xi_2)| < R^{-1/D}R_\sigma$$

at first for σ in Σ and then for all σ. Similarly by eliminating $\mu_\mathfrak{p}$ from the two (14.40), we get

$$|\xi_1 - \xi_2|_\mathfrak{p} \leq R_\mathfrak{p}$$

for all $\mathfrak{p}$. Multiplying up we get $\prod_v |\xi_1 - \xi_2|_v < 1$ by the definition of R. So (14.38) implies that $\xi_1 = \xi_2$, the desired non-overlapping property.

Also (14.40) gives $|\sigma(\xi) - \sigma(\mu_\sigma)| \leq R_\sigma$ for σ in Σ. Therefore each $B(\xi)$ lies in the ball B defined by

$$|y_\sigma - \sigma(\mu_\sigma)| < \left(1 + \frac{1}{2}R^{-1/D}\right)R_\sigma. \tag{14.42}$$

So a lot of disjoint small balls in a big ball; and now we compare volumes.

The unit ball defined by $|y_\sigma| \leq 1$ has a certain volume $\Theta > 0$ (in fact $2^r\pi^s$ in the standard notation). Thus each small ball $B(\xi)$ has volume

$$\Theta \prod_{\sigma\in\Sigma}\left(\frac{1}{2}R^{-1/D}R_\sigma\right)^{[K_\sigma:\mathbb{R}]} = \Theta\left(\frac{1}{2}R^{-1/D}\right)^D \prod_{\sigma\in\Sigma} R_\sigma^{[K_\sigma:\mathbb{R}]}.$$

And likewise the big ball B has volume

$$\Theta\left(1 + \frac{1}{2}R^{-1/D}\right)^D \prod_{\sigma\in\Sigma} R_\sigma^{[K_\sigma:\mathbb{R}]},$$

so the lemma follows by taking the quotient. □

The upper bound S in (14.39) is not so bad. For example with $K = \mathbb{Q}$ and $R_p = 1, \mu_p = 0$ for all p and $R_\sigma = R_\infty$, we are counting all rational ξ with $|\xi|_p \leq 1$ for all p and $|\xi - \mu_\sigma| \leq R_\infty$. These are equivalent to ξ being an integer in some interval of length $2R_\infty$. So at most $2R_\infty + 1 = S$.

Or if we throw in $|\xi - 1|_2 \leq 1/2$, then we are counting odd integers, so now at most $R_\infty + 1$ also exactly S.

Now the Siegel Lemma in Proposition 13.2 involves equations $\Lambda = 0$, where Λ is a linear form in $K[X_1, \ldots, X_N]$ (in fact with coefficients in $\mathbb{Z}_K$). We define a height $H(\Lambda)$ by writing

$$\Lambda = \alpha_1 X_1 + \cdots + \alpha_N X_N$$

and

$$|\Lambda|_v = \max\{|\alpha_1|_v, \ldots, |\alpha_N|_v\}, \tag{14.43}$$

somewhat as in Gauss's Lemma used in the proof of Proposition 14.4, and then

$$H(\Lambda)^{[K:\mathbb{Q}]} = \prod_v |\Lambda|_v,$$

somewhat as in (14.37). One checks as in the proof of Proposition 14.4 (also using Exercise 14.35) that this is independent of the choice of K. However the 1 in (14.37) is missing. This means that for any non-zero α in K we have

$$H(\alpha\Lambda)^{[K:\mathbb{Q}]} = \prod_v |\alpha\Lambda|_v = \prod_v |\alpha|_v |\Lambda|_v = (\prod_v |\alpha|_v) H(\Lambda)^{[K:\mathbb{Q}]} = H(\Lambda)^{[K:\mathbb{Q}]}$$

so that $H(\alpha\Lambda) = H(\Lambda)$. This is not unnatural, as the equation $\alpha\Lambda = 0$ is equivalent to $\Lambda = 0$. But for $N = 1$ it also means that $H(\Lambda) = 1$ if Λ is not identically zero!

And for general N if Λ is not identically zero, then for example $\alpha_1 \neq 0$ so that

$$H(\Lambda)^{[K:\mathbb{Q}]} \geq \prod_v |\alpha_1|_v = 1.$$

In $K = \mathbb{Q}$ one can always multiply Λ by a non-zero rational, much as for A at the beginning of this chapter, so that $\alpha_1, \ldots, \alpha_N$ are coprime integers. Now each $\max\{|\alpha_1|_p, \ldots, |\alpha_N|_p\} = 1$, so that $H(\Lambda) = \max\{|\alpha_1|, \ldots, |\alpha_N|\}$.

Here is our long-heralded version of Siegel's Lemma for heights. Compared with Proposition 13.2, the algebraic integer coefficients have vanished from the picture as well as the basis of $\mathbb{Z}_K$. In fact it is nearer in form to the original Proposition 8.3.

Proposition 14.12 *Suppose* $[K : \mathbb{Q}] = D$. *For positive integers* M, N *with* $DM < N$ *and real* $\mathcal{H} \geq 1$ *let* Λ_m *be linear forms with coefficients in* K *and*

$H(\Lambda_m) \leq \mathcal{H}$ $(m = 1, \ldots, M)$. Then there are rational integers $x_1, \ldots, x_N$, not all zero, with

$$\Lambda_m(x_1, \ldots, x_N) = 0 \quad (m = 1, \ldots, M)$$

and

$$|x_n| \leq (N\mathcal{H})^{\frac{DM}{N-DM}} \qquad (n = 1, \ldots, N).$$

Proof We try to imitate that of Proposition 8.3. We have to solve $u(\mathbf{x}) = 0$, where u is the linear map from $\mathbb{Q}^N$ to K^M defined by

$$u(\mathbf{x}) = (\Lambda_1(\mathbf{x}), \ldots, \Lambda_M(\mathbf{x})) \tag{14.44}$$

for $\mathbf{x} = (x_1, \ldots, x_N)$.

For real $\mathcal{X} > 0$ to be chosen shortly, write $\mathcal{S}$ for the set of all $\mathbf{x}$ in $\mathbb{Z}^N$ with $0 \leq x_1, \ldots, x_N \leq [\mathcal{X}]$ for the Gauss brackets. The cardinality is $|\mathcal{S}| = ([\mathcal{X}] + 1)^N$. The set $u(\mathcal{S})$ is also finite, and we can estimate its cardinality by imitating "cube centres" as follows.

Fix m and define $\mu_\mathfrak{p} = 0$ for every prime ideal of K and $\mu_\sigma = \Lambda_m(\mathbf{x}_0)$ for every complex embedding of K, where $\mathbf{x}_0 = ([\mathcal{X}]/2, \ldots, [\mathcal{X}]/2)$.

Then with $\xi = \Lambda_m(\mathbf{x})$ we have

$$|\xi - \mu_\mathfrak{p}|_\mathfrak{p} = |\xi|_\mathfrak{p} \leq |\Lambda_m|_\mathfrak{p}$$

as in (14.43). And as usual we get extra factors for non-ultrametric, namely

$$|\xi - \mu_\sigma|_\sigma = |\Lambda_m(\mathbf{x}) - \Lambda_m(\mathbf{x}_0)|_\sigma = |\Lambda_m(\mathbf{x} - \mathbf{x}_0)|_\sigma \leq N([\mathcal{X}]/2)|\Lambda_m|_\sigma.$$

Thus by Lemma 14.11 the number of ξ is at most $(2R^{1/D} + 1)^D$, where

$$R = (N[\mathcal{X}]/2)^D H(\Lambda_m)^D \leq (N\mathcal{H}[\mathcal{X}]/2)^D.$$

So the number of ξ is at most

$$(N\mathcal{H}[\mathcal{X}] + 1)^D \leq (N\mathcal{H}([\mathcal{X}] + 1))^D.$$

Therefore $u(\mathcal{S})$ by (14.44) has cardinality $|u(\mathcal{S})| \leq (N\mathcal{H}([\mathcal{X}] + 1))^{DM}$.

We now choose $\mathcal{X}$ so that $|u(\mathcal{S})| < |\mathcal{S}|$. This is ensured by

$$(N\mathcal{H}([\mathcal{X}] + 1))^{DM} < ([\mathcal{X}] + 1)^N,$$

that is, $[\mathcal{X}] + 1 > (N\mathcal{H})^{\frac{DM}{N-DM}}$; and so $\mathcal{X} = (N\mathcal{H})^{\frac{DM}{N-DM}}$ is just fine.

Hence u on $\mathcal{S}$ is not injective. Consequently there are distinct $\mathbf{x}', \mathbf{x}''$ in $\mathcal{S}$ with $u(\mathbf{x}') = u(\mathbf{x}'')$. Thus $u(\mathbf{x}) = 0$ for $\mathbf{x} = \mathbf{x}' - \mathbf{x}'' \neq 0$ with components x_n satisfying $|x_n| \leq [\mathcal{X}] \leq \mathcal{X}$ $(n = 1, \ldots, N)$. □

Finally we record a version of Proposition 11.1 for heights.

Proposition 14.13 *Suppose K is in $\mathbb{C}$ with $[K:\mathbb{Q}]=D$ and $\alpha \neq 0$ is in K. Then $|\alpha| \geq H(\alpha)^{-D}$.*

Proof Now one of the σ is the identity in (14.19), and if we change α to $1/\alpha$ the left-hand side is not changed, by (14.24). We deduce $H(\alpha)^D \geq 1/|\alpha|$ as required. □

Such an estimate is also known as a "Liouville inequality". And here too the exponent D (not $D-1$) is best possible (see Exercise 14.15, and compare Exercise 11.6). We have come a long way from the Fundamental Theorem!

The result in Theorem 14.8 has generated a lot of subsequent work, associated with the name of Bogomolov. Considering $(\alpha, 1-\alpha)$ as a point on the line $x+y=1$, we can generalize the situation to points on algebraic varieties in any number n of dimensions. An interesting example concerns the quantity $H(\alpha)H(\beta)H(1-\alpha-\beta)$ corresponding to the plane $x+y+z=1$ (see Exercise 14.63). One can consult Bombieri and Gubler (2006) for more references.

It fails for the hyperbola $xy=1$ though, as $H(\alpha)=H(1/\alpha)$ can be arbitrarily small. This suggests a multiplicative aspect; and indeed the thing is better interpreted in $\mathbb{G}_{\mathrm{m}}^n$. And then it is a knee-jerk reaction to extend further to arbitrary commutative group varieties. This goes as well for the the special case of Manin–Mumford problems, where the points on the group varieties have finite order; for example both α and $1-\alpha$ are roots of unity.

Similarly for the result in Theorem 14.9, which involves particular points on $x+y=1$ whose coordinates are "multiplicatively dependent"; that is, there are integers r, s, not both zero, with

$$x^r y^s = 1. \tag{14.45}$$

So we had better restrict to $x \neq 0, y \neq 0$; and we are now at once in $\mathbb{G}_{\mathrm{m}}^2$. Indeed the group subvarieties of this are defined precisely by equations (14.45), for example $xy=1$ (for which the height is not bounded above, since $H(\alpha)=H(1/\alpha)$ can be arbitrarily large). This fact gives strong hints on how to generalize (see Bombieri et al. (1999) for more) and extend to arbitrary commutative group varieties. An interesting example concerns the equation $\alpha^r(1-\alpha)^s(1+\alpha)^t=1$ corresponding to the line $x+y=z-x=1$ in $\mathbb{G}_{\mathrm{m}}^3$ (see Exercise 14.75). And go to Cohen and Zannier (2000) for the last word on $\alpha^r(1-\alpha)^s=1$.

And while we are in this expansive mood, we should mention that equations like (12.46), itself also in the form $x+y=1$, can similarly be generalized and extended.

But we have no room to say anything about likely or unlikely intersections or anything associated with the nametag Zilber–Pink (see Zannier (2012) for a comprehensive account).

Finally we should mention some other versions of Siegel's Lemma, here stated as Proposition 14.12.

Sometimes it is necessary to have a solution with $x_1, \ldots, x_N$ in K, most commonly integers of $\mathbb{Z}_K$. Now $M < N$ suffices. But the estimates for the (heights of the) x_n must then involve the discriminant Δ_K of K. In Exercises 14.87, 14.88 we state such a version, due to Bombieri and Vaaler. In fact this version has an additional feature that it is invariant under taking general linear combinations of $\Lambda_1, \ldots, \Lambda_M$, not just merely replacing them with non-zero multiples.

Quite often the discriminant can be avoided in such situations by passing from K to the field $\overline{\mathbb{Q}}$ of all algebraic numbers. This appears to lose control of degrees over $\mathbb{Q}$ and so pose a danger for Liouville estimates like Proposition 14.13; but that need not always matter. The result is known as the Absolute Siegel Lemma, associated with the names of Zhang as well as Roy and Thunder, and David and Philippon. See David and Philippon (1999) for references.

Finally, as mentioned near the end of Chapter 8, it is sometimes useful to know about more than one solution. All the versions above can be extended to give such information (in the precise sense of Exercise 8.16). For example, Theorem 2.9.19 of Bombieri and Gubler (2006) includes all the aspects above apart from absoluteness. And Proposition 4.8 of David and Philippon (1999) includes even that. But not the Thunder (1995) version for function fields.

Exercises

Here $\zeta_n = \exp(2\pi i/n)$.

14.1 Let a, b, c be coprime in $\mathbb{Z}$ with $b^2 < 4ac$, and let $a\alpha^2 + b\alpha + c = 0$. Show that $H(\alpha) = \sqrt{\max\{|a|, |c|\}}$.

14.2 Let α be algebraic. Show that $H(\alpha)$ is also algebraic.

14.3 Let d be in $\mathbb{N}$ and $\mathcal{H} \geq 1$ in $\mathbb{R}$. Show that there are at most $(12\mathcal{H})^{d(d+1)}$ algebraic numbers α in $\mathbb{C}$ with $[\mathbb{Q}(\alpha) : \mathbb{Q}] = d$ and $H(\alpha) \leq \mathcal{H}$.

14.4 Guess $\lim_{d \to \infty} H(1 + 2^{1/d})$ as an integral.

14.5 For $\alpha \neq 0$ in a number field K and a prime ideal $\mathfrak{p}$ of K one defines the order

$$\mathrm{ord}_{\mathfrak{p}}(\alpha) = -\frac{\log |\alpha|_{\mathfrak{p}}}{\log N(\mathfrak{p})}$$

and abbreviates it to ord_p for $p\mathbb{Z}$ ($p \in \mathbb{N}$).

(a) Let $a_0 X^d + \cdots + a_d$ be the minimal polynomial of α in $\mathbb{Z}[X]$ with $a_0 = a_0(\alpha) > 0$. Show that

$$\mathrm{ord}_p a_0 = \frac{1}{[K:\mathbb{Q}(\alpha)]\log p}\sum_{\mathfrak{p}|p\mathbb{Z}} \max\{0, -\mathrm{ord}_{\mathfrak{p}}(\alpha)\}\log N(\mathfrak{p}).$$

(b) Let $a_{\min} = a_{\min}(\alpha)$ be the smallest a in $\mathbb{N}$ such that $a\alpha$ is integral. Show that $\mathrm{ord}_p a_{\min}$ is the smallest $e \geq 0$ in $\mathbb{Z}$ with

$$e \geq \max_{\mathfrak{p}|p\mathbb{Z}}\left\{-\frac{\mathrm{ord}_{\mathfrak{p}}(\alpha)}{\mathrm{ord}_{\mathfrak{p}}(p)}\right\}$$

[*Hint*: $p^e\alpha$].

14.6 For $A = \sum_i \alpha_i X^i$ in $\mathbb{C}[X]$ define $|A| = \max_i |\alpha_i|$. Show that

$$|AB| \geq c|A||B| \qquad (*)$$

for all A of degree 1 and all B of degree 1, where $c = \theta = \frac{1}{2}(\sqrt{5}-1)$ [*Hint*: if $|\alpha_0| = |A|, |\beta_1| = |B|$ and $|\alpha_1| < \theta|A|, |\beta_0| < \theta|B|$ use $\alpha_0\beta_1 + \alpha_1\beta_0$ and $1 - \theta^2 = \theta$].

14.7 Let $\mathcal{K}$ be a field and let $|\cdot| : \mathcal{K} \to [0, \infty)$ with the properties

$$|\alpha\beta| = |\alpha||\beta|, \quad |\alpha+\beta| \leq \max\{|\alpha|, |\beta|\}, \quad |\alpha| = 0 \text{ if and only if } \alpha = 0. \quad (**)$$

If $\beta, \alpha_1, \ldots, \alpha_n$ are in $\mathcal{K}$ with $|\alpha_1| \leq 1, \ldots, |\alpha_n| \leq 1$ and $\beta^n + \alpha_1\beta^{n-1} + \cdots + \alpha_n = 0$, show that $|\beta| \leq 1$.

14.8 Let $\mathcal{K}$ be a number field K in Exercise 14.7, and assume that $|\cdot|$ is non-trivial in the sense that there is at least one $\alpha \neq 0$ with $|\alpha| \neq 1$.

(a) Show that $|\alpha| \leq 1$ for each α in $\mathfrak{o} = \mathbb{Z}_K$ [*Hint*: $|1 + \cdots + 1|$ and Exercise 14.7].

(b) Show that $\mathfrak{p} = \{\alpha \in \mathfrak{o};\ |\alpha| < 1\}$ is a prime ideal.

(c) Let β be in K with $|\beta|_{\mathfrak{p}} = 1$. Show that $|\beta| = 1$ [*Hint*: $\beta\mathfrak{o} = \mathfrak{c}/\mathfrak{d}$].

(d) Show that there exists $\lambda > 0$ such that $|\alpha| = |\alpha|_{\mathfrak{p}}^{\lambda}$ for each α in K.

14.9 Let $M(\mathcal{D})$ be the cardinality $\#\{\beta : [\mathbb{Q}(\beta):\mathbb{Q}] \leq \mathcal{D},\ H(\beta) \leq 2\} \geq 1$. If $H(\alpha) \leq 2^{1/M(d)}$ for some $\alpha \neq 0$ with $[\mathbb{Q}(\alpha):\mathbb{Q}] = d$, show that α is a root of unity (so $H(\alpha) = 1$).

14.10 By drawing two circles show that if α, β are roots of unity with $\alpha + \beta = 1$, then the set $\{\alpha, \beta\} = \{\zeta_6, \zeta_6^5\}$.

14.11 Let p be prime. Show that

$$(X_1 + \cdots + X_n)^p - (X_1^p + \cdots + X_n^p) = pQ$$

where Q in $\mathbb{Z}[X_1, \ldots, X_n]$ has total degree at most p.

14.12 Let α, β be algebraic with $\alpha^2 + \alpha\beta + \beta^2 = 1$.

(a) Let p be an odd prime with $\alpha^{2p} + \alpha^p\beta^p + \beta^{2p} \neq 1$. Show that $H(\alpha)H(\beta) \geq (p/4)^{\frac{1}{2p}}$ [*Hint*: Product Formula on $\gamma = \alpha^{2p} + \alpha^p\beta^p + \beta^{2p} - 1$].

(b) If $\beta \neq 0, 1, -1, \zeta_8, \zeta_8^3, \zeta_8^5, \zeta_8^7$ show that

$$H(\alpha)H(\beta) \geq \left(\frac{5}{4}\right)^{1/10} = 1.0225651825\ldots$$

[*Hint*: the resultant of $X^2 + \beta X + \beta^2 - 1$ and $X^{10} + \beta^5 X^5 + \beta^{10} - 1$ is $25\beta^2(\beta - 1)^3(\beta + 1)^3(\beta^4 + 1)^2$].

(c) Find all roots of unity α, β with $\alpha^2 + \alpha\beta + \beta^2 = 1$.

14.13 Let K be a number field of degree D with complex embeddings σ, and let $\mathcal{X} \geq 1$. Show that $\#\{\alpha \in \mathbb{Z}_K;\ |\sigma(\alpha)| \leq \mathcal{X} \text{ all } \sigma\} \leq (2\mathcal{X} + 1)^D$.

14.14 Let $m < 0$ be in $\mathbb{Z}$ squarefree with $m \not\equiv 1 \pmod 4$, and let $K = \mathbb{Q}(\sqrt{m})$ with complex embeddings σ (so that for example the discriminant $\Delta_K = 4m$). Show that

$$\#\{\alpha \in \mathbb{Z}_K;\ |\sigma(\alpha)| \leq \mathcal{X} \text{ all } \sigma\} \leq (2\mathcal{X} + 1)\left(\frac{2\mathcal{X}}{\sqrt{|m|}} + 1\right).$$

14.15 For an integer $D \geq 2$ and an even integer $t \geq 4$ let $P(X) = (X - t)(X^{D-1} - 2) - 2$.

(a) Show that $P(X)$ is irreducible over $\mathbb{Q}$ [*Hint*: Battleship Potemkin].

(b) Show that there is real β with $P(\beta) = 0$ and $t < \beta < t + 1$.

(c) Show that $H(\beta)^D \leq 8.4^D t$ [*Hint*: Proposition 14.1].

(d) Deduce that Proposition 14.13 cannot be essentially improved for any fixed $D \geq 2$ [*Hint*: $\alpha = 1/\beta$].

(e) Show that $H(\beta)^D \geq t$ [*Hint*: Proposition 14.13].

****

14.16 Guess $\lim_{n\to\infty} H(2\cos(2\pi/n))$ as an integral.

14.17 Show that $H(1 + \zeta_n) < 2$ if $n \geq 2$.

14.18 Calculate $H(\sqrt{5} + \sqrt{7 - \sqrt{11}})$.

14.19 Calculate the heights of the solutions of the equation in the excerpt below from Lewis Carroll's First Riddle.

Yet what are all such gaieties to me,
Whose thoughts are full of indices and surds,
$x^2 + 7x + 53$
$= 11/3$.

14.20 If α in Exercise 14.2 has degree d show that $H(\alpha)$ has degree at most $(2d^2)^{d^2}$.

14.21 In Exercise 14.3 show that there are constants $c = c(d) > 0$ and $\mathcal{H}_0 = \mathcal{H}_0(d)$ such that for $\mathcal{H} \geq \mathcal{H}_0$ there are at least $c\mathcal{H}^{d(d+1)}$ algebraic numbers α in $\mathbb{C}$ with $[\mathbb{Q}(\alpha) : \mathbb{Q}] = d$ and $H(\alpha) \leq \mathcal{H}$ [*Hint*: Odessa Steps].

14.22 Show that

$$\mathcal{M}(A) = \exp\left(\frac{1}{2\pi}\int_0^{2\pi} \log|A(e^{i\theta})|d\theta\right)$$

for any non-zero A in $\mathbb{C}[X]$.

14.23 Let A be in $\mathbb{R}[X]$ with $A(X) = a_0(X-\alpha_1)\cdots(X-\alpha_d)$ over $\mathbb{C}$. Calculate $\lim_{n\to\infty} \mathcal{L}(A^n)^{1/n}$.

14.24 Let $A(X)$ in Exercise 14.23 be in $\mathbb{Z}[X]$. Show that

$$u_n = a_0^n(\alpha_1^n - 1)\cdots(\alpha_d^n - 1)$$

is in $\mathbb{Z}$ $(n = 0, 1, 2, \ldots)$.

14.25 If A in Exercise 14.24 is irreducible in $\mathbb{Z}[X]$ with $A(\alpha) = 0$, show that $\lim_{n\to\infty} |u_n|^{1/n} = H(\alpha)^d$.

14.26 Show that

$$A(X) = X^{10} + X^9 - X^7 - X^6 - X^5 - X^4 - X^3 + X + 1$$

has exactly eight different zeroes α with $|\alpha| = 1$ [*Hint*: $A(X) = X^5B(X + \frac{1}{X})$].

14.27 In Exercise 14.26 show that all $H(\alpha) = 1.016368277\ldots$ are equal.

14.28 Conway proved that in the "look-and-say" sequence

$$1, 11, 21, 1211, 111221, 312211, 13112221, 1113213211, 31131211131221, \ldots$$

the nth term has roughly α^n digits, where

$$\alpha = 1.303577269034296391257099112152551890730702504659404 87\ldots$$

is algebraic of degree 71 whose minimal polynomial has no coefficient of absolute value bigger than 14 (see Conway and Guy (1996) p. 209). Show that $H(\alpha) \leq 5$.

14.29 Show that

$$a_{\min}\left(\frac{63}{25}\left(\frac{17+15\sqrt{5}}{7+15\sqrt{5}}\right)\right) = 6725.$$

14.30 Find α of degree 3 with $a_{\min}(\alpha) < a_0(\alpha)$.

14.31 Show that $(\mathfrak{a}+\mathfrak{b})\tilde{\mathfrak{o}} = \mathfrak{a}\tilde{\mathfrak{o}} + \mathfrak{b}\tilde{\mathfrak{o}}$ and $(\mathfrak{a}\mathfrak{b})\tilde{\mathfrak{o}} = (\mathfrak{a}\tilde{\mathfrak{o}})(\mathfrak{b}\tilde{\mathfrak{o}})$ for any ideals $\mathfrak{a}, \mathfrak{b}$ of $\mathfrak{o} = \mathbb{Z}_K$, where $\tilde{\mathfrak{o}} = \mathbb{Z}_{\tilde{K}}$ for a number field $\tilde{K}$ containing K.

14.32 With $K, \tilde{K}$ as in Exercise 14.31 show that

$$\mathrm{Nm}_{\tilde{K}/\mathbb{Q}}(\alpha) = \mathrm{Nm}_{K/\mathbb{Q}}(\mathrm{Nm}_{\tilde{K}/K}(\alpha))$$

for every α in $\tilde{K}$.

14.33 With $\mathfrak{a}, \mathfrak{o}, \tilde{\mathfrak{o}}, K, \tilde{K}$ as in Exercise 14.31 show that $N(\mathfrak{a}\tilde{\mathfrak{o}}) = N(\mathfrak{a})^{[\tilde{K}:K]}$ [*Hint*: class number and Exercise 14.32].

14.34 Let $\mathfrak{a}, \tilde{\mathfrak{a}}, \mathfrak{o}, \tilde{\mathfrak{o}}$ be as in Exercise 14.31.

(a) Show that $(\mathfrak{a}\tilde{\mathfrak{o}}) \cap \mathfrak{o} = \mathfrak{a}$ (going up and coming down).

(b) Is it true that $(\tilde{\mathfrak{a}} \cap \mathfrak{o})\tilde{\mathfrak{o}} = \tilde{\mathfrak{a}}$ (coming down and going up)?

14.35 For any $\alpha_1, \ldots, \alpha_N$ in K let

$$E_K(\alpha_1, \ldots, \alpha_N) = \prod_{\mathfrak{p}} \max\{|\alpha_1|_{\mathfrak{p}}, \ldots, |\alpha_N|_{\mathfrak{p}}\}^{1/[K:\mathbb{Q}]}$$

taken over all $\mathfrak{p}$ of K. Show that $E_{\tilde{K}}(\alpha_1, \ldots, \alpha_N) = E_K(\alpha_1, \ldots, \alpha_N)$ for any number field $\tilde{K}$ containing K [*Hint*: do it first for $\alpha_1, \ldots, \alpha_N$ in $\mathfrak{o}$ using Exercises 14.31, 14.33, 14.34 and then divide by some $E_K(\alpha, \ldots, \alpha)$].

14.36 Let $|\cdot|$ be an ultrametric valuation on a field $\mathcal{K}$. For $A = \sum_{ij} \alpha_{ij} X^i Y^j$ in $\mathcal{K}[X, Y]$ define $|A| = \max_{ij} |\alpha_{ij}|$. Show that $|AB| = |A||B|$.

14.37 Show that $(*)$ in Exercise 14.6 would be false with any $c > \theta$.

14.38 For $d, e \geq 0$ in $\mathbb{Z}$ show that there is $c = c(d, e) > 0$ such that $(*)$ in Exercise 14.6 holds for all A of degree d and all B of degree e.

14.39 In Exercise 14.38 show that $c(1, 1291) < 2/1293$ (Masser and Wolbert, 1993). The biggest value of $c(1, 1291)$ is algebraic, probably of degree 1292.

14.40 In Exercise 14.38 what is the biggest value of $c(2, 2)$? I don't know.

14.41 Let x, y be in $\mathbb{Q}$ with

$$y(y^2 - x^2) = 1. \qquad (*)$$

(a) Show that $\max\{1, |y|_v\} \le \sqrt{2}\max\{1, |x|_v\}$ if $v = \infty$, with $\sqrt{2}$ omitted if $v = p$.

(b) Show that there are at most finitely many x, y in $\mathbb{Z}[\frac{1}{2}]$ ("2-integers") with $(*)$ [*Hint*: as in Exercise 4.13 with $|x|_\infty, |x|_2$ in place of $|x|$, noting that $H(x) = \max\{1, |x|_\infty\}\max\{1, |x|_2\}$]. This amounts to solving $y_1(y_1^2 - x_1^2) = 2^{3u}$ in integers x_1, y_1 and $u \ge 0$. By the way, there are exactly three (x, y) in $\mathbb{Q}^2$ with $(*)$; this follows from a remarkable result of Euler which can be interpreted as the finiteness of a certain Mordell–Weil group.

14.42 Let L be a finite extension of K in Exercise 14.8. Show that each $|\cdot|$ on K has an extension to L.

14.43 Let $|\cdot| : \mathbb{Q} \to [0, \infty)$ with the properties $|\alpha\beta| = |\alpha||\beta|$, $|\alpha + \beta| \le |\alpha| + |\beta|$, $|\alpha| = 0$ if and only if $\alpha = 0$. If $|\cdot|$ is neither trivial nor $|\cdot|_p^\lambda$, show that there is λ with $0 < \lambda \le 1$ such that $|\alpha| = |\alpha|_\infty^\lambda$ for every α in $\mathbb{Q}$.

14.44 Let k be algebraically closed, let $\mathcal{K} = k(t)$, let v be in $V = V_{\mathcal{K}} = \mathbb{P}_1(k) = k \cup \{\infty\}$, and let $\Xi > 1$. For each v in k show that there is a unique $|\cdot| = |\cdot|_v$ as in $(**)$ of Exercise 14.7 with $|t - v| = \Xi^{-1}$. If $v = \infty$ show that there is a unique $|\cdot| = |\cdot|_v$ as in $(**)$ with $|1/t| = \Xi^{-1}$.

14.45 In Exercise 14.44 with k as the algebraic closure of $\mathbb{F}_p$, show that the formal series $\psi(z)$ of Exercise 4.43 converges for all z in $k(t)$ with respect to $v = \infty$.

14.46 In Exercise 14.44 show that $\prod_{v\in V} |\alpha|_v = 1$ $(\alpha \ne 0)$.

14.47 In Exercise 14.44 define $\mathcal{H}(\alpha) = \prod_{v\in V}\max\{1, |\alpha|_v\}$. When $\alpha = \frac{\beta}{\gamma}$ with β, γ coprime in $k[t]$, show that $\log\mathcal{H}(\alpha) = \max\{\deg\beta, \deg\gamma\}\log\Xi$. Thus one often chooses $\Xi = e = 2.718281828\ldots$

14.48 Does each $|\cdot|_v$ in Exercise 14.44 have an extension to each finite extension of $\mathcal{K}$?

14.49 For $\beta, \alpha_1, \ldots, \alpha_n$ $(n \ge 2)$ in $\mathbb{C}$ with $|\alpha_1| \le 1, \ldots, |\alpha_n| \le 1$ and $\beta^n + \alpha_1\beta^{n-1} + \cdots + \alpha_n = 0$ show that $|\beta| \le x$, where $x = x_n$ is the unique solution of $x^{n+1} - 2x^n + 1 = 0$ with $1 < x < 2$.

14.50 Let $K = \mathbb{Q}(\sqrt{5})$ with embeddings σ, σ'. Show that there is α in K with $|\sigma(\alpha) - \pi| < 10^{-100}$, $|\sigma'(\alpha) - e| < 10^{-100}$ (but no such α in $\mathbb{Z}_K$).

14.51 For $\zeta = e^{2\pi i/24}$ let

$$\alpha = -\zeta + \frac{1}{4}\left(\sqrt{6} + \sqrt{2} + i(\sqrt{6} - \sqrt{2})\right)$$

in $K = \mathbb{Q}(\zeta, \sqrt{6}, \sqrt{2}, i) \subset \mathbb{C}$. Show that $|\alpha|_\mathfrak{p} \leq |\frac{1}{4}|_\mathfrak{p}$ for all $\mathfrak{p}$ of K; and that $|\alpha|_\sigma \leq 3$ for all σ of K.

14.52 In Exercise 14.51 one finds with Maple that $|\alpha| < 12^{-64}$ in $\mathbb{C}$. Show that $\alpha = 0$.

14.53 If $H(\alpha) \leq 1 + 144^{-d^2}$ in Exercise 14.9 show that α is a root of unity.

14.54 If $\alpha^m, \alpha^{m'}$ are conjugate for $\alpha \neq 0$ and different positive integers m, m', show that α is a root of unity.

14.55 Show that there are no constants c, C such that $H(\alpha + \beta) \leq c(H(\alpha) + H(\beta)) + C$ for all algebraic α, β.

14.56 Let $\alpha \neq 0$ be algebraic, let θ be rational with $0 \leq \theta \leq 1$, and let α^θ be any determination.

(a) Show that $H(\alpha + \alpha^\theta) \leq 2eH(\alpha)\max\{e, \log H(\alpha)\}$ (Habegger, 2005).

(b) Show that $H(\alpha^\theta)$ is independent of the determination.

14.57 Let $A \neq 0$ in $\mathbb{Z}[X]$ inside $\mathbb{Q}[X]$. Show that $A \prod_p |A|_p$ is in $\mathbb{Z}[X]$ with coprime coefficients.

14.58 Show that all exceptions for β in Exercise 14.12(b) are needed.

14.59 With at most finitely many exceptions β show that

$$H(\alpha)H(\beta) \geq \left(\frac{11}{4}\right)^{1/22} = 1.0470554168\ldots$$

in Exercise 14.12.

14.60 Show that $\mathrm{res}(X^2 + \beta X + \beta^2 - 1, X^{2p} + \beta^p X^p + \beta^{2p} - 1)$ lies in $\beta(\beta - 1)\mathbb{Z}[\beta]$.

14.61 Show that $\mathrm{res}(X^2 + \beta X + \beta^2 - 1, X^{2p} + \beta^p X^p + \beta^{2p} - 1)$ lies in $p\mathbb{Z}[\beta]$.

14.62 Show that there is $\lambda > 1$ such that $H(\alpha)H(\beta) \geq \lambda$ for all algebraic α, β with $\alpha^3 + \alpha^2\beta + \alpha\beta^2 + \beta^3 = 1$, apart from at most finitely many exceptions β.

14.63 (a) Let n in $\mathbb{N}$ be odd. Show that

$$A_n = \frac{X^n + Y^n + (1 - X - Y)^n - 1}{(X-1)(Y-1)(X+Y)}$$

is a polynomial in X and Y.

(b) Let ξ, η, ζ be in $\mathbb{C}^*$ with

$$\xi + \eta + \zeta = \xi^5 + \eta^5 + \zeta^5 = \xi^7 + \eta^7 + \zeta^7 = 1.$$

Show that $\xi = 1$ or $\eta = 1$ or $\zeta = 1$ [*Hint*: with $\mathcal{R} = \mathbb{C}[Y]$ and A_5, A_7 in $\mathcal{R}[X]$ we have $\mathrm{res}(A_5, A_7) = 30625Y^2(Y^2 - Y + 1)^2$].

(c) Show that there is $\lambda > 1$ such that $H(\alpha)H(\beta)H(\gamma) \geq \lambda$ for all algebraic α, β, γ with $\alpha, \beta, \gamma \neq 0, 1$ and $\alpha + \beta + \gamma = 1$.

(d) Find all roots of unity α, β, γ with $\alpha + \beta + \gamma = 1$.

14.64 Let $\xi_1, \xi_2, \xi_3, \xi_4$ be in $\mathbb{C}^*$ with $(\xi_i/\xi_j)^{30} \neq 1$ $(i \neq j)$. Show that $\xi_1^p + \xi_2^p + \xi_3^p + \xi_4^p \neq 0$ for at least one of the primes $p = 7, 37, 67, 97$.

14.65 Show that there are infinitely many roots of unity $\alpha, \beta, \gamma, \delta$ with $\alpha + \beta + \gamma + \delta = 1$; but only one set $\{\alpha, \beta, \gamma, \delta\}$ with no non-empty subset whose elements sum to zero (Warin Master Thesis 2012).

14.66 Show that $H(\alpha)H(3 - \alpha) \geq 2$ for all algebraic α [*Hint*: note that $\max\{1, x\}\max\{1, 3 - x\} \geq 2$ on $\mathbb{R}$].

14.67 Show that $H(\alpha)H(3 - \alpha) > 2$ for all algebraic $\alpha \neq 1, 2$.

14.68 Is there $\mu > 2$ such that $H(\alpha)H(3 - \alpha) \geq \mu$ for all algebraic $\alpha \neq 1, 2$?

14.69 Find a positive constant c such that $H(\alpha^3 - \alpha^2) \geq cH(\alpha)^3$ for all algebraic α [*Hint*: $(X^3 - X^2)(X^2 + X + 1) + X^2 = X^5$].

14.70 Find a positive constant c such that $H(\alpha^4 - \alpha^3) \geq cH(\alpha)^4$ for all algebraic α.

14.71 Is there C such that $H(\alpha^2 - \alpha) \geq H(\alpha)^2 - CH(\alpha)$ for all algebraic α?

14.72 Is there C such that $H(\alpha^2 - \alpha) \leq H(\alpha)^2 + CH(\alpha)$ for all algebraic α?

14.73 Find C such that $H(\alpha) \leq C$ for any algebraic $\alpha \neq 0, 1$ for which there are integers r, s with $\alpha^r(1 - \alpha)^s = 2$.

14.74 Given any non-zero algebraic β, not a root of unity, show that there is $C(\beta)$ such that $H(\alpha) \leq C(\beta)$ for any algebraic $\alpha \neq 0, 1$ for which there are integers r, s with $\alpha^r(1 - \alpha)^s = \beta$.

14.75 Show that there is C such that $H(\alpha) \leq C$ for all algebraic $\alpha \neq 0, 1, -1$ for which there are integers r, s, t, not all zero, with $\alpha^r(1 - \alpha)^s(1 + \alpha)^t = 1$ (Bombieri et al., 1999) [*Hint*: Exercise 12.49 with $\theta = r/t, \theta' = s/t$ assuming $|t| \geq \max\{|r|, |s|\}$].

14.76 Show that there is no C such that $H(\alpha) \leq C$ for all algebraic $\alpha \neq 0, 1$ for which there are integers r, s, t, not all zero, with $\alpha^r(1 - \alpha)^s 2^t = 1$.

14.77 Let P, Q be coprime polynomials in $\mathbb{Z}[X]$ with $\max\{\deg P, \deg Q\} = L \geq 1$. Show that there exist $A_0, B_0, A_\infty, B_\infty$ in $\mathbb{Z}[X]$ with degrees at most $L-1$ and $r \neq 0$ in $\mathbb{Z}$ such that

$$A_0(X)P(X) + B_0(X)Q(X) = r, \quad A_\infty(X)P(X) + B_\infty(X)Q(X) = rX^{2L-1}.$$

Deduce that for any algebraic ξ with $Q(\xi) \neq 0$ we have

$$H\left(\frac{P(\xi)}{Q(\xi)}\right) \geq C^{-1}H(\xi)^L,$$

where

$$C = \max\{\mathcal{L}(A_0) + \mathcal{L}(B_0), \mathcal{L}(A_\infty) + \mathcal{L}(B_\infty)\}$$

for the lengths.

14.78 For integers r, s define $\nu = \nu(r, s)$ as $|r| + |s|$ if r, s have the same sign and $\max\{|r|, |s|\}$ if not. Show that there is a positive constant c, depending only on r, s, such that $H(\alpha^r(1-\alpha)^s) \geq cH(\alpha)^\nu$ for all algebraic $\alpha \neq 0, 1$.

14.79 Show that there are positive constants λ, θ such that $H(\alpha^r(1-\alpha)^s) \geq \lambda^{\theta\nu}H(\alpha)^{\theta\nu}$ for all algebraic $\alpha \neq 0, 1$ and all integers r, s, with $\nu = \nu(r, s)$ as in Exercise 14.78.

14.80 Is there a positive constant λ such that $H(\alpha^r(1-\alpha)^s) \geq \lambda H(\alpha)^\nu$ for all algebraic $\alpha \neq 0, 1$ and all integers r, s, with $\nu = \nu(r, s)$ as in Exercise 14.78? I don't know.

14.81 Show that there is C such that $H(\alpha) \leq CH(\beta)$ for all α, β algebraic with $\alpha^2 + \alpha\beta + \beta^2 = 1$.

14.82 Let $\epsilon > 0$. In Exercise 14.13 show that for each D there is no upper bound $C(K, \epsilon)(2\mathcal{X} + 1)^{D-\epsilon}$.

14.83 In Exercise 14.14 show that the number is asymptotic to $\pi\mathcal{X}^2/\sqrt{|m|}$ as $\mathcal{X} \to \infty$.

14.84 Guess an asymptotic formula in Exercise 14.13.

14.85 Let α be in a number field K. Show that $|\alpha|_\mathfrak{p} = 1$ (all $\mathfrak{p}$) if and only if α is in the unit group U_K of K.

14.86 Let $\mathcal{D}, M, N$ be in $\mathbb{N}$ with $\mathcal{D}M < N$, let $\mathcal{H} \geq 1$ be in $\mathbb{R}$, let K_m be a field with $[K_m : \mathbb{Q}] \leq \mathcal{D}$ and let Λ_m be linear forms with coefficients in K_m and $H(\Lambda_m) \leq \mathcal{H}$ $(m = 1, \ldots, M)$. Show that there are $x_1, \ldots, x_N$ in $\mathbb{Z}$, not all zero, with $\Lambda_m(x_1,\ldots,x_N) = 0$ $(m = 1,\ldots,M)$ and $|x_n| \leq (N\mathcal{H})^{\mathcal{D}M/(N-\mathcal{D}M)}$ $(n = 1,\ldots,N)$ (the case $K_1 = \cdots = K_M$ is Proposition 14.12).

14.87 Let $\mathcal{S}$ be a finite subset of K. Show that $H(\mathcal{S}) = \prod_v \max_{s\in\mathcal{S}} |s|_v^{1/[K:\mathbb{Q}]}$ is independent of K [*Hint*: Exercise 14.35].

14.88 For $1 \leq M < N$ let Λ be a $M \times N$ matrix of rank M with entries in K. Show that there is a $N \times 1$ matrix $\mathbf{x}$ of rank 1 with entries in $\mathbb{Z}_K$ such that $\Lambda\mathbf{x} = 0$ and $H(\mathbf{x}) \leq C\sqrt{|\Delta_K|}^{1/D} H(\Lambda)^{1/(N-M)}$, where $D = [K : \mathbb{Q}]$, $H(\mathbf{x}) = H(\mathcal{E})$ for the set $\mathcal{E}$ of entries of $\mathbf{x}$, $H(\Lambda) = H(\mathcal{D})$ for the set $\mathcal{D}$ of determinants of the $M \times M$ submatrices of Λ, and $C = \binom{N}{M}^{1/(N-M)}$ (due to Bombieri and Vaaler (1983), who used l^2 norms on Λ – see Bombieri and Gubler (2006) for much more).

14.89 Show that $\#\{\alpha \in \mathbb{Q};\ H(\alpha) \leq \mathcal{H}\}$ is asymptotic to $\frac{12}{\pi^2}\mathcal{H}^2$ (compare Exercises 14.3 and 14.21).

14.90 For K in Exercise 14.14 with $m < -3$ show that $\#\{\alpha \in K;\ H(\alpha) \leq \mathcal{H}\}$ is asymptotic to $\frac{\pi^2 h_K}{2\zeta_K(2)|m|}\mathcal{H}^4$, where h_K is the class number and $\zeta_K(2) = \sum_{\mathfrak{a}} N(\mathfrak{a})^{-2}$ over all ideals $\mathfrak{a} \neq 0$ of $\mathbb{Z}_K$.

14.91 Google "Schanuel" (but not Lemma or Conjecture).

14.92 Let's pretend that we don't know the degree $d = d_n$ of $\zeta = \zeta_n = e^{2\pi i/n}$ over $\mathbb{Q}$.

(a) For L, N, S in $\mathbb{N}$ with $(L+1)^2 \geq 2S$ use Siegel's Lemma to find a non-zero P in $\mathbb{Z}[\zeta][X, Y]$ of degree at most $d-1$ in ζ and of degree at most L in X, Y such that

$$\phi\left(\frac{2\pi is}{n}\right) = 0 \quad (s = 0, 1, \ldots, S-1)$$

for $\phi(z) = P(e^z, e^{Nz})$.

(b) If $L < N$ show that ϕ is not identically zero (so e^z, e^{Nz} are "almost algebraically independent").

(c) If $L = N-1$ and $(L+1)^2 \leq n$ show that

$$\phi\left(\frac{2\pi is}{n}\right) = 0 \quad (s = 0, 1, \ldots, L^2 + 2L)$$

is impossible.

(d) By suitable choice of L, N, S conclude that $d_n \geq cn/\log n$ for some absolute positive effective c (see Loher and Masser (2004) for stronger results not using an auxiliary polynomial).

15

Equidistribution – Bilu

For an algebraic number α of degree d let $\alpha_1, \ldots, \alpha_d$ be the conjugates as in (14.2). After seeing some equidistribution results for conjugates on abelian varieties, Bilu proved the analogue for α (and even for a higher-dimensional $\boldsymbol{\alpha}$). Now the equidistribution is only relative to the unit circle; roughly speaking, as the height $H(\alpha) \geq 1$ gets closer to 1 then the conjugates cluster nearer the unit circle and if in addition d gets large then they are more evenly equidistributed around it.

This is very well illustrated by the example (14.7) with $H(2^{1/d}) = 2^{1/d}$ as $d \to \infty$, when the conjugates are the $\zeta_d^m 2^{1/d}$ $(m = 0, 1, \ldots, d-1)$ for $\zeta_d = e^{2\pi i/d}$. It is slightly less clear for the example (14.5) with say $H(\zeta_l) = 1$, as $l \to \infty$, because the conjugates are the ζ_l^m now only for the m prime to l.

It is more convenient to use the logarithmic height defined by

$$h(\alpha) = \log H(\alpha) \geq 0, \tag{15.1}$$

so that we are considering $h(\alpha) \to 0$. The radial distribution is relatively easy.

Theorem 15.1 *For any $r > 1$ the number n of conjugates α_i of α with $1/r \leq |\alpha_i| \leq r$ satisfies*

$$n \geq d\left(1 - 2\frac{h(\alpha)}{\log r}\right).$$

Proof We have $n = d - k - l$, where k is the number with $|\alpha_i| > r$ and l is the number with $|\alpha_i| < 1/r$. From (14.3), we have $H(\alpha)^d \geq r^k$ (equality if $k = 0$!) and so $k \leq dh(\alpha)/\log r$. For l, we note that $1/\alpha$ has conjugates $|1/\alpha_i| > r$ and the same height by (14.24); thus also $l \leq dh(\alpha)/\log r$ and we are done. □

For example if $h(\alpha) \leq 10^{-12}$, then at least 99% of the conjugates α_i satisfy $1 - 10^{-9} \leq |\alpha_i| \leq 1 + 10^{-9}$.

The problems begin with the angular distribution. We shall prove the following explicit version.

Theorem 15.2 *For any θ with $0 \le \theta \le 2\pi$ the number n of conjugates of α in any fixed sector, based at the origin, of angle θ satisfies*

$$\left|n - \frac{\theta}{2\pi}d\right| \le 24(d^{2/3}(\log 2d)^{1/3} + dh(\alpha)^{1/3}).$$

For example if $h(\alpha) \le 10^{-12}$ and $d \ge 10^{12}$, then between 49% and 51% of the conjugates lie in the upper half plane.

In fact our proof is not self-contained but based on a beautiful analytic result of Erdős–Turán. Our excuse for this, besides the explicit error term, is that our proof also involves an auxiliary polynomial, which further is related to one that we shall use in the next chapter to prove the famous theorem of Dobrowolski.

Here is the Erdős–Turán result, without proof. That is in Erdős and Turán (1950); see also Amoroso and Mignotte (1996) for a more modern proof of something slightly sharper.

Proposition 15.3 *For L in $\mathbb{N}$ let*

$$P(X) = p_0X^L + \cdots + p_L$$

be in $\mathbb{C}[X]$ with $p_0p_L \neq 0$. Then the number N of zeroes of P (with multiplicity) in any fixed sector, based at the origin, of angle θ satisfies

$$\left|N - \frac{\theta}{2\pi}L\right| \le 16\sqrt{L\log\left(\frac{\mathcal{L}(P)}{\sqrt{|p_0p_L|}}\right)} \tag{15.2}$$

for the length $\mathcal{L}(P)$.

Surely this powerful weapon implies some sort of version of Theorem 15.2 rightaway? We try $P = A$ as in (14.2), with $L = d$. If $\alpha \neq 0$ then $a_0a_d \neq 0$; and there is no reason to suppose a_0a_d large, so we have to be content with

$$\left|n - \frac{\theta}{2\pi}d\right| \le 16\sqrt{d\log\mathcal{L}(P)}.$$

But what about $\mathcal{L}(P)$? Using (14.10) we see that we are unlikely to get any upper bound better than

$$\sum_{i=0}^{d}\binom{d}{i} = 2^d, \tag{15.3}$$

which would be fatal for equidistribution.

Here the big $\binom{d}{i}$ is the number of terms in the ith symmetric polynomial of d things. For the first time in this book the number of terms is a problem. It will get worse in the next chapter.

The idea now is to find a better P vanishing at α and so at $\alpha_1, \ldots, \alpha_d$; naturally its degree cannot be smaller than d but we can hope that its length will involve something smaller than 2^d. Here is what we can achieve using Siegel's Lemma.

Lemma 15.4 *Let α be algebraic of degree d and let $L \geq d$ be an integer. Then there is non-zero P in $\mathbb{Z}[X]$ of degree at most L with $P(\alpha) = 0$ and coefficients of absolute values at most*

$$\left((L+1)H(\alpha)^L\right)^{d/(L+1-d)}.$$

Proof We just have to solve $p_0\alpha^L + \cdots + p_L = 0$, a single $M = 1$ equation $\Lambda = 0$ in $N = L+1$ unknowns over $K = \mathbb{Q}(\alpha)$, which by Proposition 14.12 has a non-trivial solution with

$$|p_i| \leq ((L+1)\mathcal{H})^{d/(L+1-d)} \qquad (i = 0, 1, \ldots, L).$$

Here we can take

$$\mathcal{H}^d = H(\Lambda)^d = \prod_v \max\{|\alpha|_v^L, \ldots, 1\} = \prod_v \max\{|\alpha|_v^L, 1\} = H(\alpha)^{dL}.$$

This gives the required result. □

Following our old $N = 2dM$ principle, now $L = 2d - 1$, we have $L+1$ terms in P and so we get

$$\mathcal{L}(P) \leq (L+1)(2d)H(\alpha)^{2d-1} = 4d^2H(\alpha)^{2d-1}, \qquad (15.4)$$

better than (15.3) because we are thinking that $H(\alpha)$ is near 1. And indeed (if $p_0p_L \neq 0$) the right-hand side of (15.2) becomes of order $\sqrt{d\log d + d^2h(\alpha)}$ which does give equidistribution; but only for the zeroes of P. As $P(\alpha) = 0$ we see that A divides P, and so these zeroes do include $\alpha_1, \ldots, \alpha_d$, but also possibly $L - d = d - 1$ other strangers. Here we see some connexion with our opening "chestnut" of the Prologue: there we wanted a "good" multiple of a given F, and here we want a "good" multiple of a given A.

We can reduce the number of strangers by taking L just a bit bigger than d. So we take

$$L - d = [\mathcal{X}]$$

for some $\mathcal{X} > 0$ to be determined in a moment; in view of the above we certainly need

$$L \leq 2d - 1 \qquad (15.5)$$

which amounts to $\mathcal{X} < d$. And here the proof of Theorem 15.2 begins. We abbreviate $h(\alpha)$ to h.

Lemma 15.4 gives now

$$\log \mathcal{L}(P) \le \frac{2d\log 2d + 2d^2 h}{\mathcal{X}}.$$

We assume also

$$p_0 p_L \neq 0. \tag{15.6}$$

Then Proposition 15.3 yields

$$\left| N - \frac{\theta}{2\pi} L \right| \le 16\sqrt{L \log \mathcal{L}(P)} \le \frac{\mathcal{E}}{\sqrt{\mathcal{X}}}$$

with

$$\mathcal{E} = 32d\sqrt{\log 2d + dh}.$$

To take the strangers into account we must use

$$\left| n - \frac{\theta}{2\pi} d \right| \le \left| N - \frac{\theta}{2\pi} L \right| + \left| \frac{\theta}{2\pi} L - \frac{\theta}{2\pi} d \right| + |N - n|,$$

which is at most

$$\frac{\mathcal{E}}{\sqrt{\mathcal{X}}} + (L - d) + (L - d) \le \frac{\mathcal{E}}{\sqrt{\mathcal{X}}} + 2\mathcal{X}.$$

The right-hand side is minimized at

$$\mathcal{X} = (\mathcal{E}/4)^{2/3} \tag{15.7}$$

with value $\mathcal{F} = (3/2)(2\mathcal{E})^{2/3}$. Now

$$\mathcal{F}^3 = \frac{27}{8}(2\mathcal{E})^2 = 24^3(d^2 \log 2d + d^3 h) \le 24^3(d^{2/3}(\log 2d)^{1/3} + dh^{1/3})^3 \tag{15.8}$$

so we get exactly what we want.

But we forgot (15.5) and (15.6).

Actually Theorem 15.2 is trivial unless

$$d \ge 24(d^{2/3}(\log 2d)^{1/3} + dh^{1/3}).$$

But from this (which implies $d \neq 1$) and (15.8) follows $d^3 \ge (27/8)(2\mathcal{E})^2$, so $\mathcal{X}^3 = (\mathcal{E}/4)^2 \le d^3/216$ and then $L = d + [\mathcal{X}] \le (7/6)d$ comfortably for (15.5).

And if (15.6) fails, then we just use P deprived of its factors X with its actual degree taken into account, and we end up better off. This completes the proof of Theorem 15.2.

The argument we have given goes back to Mignotte, who gave in 1980 an upper bound $cd^{2/3}(\log 2d)^{1/3}$ provided $h(\alpha) \le (\log 2)/d$. See also Mignotte (1989).

The combination of Theorem 15.1 and Theorem 15.2 leads very quickly to a weaker version of Theorem 14.8. Thus if $h(\alpha)$ is small and d is large, then the conjugates of α cluster uniformly round the unit disc centred at 0. But if $h(1-\alpha)$ is also small, then they also cluster uniformly round the unit disc centred at 1. That is patently absurd (and if d is not large then we have only finitely many α to consider, thanks to Northcott). This argument can be made to work even using just Theorem 15.1. We get many i with $1/r \leq |\alpha_i| \leq r$ and simultaneously $1/r \leq |1-\alpha_i| \leq r$; these force many α_i to be near some $\zeta = e^{\pm\pi i/3}$ (compare Exercise 14.10) and we may contradict the Product Formula (14.17) for $\alpha - \zeta$ or $\alpha^2 - \alpha + 1$ (see Exercise 15.9).

This combination can be formulated in terms of weak convergence. For example, if f is a sufficiently nice function then

$$\lim_{h(\alpha)\to 0, d\to\infty} \frac{1}{d}\sum_{i=1}^{d} f(\alpha_i) = \frac{1}{2\pi}\int_0^{2\pi} f(e^{i\theta})\mathrm{d}\theta. \tag{15.9}$$

In particular it holds for $f(z) = \log\max\{1, |1+z|\}$, and (14.6) is a very special case. For more explicit versions in this formulation see Petsche (2005) and Favre and Rivera-Letelier (2006).

For the higher-dimensional version (which makes better sense in $\mathbb{G}_{\mathrm{m}}^n$) see the original Bilu (1997) paper, and for the results on abelian varieties (which are commutative group varieties generalizing elliptic curves and products of these) see Szpiro et al. (1997).

Exercises

15.1 Show that there is $\delta > 0$ and a finite set $\mathcal{S}$ such that every algebraic α not in $\mathcal{S}$ with $h(\alpha) \leq \delta$ has at least 1000 conjugates $\sigma(\alpha)$ with

$$\left|\sigma(\alpha) - \frac{1+i}{\sqrt{2}}\right| < \frac{1}{1000} \tag{$*$}$$

[*Hint*: Bilu+Northcott].

15.2 Let $\mu < 1$ and $\kappa \geq 0$. Show that there is no $C = C(\mu, \kappa)$ such that

$$\left|n - \frac{\theta}{2\pi}d\right| \leq C(d^\mu + d^\mu h^\kappa) \qquad (h = h(\alpha))$$

in Theorem 15.2 [*Hint*: $1 + e^{2\pi i/m}$].

* * * * * * * * * * * * * * * * * * * *

15.3 Show that there is $\delta > 0$ such that every algebraic α with $0 < h(\alpha) \le \delta$ has at least 1000 conjugates $\sigma(\alpha)$ with $(*)$ in Exercise 15.1.

15.4 Assuming (15.9) for a suitable f, show that

$$\lim_{h(\alpha)\to 0,\ d\to\infty} \frac{1}{d} \log |\mathrm{Nm}_{\mathbb{Q}(\alpha)/\mathbb{Q}}(1+\alpha)| = \lim_{h(\alpha)\to 0,\ d\to\infty} \frac{1}{d} \sum_{\sigma} \log |1+\sigma(\alpha)|$$

is $\frac{1}{2\pi}\int_0^{2\pi} \log|1+e^{i\theta}|d\theta = 0$.

15.5 Let $p \ge 3$ be prime. Show that $X^p - pX^{p-1} + 1$ is irreducible over $\mathbb{Q}$.

15.6 Show that

$$\lim_{h(\alpha)\to 0, d\to\infty} \frac{1}{d} \mathrm{Tr}_{\mathbb{Q}(\alpha)/\mathbb{Q}}(\alpha) = \lim_{h(\alpha)\to 0, d\to\infty} \frac{1}{d} \sum_{\sigma} \sigma(\alpha)$$

is not always $\frac{1}{2\pi}\int_0^{2\pi} e^{i\theta}d\theta = 0$.

15.7 Show that

$$\lim_{m\to\infty} h(1+\zeta_m) = \frac{1}{2\pi}\int_0^{2\pi} \log\max\{1, |1+e^{i\theta}|\}d\theta = 0.32306594\ldots.$$

15.8 Let α be algebraic of degree d and let $L \ge 2d$ in $\mathbb{N}$. Show that there is $P \ne 0$ in $\mathbb{Z}[X]$ of degree at most L with $P(\alpha) = P'(\alpha) = 0$ and coefficients of absolute value at most $\left((L+1)^{2d}H(\alpha)^{dL}\right)^{\frac{2}{L+1-2d}}$.

15.9 (a) Show that there is c such that $|z^2 - z + 1| \le c(r-1)$ for any complex z with $1/r \le |z| \le r$ and $1/r \le |1-z| \le r$.

(b) Deduce a weaker version of Theorem 14.8 using only Theorem 15.1 and the Product Formula.

15.10 Let $P(X,Y)$ be in $\mathbb{C}[X,Y]$ with algebraic coefficients such that the set $\{(\theta,\varphi) \in \mathbb{C}^2;\ |\theta| = |\varphi| = 1, P(\theta,\varphi) = 0\}$ is finite. Show that there is $\lambda > 1$ and a finite set $\mathcal{S}$ such that $H(\alpha)H(\beta) \ge \lambda$ for each algebraic α and β not in $\mathcal{S}$ with $P(\alpha,\beta) = 0$ (for example $X + Y - 1$ as in Zhang or $X^2 + XY + Y^2 - 1$ as in Exercise 14.12, but not $X + Y$ or $X^2 + Y^2$).

15.11 Let $P(X,Y)$ be in $\mathbb{C}[X,Y]$ such that the set

$$\{(\xi,\eta) \in \mathbb{C}^2;\ \xi,\eta \text{ roots of unity},\ P(\xi,\eta) = 0\}$$

is infinite. Show that $P(X,Y)$ is divisible by $X^m - \zeta Y^n$ or $X^mY^n - \zeta$ with ζ a root of unity and $m,n \ge 0$ not both zero (this was first proved independently by Ihara, Serre and Tate in 1965 – see Lang (1983) for more details).

15.12 Let $P(X, Y)$ be in $\mathbb{C}[X, Y]$ such that the set

$$\{(\theta, \varphi) \in \mathbb{C}^2;\ |\theta| = |\varphi| = 1,\ P(\theta, \varphi) = 0\}$$

is infinite. Must $P(X, Y)$ be divisible by $X^m - \psi Y^n$ or $X^m Y^n - \psi$ with $|\psi| = 1$ and $m, n \geq 0$ not both zero?

16

Height lower bounds – Dobrowolski

In 1933 Lehmer, in connexion with the factorization of large integers, looked at integer sequences

$$u_n = (\alpha_1^n - 1)\cdots(\alpha_d^n - 1) \quad (n = 0, 1, \ldots,)$$

in the notation of (14.2) with $a_0 = 1$. It was advantageous to have as slow growth as possible. Clearly from (14.3) we have $|u_n| \le 2^d H(\alpha)^{dn}$ for the height $H(\alpha)$ and so it would be nice to make $H(\alpha)^d$ as small as possible. Of course this can be achieved with roots of unity (or zero) by (14.5), but then u_n is uninteresting. Lehmer found the value

$$H(\alpha)^d = 1.176280818259917506544070338474035050693415806564 69\ldots$$

(which I calculated by Maple with Newton's Method (12.11) after three iterations starting with Maple's first approximation 1.176280818). Here $d = 10$ and

$$\alpha^{10} + \alpha^9 - \alpha^7 - \alpha^6 - \alpha^5 - \alpha^4 - \alpha^3 + \alpha + 1 = 0$$

(as already seen in Exercise 14.26). He asked if this is the smallest possible, and to this day no-one has found a smaller value. Implicit here is the conjecture (after taking logarithms)

$$h(\alpha) \ge \frac{0.16235761200773813943219880355496580770786270030620 7\ldots}{d}.$$

Certainly this cannot be improved too much, for by (14.7) we have $h(2^{1/d}) = (\log 2)/d$.

As $a_0 = 1$ above, this discussion applies only to algebraic integers, but if $a_0 \ge 2$ then it is trivial from (14.3) that

$$h(\alpha) \ge \frac{\log 2}{d} > \frac{.69}{d}. \qquad (16.1)$$

In 1979 Dobrowolski proved in the above circumstances that there is a positive constant c such that

$$h(\alpha) \geq c\frac{(\log\log 3d/\log 3d)^3}{d}. \tag{16.2}$$

Apart from refinements of c, this still has not been improved.

Now in almost all applications of this important inequality the logarithmic terms are not significant. So we are content to prove here the following weaker result.

Theorem 16.1 *For any $\kappa > 1$ there is $c > 0$ such that*

$$h(\alpha) \geq \frac{c}{d^\kappa} \tag{16.3}$$

for any non-zero algebraic α of degree d not a root of unity.

Already we can obtain a positive lower bound from the argument of Kronecker's Theorem 14.5 together with Northcott's Proposition 14.2 – see also Exercises 14.9 and 14.53. Namely if $h(\alpha) \leq (\log 2)/M$ for some positive integer M then

$$H(\alpha^m) = H(\alpha)^m \leq 2 \quad (m = 0, 1, \ldots, M),$$

so if we choose M as the number of β with degree at most $d = [\mathbb{Q}(\alpha) : \mathbb{Q}]$ and $H(\beta) \leq 2$ we conclude as there that the $M + 1$ powers α^m cannot all be different; now a contradiction. It follows that $h(\alpha) > (\log 2)/M$. And by (14.11) we have $M \leq 36^{d^2} 2^{d(d+1)} \leq 144^{d^2}$. But this gives only

$$h(\alpha) > \frac{\log 2}{144^{d^2}}, \tag{16.4}$$

almost another atrocity.

At any rate this shows for each positive integer D that the minimum $\eta(D)$ of $h(\alpha)$, taken over all non-zero α, not roots of unity, in any number field K with $[K : \mathbb{Q}] = D$, is positive. Here are a few values (taken from Lehmer's paper). Clearly

$$\eta(1) = \log 2 = 0.6931471805 \ldots.$$

Also

$$\eta(2) = \frac{1}{2}\log\left(\frac{1+\sqrt{5}}{2}\right) = 0.2406059125 \ldots$$

and

$$\eta(3) = \frac{1}{3}\log\alpha = 0.09373319144 \ldots \tag{16.5}$$

for the unique real α with $\alpha^3 - \alpha - 1 = 0$ (see Exercise 12.1), as well as

$$\eta(4) = \frac{1}{4}\log(-1/\alpha) = 0.08057115399\ldots$$

for the unique $\alpha < 0$ with $\alpha^4 - \alpha - 1 = 0$.

We can improve (16.4) by using the trick in the proof of Theorem 14.8. It can be reformulated as follows. We have a polynomial $P(X, Y) = X + Y - 1$ with $P(\boldsymbol{\alpha}) = 0$ for $\boldsymbol{\alpha} = (\alpha, 1-\alpha)$. We then considered (in an obvious notation)

$$P(\boldsymbol{\alpha}^5) = P(\alpha^5, (1-\alpha)^5) = \beta$$

and exploited the fact that it has a factor 5 in the sense of (14.29). Only now we know nothing about $1 - \alpha$, so we have to leave it out and make do with $P(\alpha) = 0$ (fortunately already an old friend from Lemma 15.4). Now it turns out for any prime p that $\beta = P(\alpha^p)$ has a factor p in the same sense, and we will get a lower bound for $H(\alpha)$ just as we did for $H(\alpha)H(1-\alpha)$. However there is the usual vanishing snag $\beta = 0$.

Here is what comes out.

Lemma 16.2 *Let α be algebraic of degree d and let P in $\mathbb{Z}[X]$ be of degree at most L with $P(\alpha) = 0$ but*

$$P(\alpha^p) \neq 0 \tag{16.6}$$

for some prime p. Then we have

$$h(\alpha) \geq \frac{1}{pL}\log\frac{p}{\mathcal{L}} \tag{16.7}$$

with $\mathcal{L} = \mathcal{L}(P)$.

Proof In $\mathbb{F}_p$ we have $a^p = a$ and in any field of characteristic p we have $(\lambda+\mu)^p = \lambda^p + \mu^p$, so in $\mathbb{F}_p[X]$ with A as in (14.1) we deduce

$$A(X)^p = (a_0X^d + \cdots + a_d)^p = a_0^pX^{dp} + \cdots + a_d^p = a_0X^{dp} + \ldots + a_d = A(X^p).$$

Thus there is B in $\mathbb{Z}[X]$ of degree at most dp with

$$A(X)^p = A(X^p) + pB(X)$$

in $\mathbb{Z}[X]$. Further A divides P in $\mathbb{Q}[X]$ so that $P = AQ$ for Q in $\mathbb{Z}[X]$ (the usual Gauss Lemma but as a matter of fact just the p-valuation suffices here) of degree at most $L - d$. At $X = \alpha$ we get

$$\beta = P(\alpha^p) = A(\alpha^p)Q(\alpha^p) = -pB(\alpha)Q(\alpha^p).$$

To use the Product Formula we fix any K containing α. For any prime ideal $\mathfrak{p}$ we have

$$|\beta|_{\mathfrak{p}} \leq |p|_{\mathfrak{p}} \max\{1, |\alpha|_{\mathfrak{p}}\}^{pL}.$$

And for any embedding σ we have

$$|\beta|_{\sigma} \leq \mathcal{L} \max\{1, |\alpha|_{\sigma}\}^{pL}.$$

Multiplying up over all $\mathfrak{p}, \sigma$ we get

$$1 \leq C\mathcal{L}^{[K:\mathbb{Q}]} e^{h(\alpha)pL[K:\mathbb{Q}]}$$

where $C = \prod_{\mathfrak{p}} |p|_{\mathfrak{p}} = \prod_{\sigma} |p|_{\sigma}^{-1} = p^{-[K:\mathbb{Q}]}$ again by the Product Formula. And now (16.7) drops out. □

The lower bound (16.7) is positive only if $p > \mathcal{L}(P)$, and for safety it is advisable to take $p \geq 2\mathcal{L}(P)$. As in the previous chapter, we have the obvious candidate $P = A$ as in (14.1). By (14.10) we have $\mathcal{L}(A) \leq (2H(\alpha))^d$. Now we can certainly assume that $H(\alpha)^d < 2$, otherwise we get again the very strong (16.1). In this case $\mathcal{L}(A) \leq 2^{d+1}$, and we need just $p \geq 2^{d+2}$. The famous Bertrand's Postulate guarantees such a prime with $p \leq 2^{d+3}$. We conclude from Lemma 16.2 that

$$h(\alpha) \geq \frac{1}{pd} \log 2 \geq \frac{\log 2}{8d2^d}. \tag{16.8}$$

This is much better than (16.4) but still a long way from Theorem 16.1.

However we forgot the snag (16.6)! At the moment this is $A(\alpha^p) = 0$. We already prepared for this in Lemma 14.6, which shows that α is a root of unity, excluded in our present situation.

So (16.8) is proved. We see that the big 2^d comes from the bad bound for $\mathcal{L}(A)$. From the previous chapter we see a way of improving this. We already noted in (15.4) that $\mathcal{L}(P) \leq 4d^2H(\alpha)^{2d-1}$ for a certain P of degree at most $2d - 1$ with $P(\alpha) = 0$. Here we can assume as above $H(\alpha)^d < 2$. In that case $\mathcal{L}(P) \leq 16d^2$, which leads as above via Bertrand with

$$32d^2 \leq p \leq 64d^2 \tag{16.9}$$

to

$$h(\alpha) \geq \frac{1}{2pd} \log 2 \geq \frac{\log 2}{128d^3}, \tag{16.10}$$

indeed an enormous improvement.

But we forgot the snag again! This time we cannot deduce anything from $P(\alpha^p) = 0$, because as in the previous chapter there may be strangers among the zeroes of P. However suppose we know that

$$\mathbb{Q}(\alpha^p) = \mathbb{Q}(\alpha). \tag{16.11}$$

Then it would follow that α^p also has degree d and so its minimal polynomial A_p in $\mathbb{Z}[X]$ has degree d. But now P would have to be divisible by both A and A_p. Again by Lemma 14.6 we know that A, A_p are not proportional; hence they are coprime. Therefore P is divisible by AA_p. This is impossible, because the degree of P is at most $2d - 1$.

Now in connexion with (14.22) we already saw that (16.11) can fail. The following result shows that it fails for rather few p.

Lemma 16.3 *Let α be algebraic of degree d. If there are r different primes p with $\mathbb{Q}(\alpha^p) \neq \mathbb{Q}(\alpha)$ then $2^r \leq d$.*

Proof Let $p_1, \ldots, p_r$ be these primes, and consider the fields $K_i = \mathbb{Q}(\alpha^{q_i})$ with $q_i = p_1 \cdots p_i$ $(i = 1, \ldots, r)$. With $K_0 = \mathbb{Q}(\alpha)$ and $q_0 = 1$ we have a chain

$$\mathbb{Q} \subseteq K_r \subseteq \cdots \subseteq K_i \subseteq K_{i-1} \subseteq \cdots \subseteq K_1 \subseteq K_0 \tag{16.12}$$

because $\alpha^{q_i} = (\alpha^{q_{i-1}})^{p_i}$. We will show that each $K_i \neq K_{i-1}$ $(i = 1, \ldots, r)$.

If not, then some $K_i = K_{i-1}$ and so $\alpha^{q_{i-1}}$ in K_{i-1} would lie in $K_i = \mathbb{Q}(\alpha^{q_i})$. Thus also in $\mathbb{Q}(\alpha^{p_i})$. The latter contains of course α^{p_i}. Now $q_{i-1} = p_1 \cdots p_{i-1}$ and p_i are coprime and so there are rational integers u, v with $q_{i-1}u + p_i v = 1$. But then we could deduce that $\alpha = (\alpha^{q_{i-1}})^u (\alpha^{p_i})^v$ also lies in $\mathbb{Q}(\alpha^{p_i})$. Therefore $\mathbb{Q}(\alpha) \subseteq \mathbb{Q}(\alpha^{p_i})$ forcing equality which contradicts the defining property of p_i.

It now follows from (16.12) that

$$d = [K_0 : \mathbb{Q}] \geq [K_0 : K_r] = \prod_{i=1}^{r} [K_{i-1} : K_i] \geq 2^r$$

as desired. □

We can now finish the proof of (16.10). From Lemma 16.3 there are at most $\log d / \log 2$ primes p for which (16.11) fails. Surely these cannot exhaust the interval in (16.9)?

To be sure of this we need some weak form of the Prime Number Theorem; the latter would imply in the standard notation that $\pi(2x) - \pi(x)$ is asymptotically $x/\log x$, so that there are asymptotically at least $16d^2/\log d$ primes in (16.9). However the presence of error terms means that d must be taken sufficiently large. This is no problem because for small d we can use things like (16.8) or even (16.4). In fact for this chapter, for example, just the existence of $b > 0$ and $a > 1$ such that for all sufficiently large x

$$\pi(ax) - \pi(x) \geq b\sqrt{x} \tag{16.13}$$

is more than enough; and this follows easily from any version of Chebychev's estimates. But for simplicity of exposition we will carry on with $a = 2$.

So (16.10) is proved.

Let us give this estimate the name $\mathfrak{D}(3)$ (for $\mathfrak{D}$obrowolski), because the exponent is 3. We are after $\mathfrak{D}(\kappa)$ for every $\kappa > 1$. How do we close the gap?

This time we can handle strangers among the zeroes of P, so we don't need to take $L - d$ small. In the spirit of the previous chapter, let us try instead $L = \mathcal{X}d$ for some large parameter $\mathcal{X}$ independent of d. The upper bound in Lemma 15.4 becomes of order at most $(\mathcal{X}dH(\alpha)^{\mathcal{X}d})^{1/(\mathcal{X}-1)}$. Assuming as before $H(\alpha)^d < 2$ we see that this is of order at most $d^{1/(\mathcal{X}-1)}$, like a d^ϵ. Then $\mathcal{L}(P)$ is of order at most $(L+1)d^\epsilon$ or $d^{1+\epsilon}$. Now the analogues of (16.9) and (16.10) will yield $\mathfrak{D}(2+\epsilon)$. We have halved the gap (modulo the usual snag), but we are held away from the coveted $\mathfrak{D}(1+\epsilon)$ only by the factor $L+1$ which counts the number of terms! It is the fine difference between the sum norm l^1 and the sup norm l^∞.

The key is multiplicities, which we have seen to improve things for the maximum modulus principle. There we had the standard valuation on $\mathbb{C}$; but things work just as well for the valuations at $\mathfrak{p}$ used in the proof of Lemma 16.2. Here is the result, where we make the order of vanishing at least T.

Lemma 16.4 *For a positive integer T let α be algebraic of degree d and let P in $\mathbb{Z}[X]$ be of degree at most L with* $\mathrm{ord}_{X=\alpha} P \geq T$ *but $P(\alpha^p) \neq 0$ for some prime p. Then we have*

$$h(\alpha) \geq \frac{1}{pL} \log \frac{p^T}{\mathcal{L}} \tag{16.14}$$

with $\mathcal{L} = \mathcal{L}(P)$.

Proof We imitate the proof of Lemma 16.2 (for $T = 1$). As before there is B in $\mathbb{Z}[X]$ of degree at most dp with $A(X)^p = A(X^p) + pB(X)$ in $\mathbb{Z}[X]$. But now A^T divides P so that $P = A^T Q$ for Q in $\mathbb{Z}[X]$ of degree at most $L - dT$. At $X = \alpha$ we get $\beta = P(\alpha^p) = -(pB(\alpha))^T Q(\alpha^p)$. For K as before and any $\mathfrak{p}$ we have now

$$|\beta|_\mathfrak{p} \leq |p|_\mathfrak{p}^T \max\{1, |\alpha|_\mathfrak{p}\}^{pL}.$$

And for any embedding σ we have as before

$$|\beta|_\sigma \leq \mathcal{L} \max\{1, |\alpha|_\sigma\}^{pL}.$$

Multiplying up we get now $1 \leq p^{-T[K:\mathbb{Q}]}\mathcal{L}^{[K:\mathbb{Q}]}e^{h(\alpha)pL[K:\mathbb{Q}]}$, so (16.14) drops out. □

Thus (16.14) has $p^T/\mathcal{L}$ compared with only $p/\mathcal{L}$ in (16.7). This means that we need only take p about $\mathcal{L}^{1/T}$. Thus the difference between the two norms is damped.

However it may be that $\mathcal{L} = \mathcal{L}(P)$ also has to get larger due to the effect of demanding multiplicities. At any rate this effect is easily worked out.

Lemma 16.5 *For a positive integer T let α be algebraic of degree d and let $L \geq dT$ be an integer. Then there is non-zero P in $\mathbb{Z}[X]$ of degree at most L with* $\mathrm{ord}_{X=\alpha} P \geq T$ *and coefficients of absolute values at most*

$$\left((L+1)L^{T-1}H(\alpha)^L\right)^{dT/(L+1-dT)}.$$

Proof As for Lemma 15.4 (for $T = 1$) – see also Exercise 15.8 (for $T = 2$). We now have to solve

$$\sum_{i=0}^{L} p_i(L-i)(L-i-1)\cdots(L-i-j+1)\alpha^{L-i-j} = 0 \qquad (j = 0, 1, \ldots, T-1),$$

now $M = T$ equations in $N = L+1 > dM$ unknowns over $K = \mathbb{Q}(\alpha)$, which by Proposition 14.12 have a non-trivial solution with

$$|p_i| \leq ((L+1)H)^{dT/(L+1-dT)} \qquad (i = 0, 1, \ldots, L),$$

where now $H = L^{T-1}H(\alpha)^L$ has an extra term arising from the differentiation. This gives the required result. □

We already saw that with $T = 1$ we get $\mathfrak{D}(2+\epsilon)$. Let us explore in the same informal manner $T = 2$.

Again take $L = \mathcal{X}d$ for some large parameter $\mathcal{X}$. The upper bound in Lemma 16.5 becomes of order at most $(\mathcal{X}^2 d^2 H(\alpha)^{\mathcal{X}d})^{2/(\mathcal{X}-2)}$. Assuming as before $H(\alpha)^d < 2$ we see that this is of order at most $d^{4/(\mathcal{X}-2)}$, again d^ϵ. So $\mathcal{L}(P)$ is again of order at most $(L+1)d^\epsilon$ or $d^{1+\epsilon}$; the multiplicities have not made it larger. Now the analogue of (16.9) gives p^T about $d^{1+\epsilon}$ so p about $d^{(1+\epsilon)/2}$ and the analogue of (16.10) will yield $\mathfrak{D}(3/2+\epsilon)$ (modulo snag). We have again halved the gap, but we still have the option of making T large.

In fact the same discussion shows quickly that using multiplicity T will yield $\mathfrak{D}(1+1/T+\epsilon)$. So the strategy is clear: take T about $1/\epsilon$!

It is about time we did things properly. We make the routine assumption $H(\alpha)^d < 2$. Fix a positive integer T and a real $\mathcal{X} > T$, and choose $L = [\mathcal{X}d]$, so that $L + 1 > \mathcal{X}d > Td$ and thus $L \geq dT$. Lemma 16.5 gives for the coefficient bound

$$\left((L+1)L^{T-1}H(\alpha)^L\right)^{dT/(L+1-dT)} \leq ((2\mathcal{X}d)^T 2^{\mathcal{X}})^{T/(\mathcal{X}-T)} = C_1 d^\epsilon$$

with $\epsilon = T^2/(\mathcal{X} - T)$ and C_1 depending only on $T, \mathcal{X}$.

It follows that $\mathcal{L} = \mathcal{L}(P) \leq (L+1)C_1 d^\epsilon \leq C_2 d^{1+\epsilon}$ also for C_2 depending only on $T, \mathcal{X}$. We can take $C_2 \geq 1$.

We choose any p with

$$2\mathcal{L} \leq 2C_2 d^{1+\epsilon} \leq p^T \leq 2^{T+1} C_2 d^{1+\epsilon} \tag{16.15}$$

distinct from at most $\log d/\log 2$ exceptional primes regulated by Lemma 16.3. As indicated earlier (16.13) with $a = 2$ suffices, at least if d is large enough depending on $T, \mathcal{X}$. It follows from Lemma 16.4 that

$$h(\alpha) \geq \frac{1}{pL}\log 2 \geq \frac{1}{C_3 d^\kappa}$$

with

$$\kappa = 1 + \frac{1}{T} + \frac{T}{\mathcal{X} - T}$$

and C_3 depending only on $T, \mathcal{X}$. Clearly this κ can be made arbitrarily close to 1 by choosing T large and then $\mathcal{X}$ large. For example to prove $h(\alpha) \geq c/d^{1.01}$ we could take $T = 200$ and $\mathcal{X} = 40200$.

This time we only pretended to forget the snag. There might possibly be a prime with $P(\alpha^p) = 0$. In that case let A_p be the minimal polynomial of α^p in $\mathbb{Z}[X]$, of degree d thanks to the condition in Lemma 16.3. Then A_p divides P. If this happens for various p the resulting A_p are coprime because their zeroes α^p are different. So the product of all these A_p divides P. As P has degree at most $\mathcal{X}d$, it suffices to take more than $\mathcal{X}$ primes satisfying (16.15), a breeze.

That completes the proof of Theorem 16.1.

The full result (16.2) can be proved by choosing $T, \mathcal{X}$ as suitable functions of d; see for example section 3.6 of Waldschmidt (2000) or section 4.4 of Bombieri and Gubler (2006). See Exercise 16.7 for a way of avoiding things like Lemma 16.3.

The method of auxiliary polynomials does not necessarily supply the quickest proofs of these things. There is another way using big determinants (see Exercises 16.19 and 16.22), which sometimes gives better constants. Concerning these, the record in (16.2) is any $c < 9/4$ for large enough d, due to both

Louboutin and Meyer. An earlier result of Blanksby and Montgomery used Fourier analysis. And Smyth used complex function theory to show that

$$h(\alpha) \geq \frac{\log(1.32471795\ldots)}{d} \tag{16.16}$$

for any α not having $1/\alpha$ as a conjugate (which always holds when $d \geq 3$ is odd). The real number appearing here is that of Exercise 12.1, there considered as an algebraic number α; and for this α we have equality in (16.16) – see also (16.5) – so it is best possible. See Bombieri and Gubler (2006) for references to some of these.

It was Stewart (1978) who introduced auxiliary polynomials in this context. And in fact it seems vital to use them for extensions and generalizations.

One of these is to improve (16.3) to "relative" versions

$$h(\alpha) \geq \frac{c}{[\mathcal{K}(\alpha) : \mathcal{K}]^{\kappa}} \tag{16.17}$$

for some larger field $\mathcal{K}$ (still for any $\kappa > 1$). This has been done by Amoroso and Zannier (2000) with $\mathcal{K}$ as the maximal abelian extension $\mathbb{Q}_{\mathrm{ab}}$ of $\mathbb{Q}$. Also Amoroso and Dvornicich (2000) (see also Bombieri and Gubler (2006) for remarks) have proved (without using any of the above methods) that

$$h(\alpha) \geq \frac{\log 5}{12}$$

for any non-zero α in $\mathbb{Q}_{\mathrm{ab}}$ not a root of unity. Pontreau (2005) has given a simpler proof of (16.17) for the smaller maximal cyclotomic extension of $\mathbb{Q}$.

There are also higher-dimensional versions. The two-dimensional case of an important result of Amoroso and David (1999) states that for any $\kappa > 1$ there is $c > 0$ such that

$$h(\alpha)h(\beta) \geq \frac{c}{D^{\kappa}} \tag{16.18}$$

for any non-zero algebraic α, β that are multiplicatively independent (as not in (14.45) above) with $D = [\mathbb{Q}(\alpha, \beta) : \mathbb{Q}]$. If α is not a root of unity then we may apply this with either $\beta = 2$ or $\beta = 3$ to recover (16.3). Or with $\beta = 2^{1/D}$ we could deduce

$$h(\alpha) \geq \frac{c/\log 2}{D^{\kappa-1}}$$

for any α in $\mathbb{Q}(2^{1/D})$ with $\alpha, 2$ multiplicatively independent, probably much better than (16.3). See also Exercise 16.23.

And there are extensions of all these aspects to commutative group varieties; however a suitably functorial variant of the height should be taken, and this is best done for abelian varieties through the Néron–Tate height (see for example

(A.33) and especially (A.34) as well as Exercise A.18(a) in the Appendix). However the exponents κ are rarely best possible as in Theorem 16.1. They are in the case of complex multiplication, as was first proved by Laurent (1983) for elliptic curves.

Exercises

16.1 Find explicit $c > 0$ and d_0 such that $h(\alpha) \geq cd^{-5/2}$ for all $\alpha \neq 0$ of degree $d \geq d_0$ not roots of unity [*Hint*: $L = 3d - 1$ in Lemma 16.4 and $\pi(2x) - \pi(x) \geq \frac{\log x}{\log 2}$ for all $x \geq 30$].

16.2 Find explicit $c > 0$ such that $h(\alpha) \geq cd^{-3/2}$ for all $\alpha \neq 0$ of degree d large enough, apart from roots of unity [*Hint*: $\pi(2x) - \pi(x) \geq 3\frac{\log x}{\log 2}$ for all $x \geq 100$].

****

16.3 Show that $h(\alpha) \geq \frac{1}{2}\log(\frac{1+\sqrt{5}}{2})$ for all non-zero quadratic α except roots of unity.

16.4 Find explicit $c > 0$ such that $h(\alpha) \geq cd^{-5/2}$ for all $\alpha \neq 0$ of degree $d \geq 1$ not roots of unity.

16.5 Let $\epsilon > 0$. Show (without using multiplicities) that there is $c = c(\epsilon) > 0$ such that $h(\alpha) \geq cd^{-2-\epsilon}$ for all $\alpha \neq 0$ of degree $d \geq 1$ not roots of unity.

16.6 Let $\alpha \neq 0$ be of degree d, and let $p_1, \ldots, p_r$ be different primes with $\mathbb{Q}(\alpha^{p_i}) \neq \mathbb{Q}(\alpha)$ $(i = 1, \ldots, r)$. Is it true that $(p_1 - 1)\cdots(p_r - 1) \leq d$? I have set the affirmative answer as an exercise often in the past, but now I cannot reconstruct my proof (if it existed). It would imply $r \leq c\frac{\log d}{\log\log d}$ in place of $r \leq \frac{\log d}{\log 2}$ in Lemma 16.3.

16.7 (a) If α is algebraic with $\mathbb{Q}(\alpha^m) \neq \mathbb{Q}(\alpha)$ for some positive integer m, show that there is a root of unity ζ such that $\tilde{\alpha} = \zeta\alpha^e$ is in $\mathbb{Q}(\alpha^m)$ for $e = [\mathbb{Q}(\alpha) : \mathbb{Q}(\alpha^m)]$ (Rausch (1985) – see Waldschmidt (2000) for more variations) [*Hint*: "Capelli"].

(b) Give a proof of Theorem 16.1 without using things like Lemma 16.3 [*Hint*: induction on d].

16.8 Let $f(d)$ be the cardinality $\#\{\alpha \in \mathbb{C};\ [\mathbb{Q}(\alpha) : \mathbb{Q}] = d,\ h(\alpha) \leq d^{-1/2}\}$. According to the proof of Proposition 15.2 (with minimal $\mathcal{H} = 1$) there is c with $f(d) \leq c^{d^2}$. Using P of degree at most $2d - 1$ show that there is c with $f(d) \leq c^{d^{3/2}}$.

16.9 In Exercise 16.8 show that there is $c > 1$ with $f(d) \geq c^{d^{3/2}}$ ("Loher Bound" in his Doctoral Thesis 2001 – see also Masser and Vaaler (2008) for a proof).

16.10 Let $\alpha \neq 1$ in $\mathbb{C}$ be algebraic of degree d. Show that $|\alpha - 1| \geq (2H(\alpha))^{-d}$.

16.11 For $P = P(X) = a_0X^d + \cdots + a_d$ in $\mathbb{C}[X]$ define

$$|P| = \max\{|a_0|, \ldots, |a_d|\}.$$

For $L = L(X) = X - 1$ show that $|LP| \geq \frac{1}{d+1}|P|$ (compare with Exercises 14.38 and 14.39).

16.12 Suppose $H(\alpha) \leq 2^{1/d}$ in Exercise 16.10. Using P of degree at most $d+d^{1/2}$ with $P(1) \neq 0$, show that there is $c > 0$ such that $|\alpha - 1| \geq (cd)^{-cd^{1/2}}$ (Mignotte, 1979).

16.13 For a number field K of degree d and $\mathcal{H} \geq 1$ let $f_K(\mathcal{H}) = \#\{\alpha \in K;\ H(\alpha) \leq \mathcal{H}\}$. Show that

$$\lim_{\mathcal{H}\to\infty} \frac{f_K(\mathcal{H})}{\mathcal{H}^{2d}} = \frac{h_K R_K}{w_K \zeta_K(2)} \left(\frac{2^{r_K}(2\pi)^{s_K}}{\sqrt{|\Delta_K|}} \right)^2 2^{r_K+s_K-1}$$

with h_K the class number, R_K the regulator, $w_{\dot{K}}$ the number of roots of unity, Δ_K the discriminant, r_K real embeddings, $2s_K$ non-real embeddings, and $\zeta_K(z) = \sum \frac{1}{N(\mathfrak{a})^z}$ over all ideals $\mathfrak{a} \neq 0$ in $\mathbb{Z}_K$ (Schanuel (1979) – see Exercise 14.90).

16.14 In Exercise 16.13 show that there is an absolute constant C such that $f_K(\mathcal{H}) \leq C(d \log d)\mathcal{H}^{2d}$ $(d \geq 2)$ (Loher "Upper Bound", with $C = 37$, in his Doctoral Thesis 2001 – see also Loher and Masser (2004) for a proof).

16.15 Find explicit $c > 0$ such that $h(\alpha) \geq cd^{-3/2}$ for all $\alpha \neq 0$ of degree $d \geq 1$ apart from roots of unity.

16.16 Let $\epsilon > 0$. Calculate $c = c(\epsilon) > 0$ such that $h(\alpha) \geq cd^{-1-\epsilon}$ for all $\alpha \neq 0$ of degree d large enough apart from roots of unity.

16.17 Using the Prime Number Theorem show that

$$\lim_{x\to\infty} \frac{1}{x} \sum_{p \leq x} \log p = 1, \quad \lim_{x\to\infty} \frac{\log x}{x^2} \sum_{p \leq x} p = \frac{1}{2},$$

$$\lim_{x\to\infty} \frac{1}{x^2} \sum_{p \leq x} p \log p = \frac{1}{2}, \quad \lim_{x\to\infty} \frac{\log x}{x^3} \sum_{p \leq x} p^2 = \frac{1}{3}.$$

16.18 Prove Dobrowolski's Theorem (16.2).

16.19 Let $V = V(x_1, \ldots, x_n) = \prod_{1 \le s < r \le n}(x_r - x_s)$ be the Vandermonde determinant with rows $(1, x_i, \ldots, x_i^{n-1})$ $(i = 1, \ldots, n)$ Let α be in $\mathbb{Z}_{\mathbb{C}}$ (integral) of degree d with conjugates $\alpha_1, \ldots, \alpha_d$, let p be prime, and let $\beta = V(\alpha_1, \ldots, \alpha_d, \alpha_1^p, \ldots, \alpha_d^p)$. If $\beta \neq 0$ show that $H(\beta) \ge p^d$.

16.20 In Exercise 16.19 show that $H(\beta) \le (2d)^{2d} H(\alpha)^{4d^2 p}$.

16.21 If $\alpha \neq 0$ in Exercise 16.19 is not a root of unity, conclude (without an auxiliary polynomial) that there is $c > 0$ such that $h(\alpha) \ge c d^{-3}$.

16.22 Let $e_1 \ge 0, \ldots, e_n \ge 0$ be in $\mathbb{Z}$, and in V replace the single row

$$(1, x_i, \ldots, x_i^{n-1}) \qquad (i = 1, \ldots, n)$$

by the $e_i + 1$ "hyperdifferentiated" rows

$$\left(\binom{0}{f} x_i^{-f}, \binom{1}{f} x_i^{1-f}, \ldots, \binom{N-1}{f} x_i^{N-1-f}\right) \qquad (f = 0, 1, \ldots, e_i)$$

with $N = n + e_1 + \cdots + e_n$. For example with $n = 2, e_1 = 1, e_2 = 2$ one gets

$$\begin{vmatrix} 1 & x_1 & x_1^2 & x_1^3 & x_1^4 \\ 0 & 1 & 2x_1 & 3x_1^2 & 4x_1^3 \\ 1 & x_2 & x_2^2 & x_2^3 & x_2^4 \\ 0 & 1 & 2x_2 & 3x_2^2 & 4x_2^3 \\ 0 & 0 & 1 & 3x_2 & 6x_2^2 \end{vmatrix} = (x_2 - x_1)^6.$$

Show that the new determinant ("degenerate Vandermonde") is

$$\prod_{1 \le s < r \le n} (x_r - x_s)^{(e_r+1)(e_s+1)}.$$

(One can prove Dobrowolski's Theorem using also such objects.)

16.23 (a) Suppose β is non-zero algebraic such that $\beta, \sigma(\beta)$ are multiplicatively dependent for some complex embedding σ of $\mathbb{Q}(\beta)$. Show that there is $n = \pm 1$ such that $\sigma(\beta)/\beta^n$ is a root of unity.

(b) Suppose $\mathbb{Q}(\beta)$ is Galois with group G, and that $\beta, \sigma(\beta)$ are multiplicatively dependent for every σ in G. Show that there is a positive integer e with $[\mathbb{Q}(\beta^e) : \mathbb{Q}] \le 2$ [*Hint*: group theory on $n = +1$ above (thanks, Francesco!), but don't try to use Exercise 11.17(c) way back].

(c) Use (16.17) with $\mathcal{K} = \mathbb{Q}_{\text{ab}}$ and (16.18) to deduce that for any $\kappa > 1/2$ there is positive c such that $h(\alpha) > c/d^{\kappa}$ for any non-zero α of degree d, not a root of unity, with $\mathbb{Q}(\alpha)$ Galois [*Hint*: if (b) holds for $\beta = \alpha$, take e minimal and use standard lower bounds for the Euler ϕ function].

17
Height upper bounds

After the relatively easy lower bound Theorem 14.8 and the really easy upper bound Theorem 14.9, together with the distinctly tricky lower bound Theorem 16.1, it seems only fair on grounds of symmetry to present a more difficult upper bound. The Mirimanov polynomials are defined as $M_n(Y) = (Y+1)^n - Y^n - 1$ deprived of some obvious factors, and in the course of an investigation into their irreducibility properties, Beukers proved that if $n \geq 2$ then the absolute height $H(\beta) \leq 216$ for any zero β. He used hypergeometric functions. To make this look more like our previous results, we formulate it for

$$X^n + (1-X)^n - 1 = -X^n M_n(-1 + 1/X).$$

By now the reader will not be shocked by the fact that we use auxiliary polynomials. In fact there are several parallels with the proof of Thue's Theorem 12.1. Our result is as follows, an atrocity in the fine style of Chapters 9 and 10. This is justified by the wider applicability of the method, which we later briefly hint at.

Theorem 17.1 *For algebraic α suppose that there is an integer $n \geq 2$ with $\alpha^n + (1-\alpha)^n = 1$. Then we have $H(\alpha) \leq 10^{120}$.*

As for $\alpha^r(1-\alpha)^s = 1$ in Theorem 14.9, we cannot do this simply from the equation for α and the original definition (14.3) of $H(\alpha)$; for example $X^{13} + (1-X)^{13} - 1$ is

$$13X(X-1)(X^2 - X + 1)^2(X^6 - 3X^5 + 8X^4 - 11X^3 + 8X^2 - 3X + 1)$$

(some of these factors are "obvious") and so quite far from irreducible.

Our proof constructs P, Q, R of degree roughly $n/2$ such that

$$P(X)X^n + Q(X)(1-X)^n + R(X) = 0.$$

Then at $X = \alpha$ we get another relation connecting $\alpha^n, (1-\alpha)^n, 1$, and we can try to eliminate $(1-\alpha)^n$. This works if

$$P(\alpha) \neq Q(\alpha), \tag{17.1}$$

rather as in (12.21), and we eliminate this snag as there by differentiating. In fact we can use our easy zero estimate Proposition 12.2. The introduction of δ with $0 < \delta \leq 1$ keeps the coefficients under control.

Here is the analogue of Lemma 12.4 (in our situation essentially $d = 1$).

Lemma 17.2 *For any $L \geq 1, n \geq 1$ in $\mathbb{Z}$ with*

$$(2-\delta)(L+1) > (1+\delta)n \tag{17.2}$$

there are P, Q, R in $\mathbb{Z}[X]$, not all zero, of degree at most L and with coefficients of absolute values at most $(3(L+1)2^n)^{1/\delta}$, such that

$$P(X)X^n + Q(X)(1-X)^n + R(X) = 0. \tag{17.3}$$

Proof The apparently new $R(X)$ can be explained away by saying that $P(X)X^n + Q(X)(1-X)^n$ has small degree, which amounts to a condition with $\mathrm{ord}_{X=\infty}$. Anyway, we get $M = L+n+1$ equations in $N = 3(L+1)$ unknowns, and by (17.2) the Siegel exponent

$$\varsigma = \frac{M}{N-M} < \frac{1}{\delta}. \tag{17.4}$$

The coefficients in the equations have absolute values at most 2^n. Thus the prehistoric Proposition 8.3 gives the upper bound $(N2^n)^{\varsigma}$ and the lemma follows. □

In fact the P, Q of Lemma 12.4 are rather our $\tilde{P} = PX^n, \tilde{Q} = Q(1-X)^n$. Thus by analogy to (12.27) we define

$$\pi^{(m)} = \frac{\tilde{P}^{(m)}(\alpha)}{m!\alpha^n}, \quad \kappa^{(m)} = \frac{\tilde{Q}^{(m)}(\alpha)}{m!(1-\alpha)^n}, \quad \rho^{(m)} = -\frac{R^{(m)}(\alpha)}{m!} \quad (m = 0, 1, 2, \ldots), \tag{17.5}$$

where from now on we assume $\alpha \neq 0, 1$. The following is the appropriate analogue of Lemma 12.5.

Lemma 17.3 *The quantities $\alpha^m \pi^{(m)}, (1-\alpha)^m \kappa^{(m)}, \rho^{(m)}$ are polynomials in α of degree at most L with coefficients in $\mathbb{Z}$ of absolute values at most $2^{L+n+m}(3(L+1)2^n)^{1/\delta}$.*

Proof We find

$$\alpha^m \pi^{(m)} = \sum_{r=0}^{m} \frac{P^{(m-r)}(\alpha)}{(m-r)!} \binom{n}{r} \alpha^{m-r}.$$

This is clearly of degree at most L in α. The hyperderivatives introduce an extra 2^L into the coefficient estimates, so we get by Lemma 17.2 the upper bound $2^{L+n}(3(L+1)2^n)^{1/\delta}$. With $(1-\alpha)^m\kappa^{(m)}$ we get $(1-\alpha)^{m-r}$ instead of α^{m-r} and so an additional 2^m. This does it. □

The next remark overcomes the snag (17.1) as long as L is chosen minimally in (17.2); that is

$$L = \left[\frac{1+\delta}{2-\delta}n\right] \le \frac{1+3\delta}{2}n. \tag{17.6}$$

It is also handy to assume $\delta < 1/3$, so that $L < n$.

Lemma 17.4 *For any $\theta \neq -(1-\alpha)^n/\alpha^n$ there is m with*

$$0 \le m \le 2 + 3\delta n, \quad \pi^{(m)} \neq \theta\kappa^{(m)}.$$

Proof If not, then $\mathrm{ord}_{X=\alpha}(\tilde{P}(X) - \tilde{\theta}\tilde{Q}(X)) > 2+3\delta n$, where $(1-\alpha)^n\tilde{\theta} = \alpha^n\theta$. As yet we cannot use Proposition 12.2, because the degree of $\tilde{Q}$ is too big (this is connected to the infinity business). But as $\tilde{P}+\tilde{Q}+R=0$ we can replace it by R. We deduce that $\mathrm{ord}_{X=\alpha}(\tilde{P}(X) - \theta_\alpha R(X)) > 2+3\delta n$ with $\theta_\alpha = -\tilde{\theta}/(1+\tilde{\theta})$; note that $1+\tilde{\theta} \neq 0$. Also $\tilde{P}$ cannot be a constant multiple of R, else comparison of degrees using $L < n$ would give $P = 0$, then $\tilde{Q} = -R$, then similarly $Q = 0$, forcing also $R = 0$ against Lemma 17.2.

Now already $\mathrm{ord}_{X=0}(\tilde{P}(X)) \ge n$ and $\mathrm{ord}_{X=1}(\tilde{P}(X) + R(X)) \ge n$. Thus we have the situation of Proposition 12.2 with $K = 3$. But

$$w_1 + w_2 + w_3 \;>\; 2 + 3\delta n + 2n \;\ge\; 3 + (L+n) + L - 1$$

by (17.6), a contradiction. □

We can now prove Theorem 17.1. We can assume $\alpha \neq 0, 1$. For some δ to be determined with $0 < \delta < 1/3$ define L by (17.6). Choose m as in Lemma 17.4 with $\theta = 1 \neq -(1-\alpha)^n/\alpha^n$. Differentiating (17.3) m times and putting $X = \alpha$ we deduce

$$\pi^{(m)}\alpha^n + \kappa^{(m)}(1-\alpha)^n - \rho^{(m)} = 0.$$

Now we can eliminate $(1-\alpha)^n$ to get

$$\alpha^n = \frac{\kappa^{(m)} - \rho^{(m)}}{\kappa^{(m)} - \pi^{(m)}}.$$

Multiplying numerator and denominator by $\alpha^m(1-\alpha)^m$, we bring them by Lemma 17.3 into $\mathbb{Z}[\alpha]$ ready for use of Proposition 14.10. The degrees in α are at most $D = 2m + L$ and the lengths are at most

$$\mathcal{L} \;=\; 2^{m+1}(L+1)2^{L+n+m}(3(L+1)2^n)^{1/\delta}.$$

We obtain $H(\alpha)^n \leq \mathcal{L}H(\alpha)^D$. Here

$$D \leq 4 + 6\delta n + \frac{1+\delta}{2-\delta}n \leq \frac{2}{3}n$$

if δ is sufficiently small and n large; so we see some more light drudgery looming. In fact we can take $\delta = 1/50$ and $n \geq 127$. We deduce $H(\alpha)^n \leq \mathcal{L}^3$. Now as $L + 1 \leq n$ we get $\mathcal{L} \leq Cn^{51}2^{(52+6\delta)n}$ with $C = 3^{50}32$. And $n/127 < 4^{n/127}$ so

$$\mathcal{L} \leq C'2^{(52+6\delta+102/127)n}$$

with $C' = 127^{51}C$. Also $C' < 4^{\log C'} \leq 4^{\epsilon n}$ with $\epsilon = \log C'/127$. Thus $\mathcal{L} \leq 2^{\mu n}$ with

$$\mu = 52 + 6\delta + 102/127 + 2\epsilon,$$

giving $H(\alpha) \leq 2^{3\mu} < 10^{60}$, not quite so atrocious as in Theorem 17.1.

What about $2 \leq n \leq 126$? The above argument can be adjusted, but we could also use Proposition 14.1(ii) with $P(X) = X^n + (1-X)^n - 1 \neq 0$. If α has degree d we get

$$H(\alpha) \leq H(\alpha)^d \leq 4^n\mathcal{L}(P) \leq 4^n(1 + (2^n - 1)) < 10^{120}.$$

This finishes the proof.

We have made no effort to get a better upper bound. It would be easy for example to take into account the degree d above by considering the reducibility of $P(X)$. For Beukers's proof see Beukers (1997).

Some explicit upper bound as in Theorem 17.1 can also be deduced from Théorème 3.1 of Rémond (2002), which uses the sophisticated Vojta machinery. See also Bays and Habegger (2015) for more on this and further generalizations, sometimes leading to finiteness results.

The method described here extends to equations like

$$\alpha^n + (1-\alpha)^n + (1+\alpha)^n = 1$$

(see Exercise 17.7), and also to more general equations with any number of terms. One can even allow coefficients which grow at most subexponentially in n (see Exercise 17.9). This is useful in dealing with equations like

$$\alpha^r + (1-\alpha)^s = 1$$

(see Exercise 17.11). There are natural interpretations involving powers of $\mathbb{G}_m$, but at the moment no interesting extensions to abelian varieties or even powers of elliptic curves.

Exercises

17.1 Find C such that $H(\alpha) \leq C$ for any algebraic α for which there is an integer $n \geq 2$ with $\alpha^n + (1-\alpha)^n = 2$.

17.2 Given any algebraic β, show that there is $C(\beta)$ such that $H(\alpha) \leq C(\beta)$ for any algebraic α for which there is an integer $n \geq 2$ with $\alpha^n + (1-\alpha)^n = \beta$.

17.3 For an integer $n \geq 2$ define $\nu = \nu(n)$ as n if n is even and $n-1$ if not. Show that there is a positive constant c, depending only on n, such that $H(\alpha^n + (1-\alpha)^n) \geq cH(\alpha)^\nu$ for all algebraic α.

* * * * * * * * * * * * * * * * * * *

17.4 Show that there are positive constants λ, θ such that $H(\alpha^n + (1-\alpha)^n) \geq \lambda^\nu H(\alpha)^{\theta\nu}$ for all algebraic α and all integers $n \geq 2$, with $\nu = \nu(n)$ as in Exercise 17.3.

17.5 Is there a positive constant λ such that $H(\alpha^n + (1-\alpha)^n) \geq \lambda^\nu H(\alpha)^\nu$ for all algebraic α and all integers $n \geq 2$, with $\nu = \nu(n)$ as in Exercise 17.3? I don't know.

17.6 Let P, Q, R, S be in $\mathbb{C}[X]$, not all zero, of degrees at most $L \geq 1$, such that $\tilde{P}(X) + \tilde{Q}(X) + \tilde{R}(X) + S(X) = 0$ for

$$\tilde{P}(X) = P(X)X^n,\ \tilde{Q}(X) = Q(X)(1-X)^n,\ \tilde{R}(X) = R(X)(1+X)^n$$

and some integer $n \geq 1$.

(a) If $2L \leq n-3$ show that any three from $\tilde{P}(X), \tilde{Q}(X), \tilde{R}(X), S(X)$ are linearly independent over $\mathbb{C}$ [*Hint*: *ABC* Exercise 12.23 possibly after dividing out by common factors].

(b) If $T \geq 3L - n + 6$ and $x \neq 0, 1, -1$ is in $\mathbb{C}$, show that the vectors

$$(\tilde{P}^{(m)}(x), \tilde{Q}^{(m)}(x), \tilde{R}^{(m)}(x)) \quad (m = 0, 1, \ldots, T)$$

span $\mathbb{C}^3$ [*Hint*: Exercise 12.26].

17.7 Show that there is an absolute constant C such that $H(\alpha) \leq C$ whenever

$$\alpha^n + (1-\alpha)^n + (1+\alpha)^n = 1$$

for some integer $n \geq 0$ [*Hint*: choose two suitable m in Exercise 17.6(b) above].

17.8 Let F, G be in $\mathbb{Z}[X]$ of degrees $p \geq 1, q \geq 1$ respectively.

(a) Show that there is $C_1 = C_1(F, G)$ with the following property. For any $n \geq 1, L \geq 1, M \geq 1, N \geq 1$ in $\mathbb{Z}$ and δ in $\mathbb{R}$ with $0 < \delta \leq 1$ and

$$L + M + N + 3 > (1+\delta)(\max\{L + pn, M + qn, N\} + 1)$$

there are P, Q, R in $\mathbb{Z}[X]$, not all zero, of degrees at most L, M, N respectively and with coefficients of absolute values at most $((L+M+N)C_1^n)^{1/\delta}$, such that $P(X)F(X)^n + Q(X)G(X)^n + R(X) = 0$.

(b) Define $\tilde{P} = PF^n, \tilde{Q} = QG^n$. Show that there is $C_2 = C_2(F, G)$ with the following property. For any α with $F(\alpha) \neq 0, G(\alpha) \neq 0$ and

$$\pi^{(m)} = \frac{\tilde{P}^{(m)}(\alpha)}{m!F(\alpha)^n}, \quad \kappa^{(m)} = \frac{\tilde{Q}^{(m)}(\alpha)}{m!G(\alpha)^n}, \quad \rho^{(m)} = -\frac{R^{(m)}(\alpha)}{m!} \quad (m = 0, 1, 2, \ldots),$$

the $F(\alpha)^m \pi^{(m)}, G(\alpha)^m \kappa^{(m)}, \rho^{(m)}$ are polynomials in α of degrees at most $L + pm, M + qm, N$ respectively, with coefficients in $\mathbb{Z}$ of absolute values at most $C_2^{L+M+N+n+m}((L+M+N)C_1^n)^{1/\delta}$ [*Hint*: Exercise 6.15].

(c) Suppose that F, G are coprime and $N < np, N < nq$. Show that for any $\theta \neq -(G(\alpha)/F(\alpha))^n$ there is m with

$$0 \leq m \leq p + q + L + N - nq, \quad \pi^{(m)} \neq \theta\kappa^{(m)}.$$

17.9 Let F, G be as in Exercise 17.8. Show that there is $C = C(F, G)$ and $c = c(p, q)$ such that $H(\alpha) \leq C(H(\beta)H(\gamma))^{c/n}$ for any algebraic α, β, γ with

$$\beta F(\alpha)^n + \gamma G(\alpha)^n = 1$$

provided $n \geq 1$ and $\beta F(X)^n + \gamma G(X)^n \neq 1$ [*Hint*: if $\gamma \neq 0$ and $p \geq q$ choose $L = [\lambda n], M = [\mu n], N = [\nu n]$ for suitable λ, μ, ν and use Exercise 14.77].

17.10 Any two integers are "small almost multiples" of a third integer. Namely, show that for any non-negative integers r, s and any real $T > 1$ we can write

$$r = pn + r_0, \quad s = qn + s_0$$

for integers $n \geq 0, p \geq 0, q \geq 0, r_0, s_0$ with

$$0 < \max\{p, q\} \leq T, \quad \max\{|r_0|, |s_0|\} \leq \frac{n}{T} + T$$

[*Hint*: if $0 < r \leq s$ approximate r/s by p/q as in (8.2) *mutatis mutandis*, and take n as the integer nearest s/q].

17.11 Show that there is an absolute constant C such that $H(\alpha) \leq C$ whenever

$$\alpha^r + (1 - \alpha)^s = 1$$

for some non-negative integers r, s not both 1 [*Hint*: Exercise 17.10].

17.12 Is Lemma 17.4 true without the restriction $\theta \neq -(1 - \alpha)^n/\alpha^n$? I don't know.

18
Counting – Bombieri–Pila

In this short interlude we consider some results midway between the Pólya Theorem of Chapter 9 and the Schneider–Lang Theorem of Chapter 19. Some higher-dimensional versions have recently become very useful in the study of unlikely intersections in semiabelian schemes and Shimura varieties.

In Chapter 9 we considered integral values of

$$f(1), f(2), f(3), \ldots.$$

Here we treat rational values of

$$f(0), f\left(\frac{1}{n}\right), f\left(\frac{2}{n}\right), \ldots, f\left(\frac{n-1}{n}\right), f(1); \tag{18.1}$$

the values at first assumed to lie in $\mathbb{Z}/n$. How many of these $n+1$ values can there be?

As in Pólya's Theorem, polynomials in $\mathbb{Z}[z]$ play a special role: if the degree is at most 1 then all $n+1$ values are in $\mathbb{Z}/n$.

As we are restricting to the real interval $[0,1]$ we may consider $f(z) = \sqrt{1+z}$, which can be defined analytically on an open disc containing this interval. If $n = m^2$ is a square and $z = (r^2 - n)/n$ for any integer r with $n \leq r^2 \leq 2n$, then $f(z) = \pm rm/n$, so the number among (18.1) in $\mathbb{Z}/n$ can be of order at least $n^{1/2}$.

Similarly with $\sqrt[3]{1+z}$, giving order at least $n^{1/3}$. And here is a less obvious example, not so easily expressed in terms of radicals. By Lemma 4.3 with $S(z,w) = z - w - w^3$ there is a unique f in $z\mathbb{C}[[z]]$ with $f^3 + f = z$. By Lemma 4.5 it converges for all sufficiently small $|z|$, and if also $|w|$ is sufficiently small with $w^3 + w = z$ then $w = f(z)$ by Proposition 4.1 (with $F(X,Y) = Y(Y-iX)(Y+iX) - X^2$; the imaginary branches are easily ruled out). So for such $z = (r/m) + (r/m)^3$ we have $f(z) = r/m = rm^2/n$ for $n = m^3$. So again the number among (18.1) in $\mathbb{Z}/n$ can be of order at least $n^{1/3}$.

Bombieri and Pila showed that if f is not an algebraic function of z, and is analytic on an open set containing $[0, 1]$, then for any $\epsilon > 0$ there is c depending only on f and ϵ, such that there are at most cn^ϵ among (18.1) in $\mathbb{Z}/n$. They used generalized Lagrange Interpolation and Mean Value Theorems in a method reminiscent of Dörge's elementary proof of the Hilbert Irreducibility Theorem.

We will give a proof on complex analytic lines. But as the rationality of $f(z)$ implies its reality, perhaps too restrictive, we shall (perhaps whimsically) replace $\mathbb{Z}/n$ by $\mathbb{Z}[i]/n$, somewhat in the style of Gramain in Chapter 10. We use height H rather than denominators, now that we have the machinery. Also we shall consider a pair of functions (f, g) not just (f, z), for safety analytic on an open set containing $\mathbb{D}_2$ (recall that $\mathbb{D}_r$ is defined by $|z| \leq r$). Thus the points $(f(z), g(z))$ can be viewed as parametrizing an analytic curve. Our result is expressed in terms of a quantity $Z_L(f, g)$ with the flavour of a zero estimate. Its definition needs an easy lemma, which can be considered as a remark that zero estimates, in the sense that we have been using them up now, always exist.

Lemma 18.1 *Let f, g be functions analytic on an open set containing $\mathbb{D}_2$ which are algebraically independent over $\mathbb{C}$, and let $L \geq 0$ be in $\mathbb{Z}$. Then there exists $Z = Z_L(f, g)$ with the following property. For any non-zero P in $\mathbb{C}[X, Y]$ of degree at most L in each variable, the function $\phi = P(f, g)$ has at most Z zeroes in $\mathbb{D}_2$ counted with multiplicity.*

Proof If this is false, then for each integer N there would be such a polynomial $P_N \neq 0$ such that $\phi_N = P_N(f, g)$ has at least N zeroes in $\mathbb{D}_2$. We can normalize it to unit length (compare the proof of Lemma 5.2). Now by compactness the P_N have a subsequence tending to some limit $P_\infty \neq 0$ in the obvious topology. Here $\phi_\infty = P_\infty(f, g)$ is not identically zero by the algebraic independence hypothesis. We can find $r > 2$ such that $\phi_\infty \neq 0$ on the boundary of $\mathbb{D}_r$. Then the infimum of $|\phi_\infty|$ on this boundary is some $\mu > 0$. Let M be the number of zeroes of ϕ_∞ on $\mathbb{D}_r$ (and so in its interior). There is $N > M$ such that $|\phi_N - \phi_\infty|_r < \mu$ (recall that $|F|_r$ is the supremum of $|F|$ on $\mathbb{D}_r$). But by Rouché's Theorem this implies that ϕ_N, ϕ_∞ have the same number of zeroes in the interior of $\mathbb{D}_r$. So ϕ_N has M zeroes there, a contradiction. □

We note the lower bound

$$Z_L(f, g) \geq L^2 + 2L, \tag{18.2}$$

because if we are given any $L^2 + 2L$ different points in $\mathbb{D}_2$ then we can certainly find $\phi = P(f, g)$ vanishing at them.

But to find an explicit upper bound for $Z_L(f,g)$ one must eliminate the above compactness argument, which accounts for the difficulty of most zero estimates. We may remind the reader of Exercise 8.19.

Here is our version of the Bombieri–Pila result.

Theorem 18.2 *Let f, g be functions analytic on an open set containing $\mathbb{D}_2$ which are algebraically independent over $\mathbb{C}$, and let L be any positive integer. Then for any real $\mathcal{H} \geq 1$ there are at most*

$$2^{101} Z_L(f,g)(1 + \lfloor f \rfloor_2 + \lfloor g \rfloor_2)^{16/L} \mathcal{H}^{80/L}$$

different points z in $\mathbb{D}_1$ such that $f(z), g(z)$ are in $\mathbb{Q}(i)$ with

$$\max\{H(f(z)), H(g(z))\} \leq \mathcal{H}.$$

We remark that the "1+" here cannot be omitted. A counterexample would be $f(z) = z/N, g(z) = (e^z - 1)/N$ with $L = 1$; then $Z_1(f,g)$ can be taken independently of $N \geq 1$ (again by Lemma 18.1), and we need only note the single point $z = 0$ (with $\mathcal{H} = 1$) and make N large to get a contradiction.

We show that Theorem 18.2 implies the above assertion for (18.1) in $\mathbb{Z}/n$. For we can divide $[0,1]$ into fixed subintervals $[x - \delta, x + \delta]$ such that f is analytic whenever $|z - x| \leq 2\delta$, perform a simple change of scale, and note that any $f(s/n)$ in $\mathbb{Z}/n$ has absolute value bounded from above so height of order at most n. So we can take $L \geq 80/\epsilon$ minimal. This argument works even if the $f(s/n)$ in (18.1) have denominators at most n, not just dividing n (or even at most n^{100} using $L \geq 8000/\epsilon$).

The proof follows the same broad lines as for the results of Pólya and Gramain: we construct an auxiliary polynomial of degree at most L in f and g such that the corresponding function has zeroes at some of the points z in the theorem. We then deduce that it has zeroes at all the points, thus bringing in $Z_L(f,g)$. But this doesn't work so easily, and we must take clustering into account. If we work on a sufficiently small part of $\mathbb{D}_1$, then the effect of the zeroes is increased, just as the effect of a zero at $z = 0$ of order T involves a factor z^T which is smaller if z is small.

For any positive integer N we can divide $\mathbb{D}_1$ into at most $4N^2$ subsets each with diameter at most $\sqrt{2}/N$; for example we cut up the square $[-1,1]^2$ first and then intersect with $\mathbb{D}_1$. We use the above strategy on each subset and show that if N is large enough, then it contains at most $Z_L(f,g)$ points z satisfying the conditions of Theorem 18.2 on the values $f(z), g(z)$.

Fix such a subset, and let $\mathcal{S}$ be the set of its points satisfying these conditions. Now if $(L+1)^2 > 2S$ then we can solve equations over $\mathbb{Q}(i)$ to find $P \neq 0$ in $\mathbb{Z}[X, Y]$ with $\phi = P(f,g)$ vanishing on any S points z in $\mathcal{S}$; but we wish to keep the coefficients of P under control and so we take $(L+1)^2 \geq 4S$ to keep the Siegel exponent bounded above. In fact it is fine to aim for equality here, so we take at once $L \geq 1$ and

$$S = \left[\frac{(L+1)^2}{4}\right] \geq \frac{L^2}{4}. \tag{18.3}$$

Note that if $\mathcal{S}$ does not contain $S \leq (L+1)^2$ points z, then by (18.2) we are done (without using P).

As for the construction of P, we choose any S points w in $\mathcal{S}$ and we note that the vanishing of ϕ at each w leads to an equation $\Lambda_w = 0$ with a linear form Λ_w whose coefficients are the $f(w)^i g(w)^j$ $(i, j = 0, 1, \ldots, L)$. To estimate the height $H(\Lambda_w)$ we use the valuations v on $K = \mathbb{Q}(i)$ and note that

$$\max\{\ldots, |f(w)^i g(w)^j|_v, \ldots\} \leq \max\{1, |f(w)|_v^L\} \max\{1, |g(w)|_v^L\}. \tag{18.4}$$

It follows that $H(\Lambda_w) \leq H(f(w))^L H(g(w))^L \leq \mathcal{H}^{2L}$. Thus by Proposition 14.12 we can take the coefficients of P in $\mathbb{Z}$ with absolute values at most $(L+1)^2 \mathcal{H}^{2L}$.

Now choose any $z = w'$ in $\mathcal{S}$. Dividing out by the zeroes and comparing maximum moduli on the circles of radii 1 and 2, we get for $\alpha = \phi(w')$ the estimate

$$|\alpha| \leq |\phi|_2 \prod_w |w' - w| \leq |\phi|_2 \left(\frac{\sqrt{2}}{N}\right)^S \leq (L+1)^4 \mathcal{H}^{2L} M^L \left(\frac{\sqrt{2}}{N}\right)^S$$

(it is here that we see the effect of clustering) with

$$M = (1 + |f|_2)(1 + |g|_2) \leq (1 + |f|_2 + |g|_2)^2.$$

On the other hand, by Proposition 14.13 if $\alpha \neq 0$ we get $|\alpha| \geq H(\alpha)^{-2}$, and by Proposition 14.7

$$H(\alpha) \leq \mathcal{L}(P) H(f(w'))^L H(g(w'))^L \leq (L+1)^4 \mathcal{H}^{4L}$$

for the length $\mathcal{L}(P)$. So to deduce $\alpha = 0$ it suffices to take $(N/\sqrt{2})^S > (L+1)^{12} M^L \mathcal{H}^{10L}$, which by (18.3) is secured by $N = [\mathcal{X}] + 1$ with

$$\mathcal{X} = \sqrt{2}(L+1)^{48/L^2} M^{4/L} \mathcal{H}^{40/L} > 1.$$

This shows indeed that $\phi = 0$ on the whole of $\mathcal{S}$, and so this set contains at most $Z = Z_L(f, g)$ points. So after multiplying by the number of subsets $\mathcal{S}$, we get the final estimate

$$4N^2 Z \leq 4(\sqrt{2}+1)^2 (L+1)^{96/L^2} Z M^{8/L} \mathcal{H}^{80/L} \leq 2^{101} Z M^{8/L} \mathcal{H}^{80/L}$$

as claimed. This completes the proof of Theorem 18.2.

The reader will see at once that we could replace $\mathbb{Q}(i)$ here by any number field K in $\mathbb{C}$; then the estimate becomes

$$2^{16D^2+16D+5}Z_L(f,g)(1+|f|_2+|g|_2)^{8D/L}\mathcal{H}^{(16D^2+8D)/L}$$

with $D = [K : \mathbb{Q}]$, provided now $L \geq \max\{1, D-1\}$ (see Exercise 18.1).

In fact it is possible to demand only that $[\mathbb{Q}(f(z), g(z)) : \mathbb{Q}] \leq \mathcal{D}$ with no mention of K, but for this a different form of Siegel's Lemma is needed (see Exercise 14.86 and also Gramain et al. (1986) for a variation). One obtains a bound of very similar type

$$2^{C\mathcal{D}^2}Z_L(f,g)(1+|f|_2+|g|_2)^{C\mathcal{D}/L}\mathcal{H}^{C\mathcal{D}^2/L},$$

provided $L \geq C\mathcal{D}$, with some absolute C. See Masser (2011) for more details with the present approach using complex analysis and auxiliary polynomials; and for the latter with real analysis see Wilkie's paper in Jones and Wilkie (2015).

The n^ϵ in the original result of Bombieri and Pila (1989) is rather interesting. Already they had given an example to suggest that it cannot in general be improved, for example to some $(\log n)^\kappa$ as one believes in the vaguely analogous situation of Roth's Theorem (12.47). They used series of Newton Interpolation type like

$$g(z) = \sum_{k=0}^{\infty} \frac{1}{D_k!} \prod_{s=1}^{D_k} \left(z - \frac{s}{D_k}\right)$$

with $D_k = 2^{d_k}$ and suitable positive integers d_k. This even defines an entire function. If it was a polynomial, then its Newton series (10.18) at the points

$$1, \frac{1}{2}, \frac{1}{4}, \frac{3}{4}, \frac{1}{8}, \frac{3}{8}, \frac{5}{8}, \frac{7}{8}, \frac{1}{16}, \ldots \frac{15}{16}, \ldots$$

would be finite. The above sum is indeed such a series (with some coefficients zero) but infinite. So $g(z)$ is not a polynomial, and so must be transcendental over $\mathbb{C}(z)$ (see Exercise 9.4 a long way back). Truncating after $k = m$, we see that each s/D_{m+1} and $g(s/D_{m+1})$ $(s = 1, 2, \ldots, D_{m+1})$ is in $\mathbb{Z}/n$ for

$$n \;=\; D_m! D_{m+1}^{D_m} \;\leq\; (D_m D_{m+1})^{D_m}.$$

Now for example the inequality $D_{m+1} > (\log n)^{1000}$ is implied by

$$2^{d_{m+1}} \;>\; (2^{d_m} d_m + 2^{d_m} d_{m+1})^{1000},$$

which is easy to secure for all m.

The underlying reason is that here the term $Z_L(z, g)$ can be much larger than the lower bound in (18.2); the same truncation shows that for $L = D_m = 2^{d_m}$ it is at least $D_{m+1} = 2^{d_{m+1}} > D_m^{1000}$ (see Exercise 8.20 for a related but simpler construction).

See also Surroca (2006) for generalizations of this example to algebraic values of z and $g(z)$.

However if there are suitable differential equations (of a special diagonal shape) then indeed it is possible to get a power of a logarithm. In the next chapter we shall see how to eliminate the dependence on the height altogether, even for non-diagonal differential equations, by returning to entire functions as in Chapters 9 and 10. However we need extra properties such as suitable functional equations or addition theorems.

The higher-dimensional versions of the Bombieri–Pila result are much more powerful and lie deeper. For example we may consider the points at which three analytic functions $f(z,w), g(z,w), h(z,w)$ of two variables (which now parametrize an analytic surface) take algebraic values of height at most $\mathcal{H}$. Again one gets an estimate of the form $\mathcal{H}^\epsilon$ but now the algebraic independence of the functions does not suffice. For example take z, w, z^w at $(z,w) = (s/n, 1)$ $(s = 0, 1, \ldots, n)$ and $\mathcal{H} = n$. One must remove all (z,w) lying on pieces of algebraic curves like $w = 1$ or indeed $w = x$ for any rational x. The first result for two variables was proved by Pila (2004), who for an inductive argument to reduce to one variable had to use a difficult result of Gabrielov and introduce subanalytic sets. Then Pila and Wilkie (2006) treated general dimensions and introduced *o*-minimality from Model Theory.

Out of many applications we mention just two recent examples, both using only the original Bombieri–Pila result with f, g coming from hypergeometric functions.

First, one can prove that there are at most finitely many complex λ for which there are A, B in $\mathbb{C}[X]$ such that the Pell-type equation

$$A^2 - X^3(X^3 + X + \lambda)B^2 = 1, \quad B \neq 0$$

holds. See however Exercise 18.12.

Second, one can prove that there are at most finitely many complex λ for which

$$\int \frac{\mathrm{d}X}{(X-2)\sqrt{X^3 + X + \lambda}}$$

can be expressed in elementary terms; that is, using only algebraic, exponential and logarithmic functions. According to Liouville's Theorem on this topic it suffices to prove that the integral is not of the form

$$f + c_1 \log g_1 + \cdots + c_m \log g_m$$

for $f, g_1, \ldots, g_m$ in $\mathbb{C}(X, \sqrt{X^3 + X + \lambda})$ and $c_1, \ldots, c_m$ in $\mathbb{C}$. This is not always so easy, as an amazing example of Euler shows, according to which

$$\int \frac{\sqrt{1+X^4}\mathrm{d}X}{1-X^4}$$

is

$$-\frac{1}{4}\sqrt{2}\log\left(\frac{\sqrt{2}X - \sqrt{1+X^4}}{1-X^2}\right) - \frac{i}{4}\sqrt{2}\log\left(\frac{i\sqrt{2}X + \sqrt{1+X^4}}{1+X^2}\right).$$

See also (20.33) and Exercise 20.65 for another example.

Wilkie (2004) had already proved (also using auxiliary polynomials) a surprising variation about counting integer points, of which the following is a special case. Let f be a function analytic on a punctured neighbourhood of infinity and meromorphic there, so a Laurent series

$$f(z) = b_0 z^d + \cdots + b_{d-1}z + b_d + \frac{a_1}{z} + \frac{a_2}{z^2} + \cdots$$

convergent for all sufficiently large $|z|$ (we have already met these, under a slightly different job description, in Chapters 4 and 5 and especially Exercise 11.28). If f is not an algebraic function of z, then for any $\mathcal{N} \geq 3$ there are at most $c \log\log \mathcal{N}$ integers z with $|z| \leq \mathcal{N}$ such that $f(z)$ is (defined and) an integer (see also Exercise 18.16). Dörge's proof mentioned above (see for example Schinzel 1982) gives only $c\mathcal{N}^\theta$ for some fixed $\theta < 1$ (but applies also to algebraic irrational f such as $\sqrt{z+1}$, for which we need $\theta \geq 1/2$).

Wilkie observed that his dependence on $\mathcal{N}$ is best possible in a very strong sense: given any sequence of positive integers growing "at least doubly exponentially" there is such an f taking integral values on the sequence.

A result of Corvaja and Zannier (2002) implies that for any integer $q \geq 2$ there are at most finitely many m such that $f(q^m)$ is an integer, provided the coefficients in the Laurent series are algebraic. Wilkie's observation for the sequence q^{2^k} means that this proviso cannot be eliminated.

Exercises

18.1 Show that there is C with the following property. Let f, g be algebraically independent and analytic on an open set containing $\mathbb{D}_2$. Let K be in $\mathbb{C}$ with $[K : \mathbb{Q}] = D$, and let $L \geq \max\{1, D-1\}$ in $\mathbb{Z}$. Then for each $\mathcal{H} \geq 1$ there are at most

$$2^{CD^2} Z_L(f, g)(1 + |f|_2 + |g|_2)^{CD/L} \mathcal{H}^{CD^2/L}$$

different points z in $\mathbb{D}_1$ such that $f(z), g(z)$ are in K with

$$\max\{H(f(z)), H(g(z))\} \leq \mathcal{H}.$$

18.2 Show that there is C such that for every $\mathcal{H} \geq 1$ there are at most $C\mathcal{H}^{1/10}$ complex z in $\mathbb{D}_1$ with $2^z, e^z$ rational of height at most $\mathcal{H}$ (probably $z = 0$ is the only point, but no-one knows how to prove this) [*Hint*: proof of Lemma 13.3].

* * * * * * * * * * * * * * * * * * * *

18.3 Let $f_1, \ldots, f_N$ be linearly independent over $\mathbb{C}$ and analytic on an open set containing a compact set $\mathcal{K}$. Show that there is $Z = Z(f_1, \ldots, f_N, \mathcal{K}) \geq N - 1$ such that for any complex $p_1, \ldots, p_N$, not all zero, the function $p_1 f_1 + \cdots + p_N f_N$ has at most Z zeroes (with multiplicity) in $\mathcal{K}$ (the case $f_n = f^i g^j$ with $\mathcal{K} = \mathbb{D}_2$ is Lemma 18.1).

18.4 For coprime positive integers r, s let $f(z) = e^{rz}, g(z) = e^{sz}$, and let $L \geq 0$ be in $\mathbb{Z}$ with $L < \min\{r, s\}$. Show that there is $Z = Z(r, s)$ such that for any non-zero P in $\mathbb{C}[X, Y]$ of degree at most L in each variable the function $P(f, g)$ has at most Z zeroes (with multiplicity) in $\mathbb{D}_2$ (note that f, g are algebraically dependent).

18.5 Find C in Exercise 18.2 [*Hint*: if $2^z, e^z$ are real then z is real (why?), and make the necessary changes in the proof of Theorem 18.2 in order to use Exercise 13.20].

18.6 In Exercise 18.2 for $\mathcal{H} \geq 2$ show that there are at most $c(\log \mathcal{H})^2$ complex z in $\mathbb{D}_1$.

18.7 Show that $e^z - z$ has at most $4 < \frac{\log(e^3+3)}{\log 2}$ zeroes (with multiplicity) in $\mathbb{D}_1$.

18.8 Find explicit C with $Z_1(z, e^z) \leq C$. Can one in principle find the smallest possible C?

18.9 Show that there is C such that $Z_L(2^z, e^z) \leq CL^2$ for every $L \geq 1$ (Tijdeman, 1971a).

18.10 Let $\nu(n)$ be the cardinality $\#\{z \in [0, 1];\ z, \sqrt{1+z} \in \mathbb{Z}/n\}$. Show that $\nu(p) = 1$ (p prime), $|\nu(m^2) - (\sqrt{2} - 1)m| \leq 1$, and $\nu(n) = [n_0\sqrt{2}] - n_0 + 1 \leq \sqrt{2n}$ with $n_0 = \prod_p p^{[(\mathrm{ord}_p n)/2]}$.

18.11 Let ε be a function from $\mathbb{N}$ to $(0, \infty)$ such that $\varepsilon(n)/n^\epsilon \to 0$ for every $\epsilon > 0$. Show that there is an entire function f, not algebraic, together with infinitely many n such that at least $\varepsilon(n)$ of the values (18.1) are in $\mathbb{Z}/n$.

18.12 Show that there are infinitely many complex λ, for example

$$\lambda = 0,\ \pm\frac{i}{2},\ \pm\frac{i\sqrt{2}}{4},\ \pm\frac{i}{4}\sqrt{-2+2\sqrt{5}},\ \pm\frac{1}{4}\sqrt{2+2\sqrt{5}},\ \pm\frac{i}{2}\sqrt{2\pm\sqrt{2}}$$

for which there are A, B in $\mathbb{C}[X]$ such that

$$A^2 - X(X^3 + X + \lambda)B^2 = 1, \quad B \neq 0$$

(see Masser and Zannier (2010) and Masser and Zannier (2015) for hints). Find A, B for $\lambda = 0$.

18.13 Let f be a function analytic on an open set containing all z with $|z| \geq 1/2$ having a pole of order at most $d \geq 0$ at infinity. Show that for $R \geq 1/2$ and $|f|_{=R} = \sup_{|z|=R} |f(z)|$ we have $|f|_{=R} \leq (2R)^d |f|_{=1/2}$ [*Hint*: Exercise 9.5].

18.14 Let f, g be analytic on an open set containing all z with $|z| \geq 1/2$, meromorphic at ∞, and algebraically independent over $\mathbb{C}$, and let $L \geq 0$ be in $\mathbb{Z}$. Show that there is $Z = Z_L^*(f, g)$ such that $P(f, g)$ has at most Z zeroes z with $|z| \geq 1$ for all polynomials $P \neq 0$ in $\mathbb{C}[X, Y]$ with degree at most L in X and Y.

18.15 Show that there is an absolute constant c with the following property. Let f, g be analytic on an open set containing all z with $|z| \geq 1/2$, having poles of order at most $d \geq 1$ at infinity, and algebraically independent over $\mathbb{C}$. Then there is

$$L \leq c(d + \log(1 + |f|_{=1/2} + |g|_{=1/2}))$$

such that for any $R \geq 1$ there are at most $Z_L^*(f, g)$ different points z with

$$R \leq |z| \leq R^2$$

and $f(z), g(z)$ in $\mathbb{Z}$ [*Hint*: construct $\phi(z) = P(f(z), g(z))$ with zeroes at $S = [(L+1)^2/2]$ of these points, then try to use clustering on $\phi(1/u)$ to get more zeroes].

18.16 For f, g in Exercise 18.15 show that there is $C = C(f, g)$ such that for each $\mathcal{N} \geq 3$ there are at most $C \log\log \mathcal{N}$ points z with $1 \leq |z| \leq \mathcal{N}$ and $f(z), g(z)$ in $\mathbb{Z}$.

18.17 Show that there is C such that for any $R \geq 3$ in $\mathbb{R}$ and any $L \geq 1$ in $\mathbb{Z}$ and any non-zero polynomial P in $\mathbb{C}[X, Y]$ of degree at most L in each variable the function $P(z, e^z)$ has at most $C(L^2 + LR)$ zeroes in $\mathbb{D}_R$ [*Hint*: periodicity]. Thus for example $Z_L(z, e^z) \leq CL^2$.

18.18 Let $\zeta(z)$ be the Riemann zeta function. Show that there is C such that for any $R \geq 3$ in $\mathbb{R}$ and any $L \geq 1$ in $\mathbb{Z}$ and any non-zero polynomial P in $\mathbb{C}[X, Y]$ of degree at most L in each variable the function $P(z, \zeta(z))$ has at most $C(L^2 + LR \log R)$ zeroes in $\mathbb{D}_R$ (Masser, 2011). Thus $Z_L(z, \zeta(z + 4)) \leq CL^2$, for example. Compare with Exercise 22.23.

19

Transcendence III – Gelfond–Schneider–Lang

Here we prove among many other things the transcendence of $2^{\sqrt{2}}$, and more generally the Gelfond–Schneider Theorem (see also Exercise 13.6) on α^β, which is allowed to be $e^{\beta \log \alpha}$ for any interpretation of the logarithm. We illustrate the procedure with the Hermite–Lindemann Theorem of Chapter 13 on the transcendence of e^α. Namely, if α and $e^\alpha = \beta$ are both algebraic then they both lie in some number field K, and then so do the values $k\alpha, f(k\alpha) = \beta^k$ $(k = 0, 1, 2, \ldots)$ of the functions z and $f(z) = e^z$. Already we see hints of Chapters 9, 10 and 18.

Similarly for Gelfond–Schneider: if α and $\alpha^\beta = \gamma$ both lie in a number field K then so do the values $f(k \log \alpha) = \alpha^k$ and $g(k \log \alpha) = \gamma^k$ $(k = 0, 1, 2, \ldots)$ of the functions $f(z) = e^z$ and $g(z) = e^{\beta z}$. Thus we may hope to use an auxiliary polynomial in f and g as in Chapter 18. But now we have entire functions. And indeed the proof scheme of Chapter 13 for z and e^z does go through with only minor changes, since in that proof we did not use the small growth of z compared with that of e^z. The only essential thing to check is the algebraic independence of f and g as in Lemma 13.3 (which we do below).

In fact the same scheme succeeds for quite a few pairs of functions, mostly worked out by Schneider. He also gave a result for general functions. It was refined by Lang, and this we present here. Apart from the algebraic independence, we need some growth restrictions as in Chapters 9 and 10; and crucial also are appropriate differential equations, such as

$$f' = f, \quad g' = \beta g \tag{19.1}$$

for the Gelfond–Schneider example. Because the applications outside Gelfond–Schneider involve functions that are only meromorphic (see Chapter 20), we cover these too.

Let $\rho > 0$. We might say for simplicity that an entire function f has (strict) growth order at most ρ if there are constants c, C such that the maximum modulus $|f|_R \leq cC^{R^\rho}$. If we did so then Chapter 9 studied the case $\rho = 1$, and Chapter 10 the case $\rho = 2$. The applications to Hermite–Lindemann and Gelfond–Schneider involve $\rho = 1$, but in fact all other known applications involve $\rho = 2$. We need an extension to meromorphic f, for us simply quotients $f = f^*/f_*$ of entire functions. We say that f has growth order at most ρ if there are entire f^*, f_* themselves with growth order at most ρ. In fact if we want both definitions then some consistency has to be proved; this is possible but it is easier just to jettison the first definition. We then have the following result attached to the nametag Schneider–Lang. For reasons connected with the linearization of differential equations we allow several functions instead of just two.

Theorem 19.1 *Let $f_1, \ldots, f_m$ be meromorphic functions of growth order at most $\rho > 0$, with at least two among them algebraically independent over $\mathbb{C}$. Suppose that for some number field K the derivatives $f_1', \ldots, f_m'$ lie in the ring $K[f_1, \ldots, f_m]$. Then there are at most $16\rho[K : \mathbb{Q}]$ complex numbers w such that $f_1, \ldots, f_m$ are analytic at w with values in K.*

This apparently says nothing about transcendence (and nothing about heights as in Chapter 18). Nevertheless let us see how it implies the Hermite–Lindemann Theorem 13.1. As above we suppose to the contrary that $e^\alpha = \beta$ is algebraic, and we work in the finite extension $K = \mathbb{Q}(\alpha, \beta)$ of $\mathbb{Q}$. The functions $f_1(z) = z$ and $f_2(z) = e^z$ have $\rho = 1$, and by Lemma 13.3 they are algebraically independent. Also we have differential equations $f_1' = 1, f_2' = f_2$. Furthermore $f_1(k\alpha) = k\alpha$ and $f_2(k\alpha) = \beta^k$ are in K for $k = 0, 1, 2, \ldots$. As $\alpha \neq 0$ this contradicts the finiteness in Theorem 19.1 (and of course the explicit cardinality bound is irrelevant).

In a similar way we deduce Gelfond–Schneider.

Corollary 19.2 *For an irrational algebraic number β and an algebraic number $\alpha \neq 0$ let $\log\alpha$ be any non-zero determination of the logarithm. Then $\alpha^\beta = e^{\beta\log\alpha}$ is transcendental.*

Also in connexion with the zoo in Chapter 20, we remark that this result corresponds to the group variety $\mathbb{G}_m \times \mathbb{G}_m$, another commutative group surface, this time with exponential map (e^w, e^z) – see Exercise 19.29. But neither Gelfond nor Schneider knew this. However Lang knew it very well.

To prove Corollary 19.2, suppose to the contrary that $\alpha^\beta = \gamma$ is algebraic, and work in the finite extension $K = \mathbb{Q}(\alpha, \beta, \gamma)$ of $\mathbb{Q}$. The functions $f_1(z) = e^z$ and $f_2(z) = e^{\beta z}$ have $\rho = 1$. They are algebraically independent because a non-trivial polynomial relation between them would imply the linear dependence over $\mathbb{C}$ of certain $f_1(z)^{i_1}f_2(z)^{i_2} = e^{\lambda z}$ for $\lambda = i_1 + i_2\beta$; however the irrationality of β means that all the λ are different and so by the stronger result established during the proof of Lemma 13.3, the $e^{\lambda z}$ are linearly independent even over $\mathbb{C}(z)$. Also as mentioned above we have differential equations (19.1) over K. Furthermore $f_1(k\log\alpha) = \alpha^k$ and $f_2(k\log\alpha) = \gamma^k$ are in K for $k = 0, 1, 2, \ldots$. As $\log\alpha \neq 0$ this again contradicts the finiteness in Theorem 19.1.

So of course $2^{\sqrt{2}}$ is transcendental. But also $e^{\pi} = e^{-i\log(-1)}$ for $\log(-1) = \pi i$. And $e^{\pi\sqrt{163}} = e^{-i\sqrt{163}\log(-1)}$, even though our old friend

$$e^{\pi\sqrt{163}} = 262537412640768743.99999999999925007259719 81\ldots \quad (19.2)$$

is astonishingly close to an integer. So of course $\sqrt[3]{e^{\pi\sqrt{163}} - 744}$ is also transcendental, but you might not think it from the decimal expansion

$$640319.99999999999999999999999390317352319470126502835539026\ldots$$

which we have also seen before.

Further $\beta = \log 3/\log 2$ satisfies $2^{\beta} = 3$, and it is irrational because $\beta = r/s$ implies the absurd $2^r = 3^s$. So $\log 3/\log 2$ is also transcendental. The same arguments prove the transcendence of $\log\gamma/\log\alpha$ for non-zero algebraic numbers α, γ, provided this quotient is irrational. But when we calculate

$$\frac{\log(30122754096401 + 21300003689580\sqrt{2})}{\log(1607521 + 1136689\sqrt{2})} = 2.1176470588235294\ldots \quad (19.3)$$

(as in Exercise 13.11), we find a number that occurred in the context of Apéry's proof of the irrationality of $\zeta(3) = \sum_{k=1}^{\infty}\frac{1}{k^3}$ while trying to show that

$$\frac{\sum_{k=1}^{\infty}\frac{1}{k^4}}{\sum_{k=1}^{\infty}\frac{1}{k^4\binom{2k}{k}}} = 2.1176470588235294\ldots \quad (19.4)$$

is after all rational; see van der Poorten (1979). The continued fraction

$$[2; 8, 1, 1, 294117647058823]$$

of 2.1176470588235294 is much more encouraging than (12.39), and strongly suggests both ratios are $[2; 8, 1, 1] = 36/17$. Indeed this is the case, although a proof for (19.4) is much more difficult than a proof for (19.3). A hint for the former is in Exercise 19.12 and for the latter see Exercise 19.3. By the way, the irrationality of $\zeta(4) = \sum_{k=1}^{\infty}\frac{1}{k^4} = \pi^4/90$ is of course classical, but up to now there is no proof by methods like Apéry's.

In fact there are several ways of deciding whether a given $\log\gamma/\log\alpha$ is rational r/s. One is by factorizing γ, α over a suitable number field and thus determining the possible solutions of $\alpha^r = \gamma^s$. Another is to use heights and a modicum of diophantine approximation (see Exercise 19.28). But the method most amenable to generalization is to imitate the direct proof of Corollary 19.2 with auxiliary polynomials and see what turns up if β is after all rational. We shall say a few words about this at the end of the chapter.

A more symmetric version of the last transcendence result is that if α_1, α_2 are non-zero algebraic numbers, then their logarithms are linearly independent over the field of all algebraic numbers if and only if they are linearly independent over $\mathbb{Q}$ (see Exercise 19.10). In this form it generalizes without change to a famous theorem of Baker about $\alpha_1, \ldots, \alpha_n$ (corresponding to the commutative group variety $\mathbb{G}_{\mathrm{m}}^n$). In this situation he even proved the linear independence of $1, \log \alpha_1, \ldots, \log \alpha_n$ (corresponding to $\mathbb{G}_{\mathrm{a}} \times \mathbb{G}_{\mathrm{m}}^n$ as in Exercise 19.30), thus including Hermite–Lindemann. He also did not use any theory of group varieties; instead he found a magical way of increasing the number of zeroes of the auxiliary function, now defined on $\mathbb{C}^n$, using a complex line.

Before we start the proof of Theorem 19.1 we have to know how to differentiate $f_1, \ldots, f_m$ a large number of times; this is not quite as easy as in (13.5) for z and e^z. Fortunately we solved a similar problem during the proof of Lemma 4.4. Here it is necessary to formulate things in terms of derivations on algebras. For us this means just a map δ from a polynomial ring $\mathcal{K}[X_1, \ldots, X_m]$ to itself satisfying the usual rules

$$\delta(A+B) = \delta(A) + \delta(B), \quad \delta(AB) = \delta(A)B + A\delta(B)$$

with $\delta = 0$ on the field $\mathcal{K}$ (compare the endomorphism σ on $\mathbb{C}[X, Y]$ mentioned in Lemma 8.2). We can iterate it k times to get δ^k, with δ^0 the identity map.

Lemma 19.3 *Let δ be a derivation on $\mathcal{K}[X_1, \ldots, X_m]$. Then there are integers a, b and polynomials P_{lk} in $\mathcal{K}[X_1, \ldots, X_m]$ with total degrees at most $a + bk$ such that for any $k \geq 0$, we have*

$$\delta^k(X_l) = P_{lk}(X_1, \ldots, X_m) \quad (l = 1, \ldots, m).$$

Further if $\mathcal{K} = \mathbb{C}$ there is an integer c such that the coefficients of P_{lk} have absolute values at most $k!c^k$.

Proof For $m = 3$ and $\mathbb{C}[X, Y, Z]$ this implies (I_k) occurring in the proof of Lemma 4.4, because we can define

$$\delta(X) = P_1(X, Y, Z), \ \delta(Y) = Q_1(X, Y, Z), \ \delta(Z) = R_1(X, Y, Z)$$

according to (4.30), so that δ imitates $\frac{d}{dz}$ (just as σ imitated squaring z in Lemma 8.2), and then extend δ to $\mathbb{C}[X, Y, Z]$. And indeed we can use the argument there, but now for m functions instead of three. We have

$$\delta(X_l) = P_l(X_1, \ldots, X_m) \quad (l = 1, \ldots, m)$$

so $P_{l1} = P_l$ (and the obvious $P_{l0} = X_l$) and we get the recurrences

$$P_{lk} = \frac{\partial P_{l,k-1}}{\partial X_1} P_1 + \cdots + \frac{\partial P_{l,k-1}}{\partial X_m} P_m \quad (l = 1, \ldots, m;\ k = 1, 2, \ldots) \tag{19.5}$$

as in (4.31). So we just imitate the earlier proof with no problems (as the reader may have already done in Exercise 4.30). □

We now follow the strategies of Chapters 13 and 18 to prove Theorem 19.1. Pick any f, g among $f_1, \ldots, f_m$ which are algebraically independent over $\mathbb{C}$. There is no need for local considerations as in Chapter 18, and we shall use an auxiliary polynomial in f, g with zeroes in a finite set of complex w such that $f_1, \ldots, f_m$ are analytic at w with values in K. Because of the differential equations we can build in high multiplicities as in Chapter 13, and as there we try to make the multiplicities go to infinity. This will be possible if the cardinality of the set of all such w exceeds $16\rho D$ for $D = [K : \mathbb{Q}]$. So the auxiliary function must be identically zero, and the resulting contradiction proves what we want.

The proof doesn't really need all the technicalities of heights, but again now that we have done them we may as well use them.

We start with the analogue of Lemma 13.4, and we take a finite subset $\mathcal{S}$ of the above set with cardinality S. Now we will use $c_1, c_2, \ldots$ for quantities depending only on $f_1, \ldots, f_m$ and $\mathcal{S}$.

Lemma 19.4 *For any $L \geq 2, T \geq 1$ in $\mathbb{Z}$ with*

$$(L+1)^2 \geq 2DST \tag{19.6}$$

there is P in $\mathbb{Z}[X, Y]$, of degree at most L in each variable and with coefficients of absolute values at most $c_1^{L+T} L^T T^T$, such that

$$\phi(z) = P(f(z), g(z)) \tag{19.7}$$

is not identically zero and

$$\mathrm{ord}_{z=w}\, \phi(z) \geq T \quad (w \in \mathcal{S}). \tag{19.8}$$

Proof The equations (19.8) are $\Lambda = 0$, where $\Lambda = \Lambda_{tw}$ ($t = 0, 1, \ldots, T-1$; $w \in \mathcal{S}$) are linear forms in the unknown coefficients p_{ij} of P. A typical coefficient α_{twij} of Λ is $(\mathrm{d}/\mathrm{d}z)^t (f^i g^j)$ evaluated at $z = w$, and we will have to do a heights calculation as in (18.4). For this we use Leibniz as in (13.5) again to distribute the derivations over the $i + j$ factors, so that

$$(\mathrm{d}/\mathrm{d}z)^t (f^i g^j) = \sum C f^{(t_1)} \cdots f^{(t_i)} g^{(s_1)} \cdots g^{(s_j)} \tag{19.9}$$

taken over $t_1 + \cdots + t_i + s_1 + \cdots + s_j = t$, where

$$C = C(t_1, \ldots, t_i, s_1, \ldots, s_j) = \frac{t!}{t_1! \cdots t_i! s_1! \cdots s_j!}$$

is a multinomial coefficient. Now we have differential equations

$$f_l' = P_l(f_1, \ldots, f_m) \quad (l = 1, \ldots, m) \tag{19.10}$$

and these yield a derivation δ on $K[X_1, \ldots, X_m]$, so that by Lemma 19.3 all the $f^{(k)}, g^{(h)}$ are in $K[f_1, \ldots, f_m]$. Therefore the Λ are defined over K, and thus solvability is guaranteed by just $(L+1)^2 > DST$. As usual the stronger (19.6) is needed to tame the Siegel exponent.

Next we substitute $f^{(k)} = F_k(f_1, \ldots, f_m)$, $g^{(h)} = G_h(f_1, \ldots, f_m)$ arising from Lemma 19.3 into (19.9) and evaluate at $z = w$ to find

$$\alpha_{twij} = H_{tij}(f_1(w), \ldots, f_m(w)) \tag{19.11}$$

for the polynomials

$$H_{tij}(X_1, \ldots, X_m) \;=\; \sum C F_{t_1} \cdots F_{t_i} G_{s_1} \cdots G_{s_j}. \tag{19.12}$$

They have degree at most

$$\begin{aligned} &\max\{(a+bt_1) + \cdots + (a+bt_i) + (a+bs_1) + \cdots + (a+bs_j)\} \\ &\quad \le a(i+j) + bt \;\le\; 2aL + bT. \end{aligned} \tag{19.13}$$

For each valuation v corresponding to a complex embedding σ of K we get

$$|\alpha_{twij}|_v = |\sigma(\alpha_{twij})| \le \mathcal{L}(\sigma(H_{tij}))c_2^{L+T}. \tag{19.14}$$

Here the length

$$\mathcal{L}(\sigma(H_{tij})) \;\le\; \sum C\mathcal{L}(\sigma(F_{t_1})) \cdots \mathcal{L}(\sigma(F_{t_i}))\mathcal{L}(\sigma(G_{s_1})) \cdots \mathcal{L}(\sigma(G_{s_j})).$$

Now $\sum C = (i+j)^t \le (2L)^T$. Also we can apply σ to δ giving a derivation $\sigma(\delta)$ on $\sigma(K)[X_1, \ldots, X_m]$ and even on $\mathbb{C}[X_1, \ldots, X_m]$, and now Lemma 19.3 gives

$$\mathcal{L}(\sigma(F_{t_1})) \le (a + bt_1 + 1)^m t_1! c^{t_1} \le t_1! c_3^{t_1}.$$

So we get

$$\begin{aligned} |\alpha_{twij}|_v &\le (2L)^T \max\{t_1! \cdots t_i! s_1! \cdots s_j!\} c_3^t c_2^{L+T} \\ &\le (2L)^T t! c_3^t c_2^{L+T} \;\le\; c_4^{L+T} L^T T^T. \end{aligned} \tag{19.15}$$

This suffices for the analogue of (18.4) when v corresponds to a complex embedding of K.

Things are much easier for v corresponding to a prime ideal. One way is to fix some $\alpha \neq 0$ in K such that all the αP_l in (19.10) land in $\mathbb{Z}_K[X_1, \ldots, X_m]$ (a step back into the primeval world of denominators). It follows from (19.5) by simple induction that the $\alpha^k P_{lk}$ are also there, even for $k = 0$. And by (19.12) the same for $\alpha^t H_{tij}$. So by (19.11) and (19.13) we get

$$|\alpha_{twij}|_v \leq \max\{1, |\alpha^{-1}|_v\}^t \max\{1, |f_1(w)|_v\}^{2aL+bT} \cdots \max\{1, |f_m(w)|_v\}^{2aL+bT}. \tag{19.16}$$

Now taking this with (19.15), then the maximum over the i, j corresponding to the coefficients of the linear form Λ_{tw}, and then the product over all v, we get the height $H(\Lambda_{tw}) \leq c_5^{L+T} L^T T^T$. Then the present lemma follows by using Siegel's Lemma as in Proposition 14.12 (recall that f, g are algebraically independent). □

We now prepare for increasing the order of vanishing as in Lemma 13.5. But the fact that ϕ may not be entire forces us to use a sort of secondary induction involving t.

Lemma 19.5 *Suppose for some integer $T_1 \geq T$ that*

$$\mathrm{ord}_{z=w}\, \phi(z) \geq T_1 \quad (w \in \mathcal{S}).$$

Suppose further that for some w_1 in $\mathcal{S}$ and some $t = 0, 1, \ldots, 2T_1 - 1$ we have

$$\mathrm{ord}_{z=w_1}\, \phi(z) \geq t. \tag{19.17}$$

Then for any $\mathcal{Y} \geq 1$ we have

$$|\phi^{(t)}(w_1)| \leq \mathcal{Y}^{-ST_1} c_6^{L+T_1} L^{T_1} T_1^{3T_1} c_6^{L\mathcal{Y}^\rho}. \tag{19.18}$$

Proof We will divide by zeroes but also "divide by poles" to make ϕ entire. Namely we can find a non-zero function h such that h, hf, hg are entire of growth order at most ρ (another sort of denominator, this time an analytic one). We also need h minimal in the sense that $h(w_1) \neq 0$. We can secure this simply by replacing h by $h_1 = h/(z-w_1)^p$ with $p = \mathrm{ord}_{z=w_1} h$. Then of course h_1 stays entire; but also the growth does not increase because for any $R \geq |w_1| + 1$ we have $|h_1|_R \leq |h|_R$ (this trick is related to the definition-jettisoning mentioned at the beginning of this chapter).

Now with the minimal h we see that $\psi = h^{2L}\phi$ and then

$$\tilde{\phi}(z) = \frac{\psi(z)}{\prod_{w \in \mathcal{S}} (z - w)^{T_1}} \tag{19.19}$$

are entire. In particular we have

$$|\tilde{\phi}|_R \leq |\tilde{\phi}|_{R_1} \tag{19.20}$$

for any R, R_1 with $R \le R_1$. Ever since the first serious use of the Schwarz Lemma (in (9.12) years ago), we always took R_1 (there $3S$) as roughly the same order of magnitude as R (there S). Actually in the proof of Gramain's Theorem 10.1 the precise choice of ratio (in Lemma 10.8) was critical. But now we need R_1 to be large compared with R, the so-called "method of large radius" which is useful in other contexts (see for example Chapter 22 on algebraic independence). We take in fact $R_1 = (2\mathcal{Y} + 1)R$ with $\mathcal{Y} \ge 1$ as above. It will be enough to take $R = \sup_{w \in \mathcal{S}} |w| + 1$ (not itself "large").

Now the right-hand side of (19.20) is at most $(2\mathcal{Y}R)^{-ST_1}|\psi|_{R_1}$ and the left-hand side is at least $(2R)^{-ST_1}|\psi|_R$. We deduce

$$|\psi|_R \le \mathcal{Y}^{-ST_1}|\psi|_{R_1}.$$

The supremum on the right-hand side can be estimated from Lemma 19.4, and we get say $c_6^{L+T}L^T T^T c_6^{L\mathcal{Y}^\rho}$.

Next we use the Cauchy estimate (12.29) with $m = t \le 2T_1$ and $r = R - 1$ to deduce

$$|\psi^{(t)}(w_1)| \le t!|\psi|_R \le \mathcal{Y}^{-ST_1} c_7^{L+T_1} L^{T_1} T_1^{3T_1} c_7^{L\mathcal{Y}^\rho}.$$

Finally by $\psi = h^{2L}\phi$ and (19.17) we see that $\psi^{(t)}(w_1) = h(w_1)^{2L}\phi^{(t)}(w_1)$, and so the minimality of h leads the required upper bound. □

The last step is to obtain a competing lower bound as in Lemma 13.6.

Lemma 19.6 *For any integer $T_1 \ge T$ and each of the quantities*

$$\xi = \phi^{(t)}(w) \quad (w \in \mathcal{S};\ t = 0, 1, \ldots, 2T_1 - 1)$$

we have either $\xi = 0$ or

$$|\xi| \ge c_8^{-(L+T_1)} L^{-3DT_1} T_1^{-3DT_1}.$$

Proof Of course this ξ is the linear form Λ_{tw} of the proof of Lemma 19.4 evaluated at the vector whose entries are the coefficients p_{ij} of the polynomial P. Arguing somewhat as in the proof of Proposition 14.7, we get for any v

$$|\xi|_v \le C_v \max_{ij} |\alpha_{twij}|_v$$

with a suitable $C_v \ge 1$ of a form that is by now familiar. In estimating the right-hand side we note that the bounds in (19.15) and (19.16) are also ≥ 1. Thus we get an estimate also for $\max\{1, |\xi|_v\}$. As we are now differentiating up to order $2T_1$, the resulting upper bound for $H(\xi)$ comes out as

$$(L+1)^2 c_1^{L+T} L^T T^T c_9^{L+2T_1} L^{2T_1} (2T_1)^{2T_1} \le c_{10}^{L+T_1} L^{3T_1} T_1^{3T_1}.$$

Then the desired result follows from this and Proposition 14.13. The reader will see that we are sometimes estimating D by c_{11}, but we cannot always; at least not if we insist on proving $S \leq 16\rho D$. □

Now Theorem 19.1 follows quickly after choosing the parameters in the following way.

We see from the previous two lemmas that the $\mathcal{Y}^{-ST_1}$ in (19.18) has to beat powers of $e^L, e^{T_1}, L^{T_1}, T_1^{T_1}$ and $e^{L\mathcal{Y}^\rho}$, subject to $T_1 \geq T$ and $(L+1)^2 \geq 2DST$ in (19.6). We can dispose of the $e^{L\mathcal{Y}^\rho}$ by fixing $L = [\sqrt{2DST}]$ and then $\mathcal{Y}^\rho = T_1^{1/2}$, so it becomes at most a power of e^{T_1}.

Now we are left essentially with only powers of $T_1^{T_1}$, and we proceed to count and balance these.

In Lemma 19.5 we have

$$-\frac{S}{2\rho} + \frac{1}{2} + 3$$

of them, and in Lemma 19.6 we have

$$-\frac{3D}{2} - 3D$$

of them. Thus if $S > \rho(7 + 9D)$ we are in business. We fix T sufficiently large in terms of the data in Theorem 19.1, define L, and then Lemmas 19.5, 19.6 for $T_1 = T$ and corresponding $\mathcal{Y}$ lead to $\xi = 0$, at first for $t = T$. Then for $t = T + 1$, and so by induction on t we see that ϕ vanishes to order at least $2T$ on $\mathcal{S}$. Then again with $T_1 = 2T$ and corresponding $\mathcal{Y}$ we get order at least $4T$. And so on, until we get zeroes of infinite order (and at this last stage just a single w suffices), contradicting the fact that ϕ is not identically zero. Thus we cannot have been in business; that is, we must have $S \leq \rho(7 + 9D) \leq 16\rho D$, and this proves the Schneider–Lang Theorem.

For Schneider's original version see Satz 13 in his book (Schneider, 1957), and for Lang's refinement his book (Lang, 1966a). The proofs there are of course height-free, and so are those in the books of Baker (1990) and Waldschmidt (1974). The transcendence of α^β had already been proved independently by Gelfond and Schneider in 1934 after Gelfond in 1929 had done this for quadratic β using his ideas on integer-valued functions (see Chapters 9 and 10).

To get any sort of transcendence measures we would need sufficiently good zero estimates. These can be proved, but only by using techniques from commutative algebra; see the discussion near the end of Chapter 22. For the special situations of Hermite–Lindemann and Gelfond–Schneider they are much easier, and simple resultant arguments suffice. For Baker's Theorem on linear independence, even without measures, it so happens that zero estimates are also needed; Baker could get by with ingenious *ad hoc* arguments in Baker (1966, 1967a,b, 1968a); but the much later refinements to measures (see for example Baker and Wüstholz, 1993) also required more sophisticated algebraic techniques. In fact it is these linear independence measures that have the spectacular applications for which Baker obtained his Fields Medal; we have already mentioned some of these in connexion with diophantine equations like $F_0(x,y) = m$. Here one should not forget the p-adic aspect, which as indicated at the end of Chapter 12 allows $F_0(x,y) = p_1^{s_1} \cdots p_k^{s_k}$ to be solved; however this time effectively.

From Schneider–Lang we have already deduced Hermite–Lindemann and Gelfond–Schneider, and we postpone to the next chapter the applications involving elliptic functions. One can also do something for abelian functions. All these have $\rho = 2$. But it is amusing to note also that Bertrand, by refining the bound $16\rho D$, was able to squeeze out a special case of Siegel's Theorem (Siegel, 1929) on the Bessel function J with $\rho = 1/2$ (see Exercise 19.2), namely the transcendence of $J'(\alpha)/J(\alpha)$ for non-zero algebraic α (see Baker and Masser (1977) p. 192 and also Exercise 19.14).

The Schneider–Lang Theorem can be generalized from w in $\mathbb{C}$ to $\mathbf{w}$ in $\mathbb{C}^n$, most easily if the set of $\mathbf{w}$ lies in a Cartesian product $\mathcal{S}_1 \times \cdots \times \mathcal{S}_n$ (see Exercise 19.9). This has applications to abelian varieties (see Chapter 20), and it can also be used to prove Baker's Theorem (see also Exercises 19.23 and 19.24), and even its elliptic analogue (Bertrand and Masser, 1980). A famous result of Bombieri (1970) asserts that in general the set of $\mathbf{w}$ lies in a proper algebraic hypersurface; the proof (also magical) uses currents and plurisubharmonic functions. This was mentioned in the citation for his own Fields Medal.

As threatened, we now return to the problem of deciding whether a given $\log\gamma/\log\alpha$ is rational r/s. If it is, then $\alpha^\beta = \gamma$ is algebraic for $\beta = r/s$. We can now go through the Schneider–Lang proof with $f(z) = e^z$ and $g(z) = e^{\beta z}$, forgetting that these are algebraically dependent. With $D = [\mathbb{Q}(\alpha,\beta,\gamma) : \mathbb{Q}] = [\mathbb{Q}(\alpha,\gamma) : \mathbb{Q}]$, we choose any $S > 16\rho D = 16D$ and find $\phi(z) = P(f(z), g(z))$ as in Lemma 19.4. We pretend that this is not identically zero. Then the subsequent arguments show that after all $\phi = 0$ if T and $L = [\sqrt{2DST}]$ are sufficiently large depending on α, β, γ. This gives the linear dependence over $\mathbb{C}$ of certain $f(z)^i g(z)^j = e^{\lambda z}$ for $\lambda = i + j\beta$; and here $0 \le i,j \le L$. Thus β is rational, as indeed it always was, but with numerator and denominator of absolute value at most L. In particular the logarithmic height $h(\beta) \le \log L$.

We seem to be going round in circles, as L depends on β! But in fact an analysis of the various constants occurring shows that $L \le c(h(\beta)+2)^\kappa$, where c and κ depend only on the given α, γ. Therefore

$$h(\beta) \leq \log c + \kappa \log(h(\beta) + 2).$$

This implies $h(\beta) \leq C$ for C also depending only on α, γ. We can make a finite list of such $\beta = r/s$, and then test if $\log \gamma / \log \alpha = \beta$ via $\alpha^r = \gamma^s$.

I hope this sketch is not too short; I have never forgotten Littlewood's dictum: *two* trivialities omitted can add up to an *impasse*.

See also Exercise 13.15 for a (pre-height) version more explicit in its dependence on α, γ.

As mentioned earlier, this is not the easiest method. But up to now it is the only method to generalize to the elliptic and abelian analogues, and there it gives valuable information about isogenies, endomorphisms and even polarizations. See the account in the book of Baker and Wüstholz (2007).

Exercises

19.1 Show that $\pi \neq \frac{12}{\sqrt{190}} \log((2\sqrt{2} + \sqrt{10})(3 + \sqrt{10}))$.

19.2 Let $J = J(z) = \sum_{k=0}^{\infty} z^k/k!^2$ be a normalized Bessel function. Using Schneider–Lang, show that there are at most 8 rational numbers $w \neq 0$ with $J(w)$ and $J'(w)$ rational [*Hint*: $zJ'' + J' - J = 0$ and $k!^2 \geq 2^{-2k}(2k)!$ and $f_1 = z^{-1}$]. In fact $J(\alpha)$ is transcendental for all algebraic $\alpha \neq 0$, and even $J(\alpha), J'(\alpha)$ are algebraically independent (Siegel, 1929).

19.3 Show that

$$\frac{\log(30122754096401 + 21300003689580\sqrt{2})}{\log(1607521 + 1136689\sqrt{2})} = \frac{36}{17}$$

[*Hint*: for $\gamma = 30122754096401 + 21300003689580\sqrt{2}$ and $\alpha = 1607521 + 1136689\sqrt{2}$ we have $\text{Nm}(\gamma) = 1$ and $\text{Nm}(\alpha) = -1$; further $\left|\frac{\log \gamma}{\log \eta} - 36\right| < 1$ and $\left|\frac{\log \alpha}{\log \eta} - 17\right| < 1$ with $\eta = 1 + \sqrt{2}$]. See Exercise 13.11.

19.4 Show that

$$\frac{\log(30122754096402 + 21300003689580\sqrt{2})}{\log(1607521 + 1136689\sqrt{2})} > \frac{36}{17}$$

is transcendental. See Exercise 13.12.

19.5 Let $f(z) = e^z$, $g(z) = e^{e^z - 1}$.

(a) Show that f', g' are in $\mathbb{Q}[f, g]$.

(b) Show that f, g are algebraically independent over $\mathbb{C}$.

(c) Show that $\{z \in \mathbb{C};\ f(z), g(z) \in \mathbb{Q}\}$ is infinite [*Hint*: $f = 1$].

(d) Why no contradiction with Schneider–Lang?

19.6 Let f be entire.

(a) If f has order at most ρ show that f' also has order at most ρ [*Hint*: Cauchy].

(b) If f' has order at most $\rho > 0$ show that f also has order at most ρ.

19.7 For the derivatives show that

$$(f_1 \cdots f_n)^{(t)} \quad = \quad \sum_{t_1+\cdots+t_n=t} \frac{t!}{t_1! \cdots t_n!} f_1^{(t_1)} \cdots f_n^{(t_n)} \qquad (*)$$

with

$$\sum_{t_1+\cdots+t_n=t} \frac{t!}{t_1! \cdots t_n!} = n^t.$$

19.8 Let G and $H \neq 0$ be entire functions and let $T \geq 0$ be in $\mathbb{Z}$ such that $\phi = G/H$ has a zero at $z = 0$ of order at least T, and let $R \geq 1$ be in $\mathbb{R}$.

(a) If $H(0) \neq 0$ show that

$$|\phi^{(T)}(0)| \ \leq \ \frac{1}{|H(0)|} \frac{T!}{R^T} |G|_R$$

[*Hint*: $(H\phi)^{(T)}$].

(b) If $H(0) = 0$ show that

$$|\phi^{(T)}(0)| \ \leq \ \frac{1}{|H^{(p)}(0)|} \frac{p!T!}{R^{p+T}} |G|_R$$

with $p = \mathrm{ord}_0 H$.

Thus "$\mathrm{ord}_0\, \phi \geq T$ implies $\phi^{(T)}(0)$ small", even when ϕ is not entire.

19.9 A two-dimensional version of Schneider–Lang implies the following. If $f_1, \ldots, f_m$ are entire on $\mathbb{C}^2$ with finite order (for example polynomials P in $\mathbb{C}[z_1, z_2]$ or $\exp(P)$ too), with at least three of $f_1, \ldots, f_m$ algebraically independent over $\mathbb{C}$, and K is a number field with

$$\frac{\partial f_1}{\partial z_1}, \frac{\partial f_1}{\partial z_2}, \ldots, \frac{\partial f_m}{\partial z_1}, \frac{\partial f_m}{\partial z_2}$$

in $K[f_1, \ldots, f_m]$, then there do not exist $\mathbf{u}, \mathbf{v}$ in $\mathbb{C}^2$, linearly independent over $\mathbb{C}$, such that all values of $f_1, \ldots, f_m$ on $\mathbb{Z}\mathbf{u} + \mathbb{Z}\mathbf{v}$ are in K. Deduce that

$$\int_0^1 \frac{dX}{X^3+1} \ = \ \frac{\pi\sqrt{3}}{9} + \frac{\log 2}{3}$$

is transcendental (originally Baker (1967b) and later by Bertrand and Masser in 1981 – see Masser (1981) for details) [*Hint*: $f_1 = e^{z_1}, f_2 = e^{z_2}, f_3 = z_1 + i\sqrt{3}z_2$ with $\mathbf{u} = (\pi i, \log 2), \mathbf{v} = (-3\log 2, \pi i)$. Don't forget the algebraic independence!].

* * * * * * * * * * * * * * * * * * *

19.10 Let α_1, α_2 be non-zero algebraic numbers with logarithms $\log\alpha_1, \log\alpha_2$. Using Schneider–Lang, show that these logarithms are linearly independent over the field of algebraic numbers if and only they are linearly dependent over $\mathbb{Q}$.

19.11 For $B = B(z) = \sum_{k=0}^{\infty} z^k/k!^{17}$ and rational $w \neq 0$ show that at least one of $\frac{d^tB}{dz^t}(w)$ $(t = 0, 1, \ldots, 16)$ is irrational.

19.12 Show that

$$\frac{\sum_{k=1}^{\infty}\frac{1}{k^4}}{\sum_{k=1}^{\infty}\frac{1}{k^4\binom{2k}{k}}} = \frac{\sum_{k=1}^{\infty}\frac{1}{k^4}}{2\int_0^{\pi/3} x(\log(2\sin\frac{x}{2}))^2 dx} = \frac{36}{17}.$$

19.13 Let f, g be entire, with f/g also entire. Show that $|f/g|_r|fg|_r^6 \leq |f|_{5r}^7|g|_{5r}^5$ (Masser, 1983). Thus if f, g have order at most ρ, then so does f/g.

19.14 Show that $J(1)/J'(1) = [1; 2, 3, 4, 5, 6, 7, 8, \ldots]$ (thus the continued fraction is transcendental).

19.15 Let

$$E(z) = 1 - 24\sum_{k=1}^{\infty}\frac{ke^{kz}}{1-e^{kz}},\quad F(z) = 1 + 240\sum_{k=1}^{\infty}\frac{k^3e^{kz}}{1-e^{kz}},\quad G(z) = 1 - 504\sum_{k=1}^{\infty}\frac{k^5e^{kz}}{1-e^{kz}}.$$

Show that

$$E' = \frac{1}{12}(E^2 - F),\quad F' = \frac{1}{3}(EF - G),\quad G' = \frac{1}{2}(EG - F^2)$$

[*Hint*: modular forms]. Unfortunately E, F, G are not meromorphic on $\mathbb{C}$.

19.16 If f has order at most ρ, must f'/f have order at most ϵ for each $\epsilon > 0$?

19.17 Show that $f(z) = \prod_{k=1}^{\infty}(1+\frac{z}{k})e^{-z/k}$ has order at most $1+\epsilon$ for each $\epsilon > 0$. This is false for $\epsilon = 0$. (With Euler's $\gamma = \lim_{m\to\infty}(1 + \frac{1}{2} + \cdots + \frac{1}{m} - \log m)$ the classical gamma function is $\Gamma(z) = \frac{1}{ze^{\gamma z}f(z)}$, and $\Gamma(n) = (n-1)!$ for all n in $\mathbb{N}$; further $\frac{\Gamma(p)\Gamma(q)}{\Gamma(p+q)} = \int_0^1 x^{p-1}(1-x)^{q-1}dx$ for all real $p, q > 0$).

19.18 Show that

$$f(z) = \prod_{k=1}^{\infty}\prod_{l=1}^{\infty}\left(1 - \frac{z}{k+li}\right)e^{\frac{z}{k+li}+\frac{1}{2}\frac{z^2}{(k+li)^2}}$$

has order at most $2+\epsilon$ for each $\epsilon > 0$. Is this false for $\epsilon = 0$? Compare with (10.1) a long time ago.

19.19 For $f_1, \ldots, f_n$ in $\mathbb{F}_p[x]$ the claim $(*)$ in Exercise 19.7 is trivial if $t > pn$, because $f^{(p)} = 0$. How can one restore non-triviality?

19.20 There is a quotient $\wp = \wp(z)$ (a Weierstrass elliptic function, not constant) of entire functions with differential equation $\wp'^2 = 4\wp^3 - 4\wp$. For $t = 0, 1, 2, \ldots$ show that there is P_t in $\mathbb{Z}[x, y]$, with total degree at most $t+1$ and coefficients of absolute value at most $t!9^t$, such that $\wp^{(t)} = P_t(\wp, \wp')$.

19.21 Find a differential equation for $Q = -1/\wp$ in Exercise 19.20; and explain. Compare also Exercise 4.29(d).

19.22 For D in $\mathbb{N}$ show that there are C, κ depending only on D with the following property. If $\alpha \neq 0, \gamma \neq 0$ in $\mathbb{C}$ are algebraic with $D = [\mathbb{Q}(\alpha, \gamma) : \mathbb{Q}]$, $\log\alpha \neq 0$ and $\frac{\log\gamma}{\log\alpha}$ algebraic, then there are r, s in $\mathbb{Z}$ with $\alpha^r = \gamma^s$ and

$$0 < \max\{|r|, |s|\} \leq C\max\{h(\alpha), h(\gamma)\}^{\kappa}$$

(compare Exercise 13.15).

19.23 Let $\alpha_1 \neq 0, \alpha_2 \neq 0, \beta$ be algebraic with $[\mathbb{Q}(\beta) : \mathbb{Q}] = 2$. Show that $\log\alpha_1 + \beta\log\alpha_2$ is transcendental [*Hint*: Exercise 19.9].

19.24 Guess some Schneider–Lang in $\mathbb{C}^3$ and deduce that

$$\int_0^1 \frac{\mathrm{d}X}{X^3 - 3X - 1} = \beta_1\log\alpha_1 + \beta_2\log\alpha_2 + \beta_3\log\alpha_3$$

is transcendental, where

$$\alpha_1 = 4 - \alpha^2,\ \alpha_2 = 2 + \alpha,\ \alpha_3 = -\alpha + \alpha^2,$$

$$\beta_1 = \frac{-4 - \alpha + 2\alpha^2}{9},\ \beta_2 = \frac{2 - \alpha - \alpha^2}{9},\ \beta_3 = \frac{2 + 2\alpha - \alpha^2}{9}$$

and $\alpha^3 - 3\alpha - 1 = 0$ with

$\alpha = 1.879385241571816768108218554649462939872416268528929266 1\ldots$

19.25 Prove Schneider–Lang with $2\rho[K : \mathbb{Q}]$ instead of $16\rho[K : \mathbb{Q}]$ (Waldschmidt, 1974).

19.26 Let f, g be entire functions, algebraically independent over $\mathbb{C}$, with f', g' in $\mathbb{C}[f, g]$. Show that there is C such that $\text{ord}_{z=w} P(f, g) \leq CL^2$ for each w in $\mathbb{C}$ and each non-zero P in $\mathbb{C}[X, Y]$ with degree at most L in each variable (Brownawell and Masser, 1980).

19.27 Let $\alpha \neq 0, \beta \neq 0$ be algebraic with any determination $\log \alpha$. Show that $\alpha^\beta = \exp(\beta \log \alpha)$ is rational if and only if there are integers $n \geq 1, m, r, s \geq 1$ and rational $\rho > 0$ with real $\log \rho$ such that $s\beta = r, n \log \alpha = \log \rho + 2\pi im$ (so that α^n is in $\mathbb{Q}$ – see Exercise 11.17(b) about that) with ns dividing $2mr$ and also $r(\text{ord}_p \rho)$ for every prime p.

19.28 We have seen in Chapter 16 that for each positive integer D, there is $\eta = \eta(D) > 0$ such that $h(\gamma) \geq \eta$ for any non-zero γ, not a root of unity, in any number field K with $[K : \mathbb{Q}] = D$. Suppose α, β are non-zero in such a K and there are n, m in $\mathbb{Z}$, not both zero, with $\alpha^n \beta^m = 1$ (that is, α and β are multiplicatively dependent). Show that there are such n, m with

$$|n| \leq \omega \max\left\{1, \frac{h(\beta)}{\eta}\right\}, \quad |m| \leq \omega \max\left\{1, \frac{h(\alpha)}{\eta}\right\}$$

where $\omega = \omega(D)$ as in Exercise 11.17(a) [*Hint*: if $n \neq 0$ take $\phi = m/n$ in (8.2), note $(\alpha^s \beta^r)^n = \beta^{rn-sm}$ and compare heights, with group theory in $\mathbb{Z}^2$]. Compare Exercise 19.22.

19.29 Show that $\exp(w, z) = (e^w, e^z)$ defines a homomorphism from the group $\mathbb{C} \times \mathbb{C}$ to the group $G = \mathbb{C}^* \times \mathbb{C}^*$ whose kernel is $2\pi i\mathbb{Z} \times 2\pi i\mathbb{Z}$ (this G is the group variety $\mathbb{G}_\text{m} \times \mathbb{G}_\text{m}$ with its exponential map).

19.30 Show that $\exp(w, z_1, \ldots, z_n) = (w, e^{z_1}, \ldots, e^{z_n})$ defines a homomorphism from the group $\mathbb{C} \times \mathbb{C}^n$ to the group $G = \mathbb{C} \times \mathbb{C}^{*n}$ whose kernel is $\{0\} \times (2\pi i\mathbb{Z})^n$ (this G is the group variety $\mathbb{G}_\text{a} \times \mathbb{G}_\text{m}^n$ with its exponential map).

20
Elliptic functions

We take another break from estimating; however the reader will have to put up with a lot of formulae, and especially the algebraic geometers are warned that some of them are of a rather explicit nature. Many of them can be found in Whittaker and Watson (1965), a classical volume here abbreviated to WW (but their omegas are half ours). We will say something later about Greenhill (1892), but nothing about Eagle (1958) (where one can see $!n$ and learn the real meaning of τ).

In Chapter 10 we saw a Weierstrass sigma function, which now we write as

$$\sigma(z) = z\prod_{\omega}\left(1-\frac{z}{\omega}\right)\exp\left\{\frac{z}{\omega}+\frac{1}{2}\left(\frac{z}{\omega}\right)^2\right\}.$$

There the product was taken with regard to $\mathbb{Z}+i\mathbb{Z}$. But more generally one can fix a lattice Ω in $\mathbb{C}$, that is, an additive subgroup of rank 2 which is discrete, and take the product over all $\omega \neq 0$ in Ω. The resulting function $\sigma = \sigma_\Omega$ (WW 20.42) is independent of the ordering of terms in the product, but to prove things it is necessary to fix an ordering, often by means of $|\omega|$ or by choosing basis elements ω_1, ω_2, so that $\Omega = \mathbb{Z}\omega_1 + \mathbb{Z}\omega_2$. In fact we are not going to prove many of the classical things about elliptic functions here. Anyway, we get an entire function. We will show later that it grows at worst like $cC^{|z|^2}$, so that $\rho = 2$ in the terminology of the previous chapter. In fact the correct value of C is something at least $\exp(\pi/2A)$, where A is the so-called determinant $\det\Omega$ of Ω, the area of a so-called fundamental parallelogram, given by $2A = |\overline{\omega}_1\omega_2 - \omega_1\overline{\omega}_2|$ (see Exercise 20.98).

There is a differential equation but of order 3 (see also Exercise 20.20(c) for a "universal" one of order 5) which is too big to lead to anything interesting in Theorem 19.1. We can reduce the order by introducing the Weierstrass zeta function $\zeta = \zeta_\Omega$ defined by

$$\zeta(z) = \frac{\sigma'(z)}{\sigma(z)} = \frac{1}{z} + \sum_{\omega}\left(\frac{1}{z-\omega} + \frac{1}{\omega} + \frac{z}{\omega^2}\right)$$

(WW 20.4) and clearly analytic everywhere except at lattice points of Ω. Then the order is 2, still too big and in fact we know nothing at all about the transcendence of $\zeta(\alpha)$ for algebraic α.

One step further gives the Weierstrass function $\wp = \wp_\Omega$

$$\wp(z) = -\zeta'(z) = \frac{1}{z^2} + \sum_{\omega}\left(\frac{1}{(z-\omega)^2} - \frac{1}{\omega^2}\right) \qquad (20.1)$$

(WW 20.2). Then the order is 1, and we have the famous differential equation

$$\wp'(z)^2 = 4\wp(z)^3 - g_2\wp(z) - g_3 \qquad (20.2)$$

involving the so-called invariants

$$g_2 = 60\sum_{\omega}\frac{1}{\omega^4}, \quad g_3 = 140\sum_{\omega}\frac{1}{\omega^6}$$

(WW 20.22) but not directly

$$\sum_{\omega}\frac{1}{\omega^8} = \frac{3}{7}\left(\sum_{\omega}\frac{1}{\omega^4}\right)^2$$

(see Exercise 20.9).

The function $\wp(z)$ can be considered as a natural analogue of e^z (see Exercise 20.100). As hinted at the end of Chapter 13, the latter is the exponential map on the group variety $\mathbb{G}_{\mathrm{m}}$. The former is the essential part of the exponential map on the commutative group variety E associated with (20.2) and $Y^2 = 4X^3 - g_2X - g_3$ (under an addition which will soon be described analytically). Of course this is a commutative group curve, an elliptic curve.

We will not need the following until the next chapter, but it is easy to deduce from (20.2) that the coefficients of the Laurent expansion of $\wp(z)$ at $z = 0$ lie in $\mathbb{Q}[g_2, g_3]$. For example the first few terms are

$$\frac{1}{z^2} + \frac{g_2}{20}z^2 + \frac{g_3}{28}z^4 + \frac{g_2^2}{1200}z^6 + \frac{3g_2g_3}{6160}z^8 + \frac{49g_2^3 + 750g_3^2}{7644000}z^{10} + \cdots \qquad (20.3)$$

(WW 20.22). One can also show that $g_2^3 \neq 27g_3^2$ (WW 20.32). This means that the cubic polynomial $4X^3 - g_2X - g_3$ has three different complex zeroes. One can show that these are the values $\wp(\omega/2)$ as ω runs through all elements of Ω not in 2Ω (WW 20.32).

In this respect the example $\Omega = \omega_0(\mathbb{Z}+i\mathbb{Z})$ is significant, because by pairing ω with $i\omega$, which satisfies $(i\omega)^6 = -\omega^6$, we see that $g_3 = 0$. Thus $g_2 \neq 0$;

and by fixing ω_0 we can make g_2 take any desired value. A traditional choice is $g_2 = 4$. Similarly with $\Omega = \omega_0^*(\mathbb{Z} + i^*\mathbb{Z})$ for $i^* = e^{2\pi i/3}$ and $\omega^4 + (i^*\omega)^4 + (i^{*2}\omega)^4 = 0$ we get $g_2 = 0$ and we can take $g_3 = 4$.

Regarding (20.2) as an analogue of $(e^z)' = e^z$, we can aim for the elliptic analogue of Hermite–Lindemann by taking one of the $f_i(z)$ in Schneider–Lang as z and another as say $f(z) = \wp(z)$. We linearize with the help of $g(z) = \wp'(z)$, so that $f' = g$ and

$$g' = 6f^2 - \frac{1}{2}g_2;$$

in fact to deduce the second of these we must know that $\wp' \neq 0$, which is pretty clear from the poles in (20.1).

But we have to check the analytic growth of the new functions. From the above we get

$$\wp = \frac{\sigma'^2 - \sigma\sigma''}{\sigma^2}, \quad \wp' = \frac{3\sigma\sigma'\sigma'' - 2\sigma'^3 - \sigma^2\sigma'''}{\sigma^3}$$

and we can settle the numerators using Cauchy's (12.29) to get $|F'|_r \leq |F|_{r+1}$ and so on. Thus $\rho = 2$ as before.

To deduce transcendence results for $\wp$ using Schneider–Lang we will as before consider the functions at nz. Already $n = 0$ makes trouble in $\wp(nz)$. In fact we can squeeze by with just $n = 2$ and the classical duplication formula

$$\wp(2z) = \frac{16\wp(z)^4 + 8g_2\wp(z)^2 + 32g_3\wp(z) + g_2^2}{16(4\wp(z)^3 - g_2\wp(z) - g_3)} \tag{20.4}$$

amazingly not explicitly in Whittaker and Watson (but see (20.10) below and Exercise 20.6). We may remark that the resultant of the polynomials in the numerator and denominator is

$$2^{28}(g_2^3 - 27g_3^2)^2 \neq 0.$$

This implies that if the denominator vanishes at some $z = z_0$, then the numerator does not, and so $2z_0$ is in the set Ω of poles of $\wp$. Thus (20.4) holds for all z not in $\Omega/2$.

By differentiation we deduce that $\wp'(2z)/\wp'(z)$ is

$$\frac{64\wp(z)^6 - 80g_2\wp(z)^4 - 320g_3\wp(z)^3 - 20g_2^2\wp(z)^2 - 16g_2g_3\wp(z) + g_2^3 - 32g_3^2}{32(4\wp(z)^3 - g_2\wp(z) - g_3)^2}, \tag{20.5}$$

which I never saw before in my life (but see Greenhill (1892) p. 174). Here the resultant is $2^{66}(g_2^3 - 27g_3^2)^6$, so a similar remark applies (note that $\wp'(z) \neq 0$ for all z not in $\Omega/2$).

In fact the trouble with $n = 0$ above can be avoided by taking a hint from the formal group to use instead $f = 1/\wp$ which is analytic at $z = 0$, even with a double zero there. Then with $g = f'$ we have

$$g' = -2g_3f^3 - \frac{3}{2}g_2f^2 + 2.$$

These preparations are almost enough to do transcendence. We need just

Lemma 20.1 *The functions $z, \wp(z)$ are algebraically independent over $\mathbb{C}$.*

Proof As in the proof of the analogous Lemma 13.3 on z, e^z there are many ways. It is intuitive that an algebraic function cannot have infinitely many poles. Or there we suggested using the periodicity $e^{z+2\pi i} = e^z$ as we used $f(z^2) = f(z) - z$ in the proof of Lemma 3.2 on the Mahler function. Actually here we have $\wp(z + \omega) = \wp(z)$ for all ω in Ω (WW 20.21) and so rather double periodicity (in this chapter I am a bit faster than Greenhill, in which Greenhill (1892), according to Hardy in connexion with Ramanujan, "complex variable and double periodicity are not mentioned until p. 254, and the double periodicity is deduced somehow from properties of Cartesian ovals"). Thus we can leave the proof to the reader (see also Exercise 20.7). We will anyway give the details for something a bit more delicate when we come to the elliptic analogue of Gelfond–Schneider. □

We can now prove elliptic Hermite–Lindemann, which corresponds to the commutative group surface $\mathbb{G}_a \times E$, with $(w, \wp(z))$ the essential part of the exponential map.

Theorem 20.2 *Suppose that g_2, g_3 are algebraic, and that $\alpha \neq 0$ is algebraic in $\mathbb{C}$. Then α is not in Ω and $\wp(\alpha)$ is transcendental.*

Proof If α were in Ω then we use $f_1 = z, f_2 = 1/\wp, f_3 = f_2'$, which for every integer n take the values $n\alpha, 0, 0$ at $n\alpha$, all in the number field $K = \mathbb{Q}(g_2, g_3, \alpha)$. This contradicts Theorem 19.1. Of course replacing α by a rational submultiple, we deduce that it is not even a rational multiple of a period.

Similarly if $\wp(\alpha)$ were algebraic then we could use $f_1 = z, f_2 = \wp, f_3 = f_2'$ and their values at $n\alpha$ for any $n = 2^r$. By repeated use of (20.4) we see that the $f_2(n\alpha) = \wp(2^r\alpha)$ all lie in $K_1 = \mathbb{Q}(g_2, g_3, \alpha, \wp(\alpha))$. Also

$$f_3(n\alpha) = \wp'(2^r\alpha) = \wp'(\alpha)\prod_{s=0}^{r-1}\frac{\wp'(2^{s+1}\alpha)}{\wp'(2^s\alpha)},$$

as $\wp'(2^s\alpha) \neq 0$ because also $2^s\alpha$ is not in $\Omega/2$. So using (20.5) we end up in $K = K_1(\wp'(\alpha))$ which is a number field by (20.2). □

Already we have some easy consequences, such as the transcendence of any non-zero period, or the transcendence of any u not in Ω such that $\wp(u)$ is algebraic, a so-called elliptic logarithm of an algebraic number.

Classically such numbers arise from integrals. Formally the differential equation gives

$$\int_{\wp(u_*)}^{\wp(u)} \frac{\mathrm{d}X}{\sqrt{4X^3 - g_2X - g_3}} = u - u_* \tag{20.6}$$

after the substitution $X = \wp(z)$. However one has to specify the path as well as the choice of square root, and we shall be vague about this. Formally with u, u_* as half-periods, and for example $\Omega = \omega_0(\mathbb{Z}+i\mathbb{Z})$ as above with $g_2 = 4, g_3 = 0$, we find the transcendence of

$$\int_0^1 \frac{\mathrm{d}X}{\sqrt{X - X^3}} = \frac{1}{2}\int_0^1 \frac{\mathrm{d}t}{t^{3/4}(1-t)^{1/2}} = \frac{1}{2}B\left(\frac{1}{4}, \frac{1}{2}\right)$$

for the Euler beta function. Actually this can be taken as ω_0 (and then $\wp(\omega_0/2) = 1$ – see Exercise 20.52). In terms of the Euler gamma function it is

$$\omega_0 = \frac{\Gamma(\frac{1}{4})\Gamma(\frac{1}{2})}{2\Gamma(\frac{3}{4})} = \frac{\Gamma(\frac{1}{4})^2}{2\sqrt{2\pi}} \tag{20.7}$$

and so we get the transcendence of

$$\frac{\Gamma(\frac{1}{4})}{\pi^{1/4}}. \tag{20.8}$$

Or formally with $u_* = 0$ we get

$$\int_{\infty}^{\wp(u)} \frac{\mathrm{d}X}{\sqrt{4X^3 - g_2X - g_3}} = u. \tag{20.9}$$

This suggests two things. Suppose we start off with any complex numbers g_2, g_3 with $g_2^3 \neq 27g_3^2$, instead of a lattice Ω. Then $4X^3 - g_2X - g_3$ is squarefree and the integral has no convergence problems. Thus we may be able to define $\wp(u)$ as the solution of (20.9). And then we may be able to solve $\wp(u) = x$ for u just by integrating. Indeed both these things are known to be true; see for example Silverman (1992) chapter 6, especially section 5, and also a hint in our own Chapter 21.

Thus we deduce the transcendence of integrals like

$$\int_2^{\infty} \frac{\mathrm{d}X}{\sqrt{X^3 - X}},$$

which are not named after anyone. As mentioned, one can think of $\wp$ as an elliptic analogue of the exponential function, which is why it is called an elliptic logarithm (of 2).

And what about integrals like

$$I = \int_2^3 \frac{\mathrm{d}X}{\sqrt{X^3 - X}}?$$

Here we need a subtler version of the duplication formula, namely the addition formula

$$\wp(z_1 + z_2) = -\wp(z_1) - \wp(z_2) + \frac{1}{4}\left(\frac{\wp'(z_1) - \wp'(z_2)}{\wp(z_1) - \wp(z_2)}\right)^2 \qquad (20.10)$$

(WW 20.31). This corresponds to the addition law on the group curve E. For example to get the transcendence of $I/2$ above, we note that by (20.6) it is $u - u_*$ with $\wp(u) = 3$ and $\wp(u_*) = 2$. But then we find $\wp(-u_*) = 2$ as $\wp$ is even, then $\wp'(u_*) = \pm\sqrt{24}$, then $\wp'(u) = \pm\sqrt{96}$, and finally $\wp(u - u_*) = 1$ or 49 (see Exercise 20.53). Thus indeed I is transcendental.

Similarly taking $g_2 = 28, g_3 = -24$ we get the transcendence of

$$\int_1^2 \frac{\mathrm{d}X}{\sqrt{X^3 - 7X + 6}}, \quad \int_1^{-3} \frac{\mathrm{d}X}{\sqrt{X^3 - 7X + 6}}, \quad \int_4^5 \frac{\mathrm{d}X}{\sqrt{X^3 - 7X + 6}}.$$

As the cubic is $(X - 1)(X - 2)(X + 3)$ the first two correspond to periods, so we need only check (for example numerically) that they are non-zero. The third might conceivably be a period, but then again it suffices to check non-vanishing.

This completes our discussion of the elliptic analogue of Hermite–Lindemann.

For elliptic Gelfond–Schneider we have to investigate the algebraic dependence of $\wp(z), \wp(\beta z)$. We saw in the previous chapter that if $e^z, e^{\beta z}$ are algebraically dependent then β is rational. This fails in the elliptic case. For example with $\Omega = \omega_0(\mathbb{Z} + i\mathbb{Z})$ as above we have $i\Omega = \Omega$. Thus we can replace ω by $i\omega$ in (20.1) to see that $\wp(iz) = -\wp(z)$. Or for $\Omega = \omega_0^*(\mathbb{Z} + i^*\mathbb{Z})$ we get $\wp(i^* z) = i^*\wp(z)$.

Similar examples can be constructed if there is non-zero μ with $\mu\Omega \subseteq \Omega/m$ for some positive integer m. It can be shown (see for example Silverman (1992) chapter 6, p.164) that the set of all such multipliers μ together with 0 is either $\mathbb{Q}$ or a complex quadratic extension of $\mathbb{Q}$ (the so-called case of complex multiplication CM for short). In the examples above this is $\mathbb{Q}(i)$ or $\mathbb{Q}(i^*)$, but it can be arbitrary. In general we write $\mathfrak{K} = \mathfrak{K}_\Omega$ for this field.

Lemma 20.3 *For any β not in $\mathfrak{K}$ the functions $\wp(z), \wp(\beta z)$ are algebraically independent over $\mathbb{C}$.*

Proof Here we shall have to use the dreaded double periodicity. We work over the field $\mathcal{K}$ of all meromorphic functions. Suppose there is $P \neq 0$ in $\mathbb{C}[X, Y]$ with $P(\wp(z), \wp(\beta z)) = 0$. Because $\wp$ is not constant, the polynomial $Q(Y) = P(\wp(z), Y)$ in $\mathcal{K}[Y]$ is not identically zero. Pick any ω in Ω. Then for any integer m

$$0 = P(\wp(z + m\omega), \wp(\beta(z + m\omega))) = Q(\wp(\beta z + \beta m\omega)).$$

Thus there are different integers m', m'' with $\wp(\beta z+\beta m'\omega) = \wp(\beta z+\beta m''\omega)$. Comparing poles we see that $\beta(m' - m'')\omega$ lies in Ω. Now taking ω as basis elements of Ω we conclude that β is in $\mathfrak{K}$, a contradiction. □

Although we don't need it in the present chapter, we prove here a converse to the above.

Lemma 20.4 *For any $\beta \neq 0$ in $\mathfrak{K}$ the functions $\wp(z), \wp(\beta z)$ are algebraically dependent over $\mathbb{C}$.*

Proof This is also with an auxiliary polynomial, somewhat in the style of the second example in the Prologue. There is b in $\mathbb{N}$ with $\beta\Omega \subseteq \Omega/b$. It follows that $\wp(bz), \wp(b\beta z)$ are both doubly periodic on Ω. Pick any z_0 at which both of these are analytic. There is $c_0 > 0$ such that, for all R sufficiently large, at least c_0R^2 of the elements w of $z_0 + \Omega$ satisfy $|w| \leq R$.

For any L in $\mathbb{N}$ we can find a non-zero P in $\mathbb{C}[X, Y]$, of degree at most L in each variable, such that $\phi(z) = P(\wp(bz), \wp(b\beta z))$ has a zero of order at least $T = L^2 + 2L$ at z_0. By double periodicity it also has a zero of order at least T at each point w above. Now with $\Sigma(z) = \sigma(bz)^{2L}\sigma(b\beta z)^{2L}$ we see that $\Phi = \Sigma\phi\prod_w(z - w)^{-T}$ is entire. From $|\Phi|_R \leq |\Phi|_{5R}$ we get in the usual way

$$|\Sigma\phi|_R \leq 2^{-c_0TR^2}|\Sigma\phi|_{5R} \leq 2^{-c_0L^2R^2}2^{cLR^2}$$

for c independent of L and R. Fixing $L \geq c/c_0$ and making $R \to \infty$ we see that $\Sigma\phi$ must be constant. In view of its zeroes it must be zero, and now $\phi = 0$ gives the required algebraic dependence. □

We can now prove elliptic Gelfond–Schneider, which corresponds to the commutative group surface $E \times E$, with $(\wp(w), \wp(z))$ the essential part of the exponential map.

Theorem 20.5 *Suppose that g_2, g_3 are algebraic, and that β is algebraic in $\mathbb{C}$ but not in $\mathfrak{K}$. If u is a complex number not in Ω such that $\wp(u)$ is algebraic, then βu is not in Ω and $\wp(\beta u)$ is transcendental.*

Proof We start by showing by contradiction that βu is not in Ω.

First suppose that u is a rational multiple of a period, say ω/m for some period ω not in $m\Omega$. We use

$$f_1 = \frac{1}{\wp(z)},\ f_2 = \frac{1}{\wp(\beta z)},\ f_3 = f_1',\ f_4 = f_2'$$

(with f_1, f_2 algebraically independent by Lemma 20.3) at $z = nu$ with n divisible by m; all the values are then zero and we contradict Theorem 19.1.

If on the other hand u is not a rational multiple of a period, then we use $f_1 = \wp(z)$, $f_2 = 1/\wp(\beta z)$ and their derivatives at $z = 2^r u$ to get a contradiction.

Thus indeed βu is not in Ω; and as above not even a rational multiple of a period.

To finish the proof we use again the two cases above. If $u = \omega/m$ then we use $f_1 = 1/\wp(mz)$, $f_2 = \wp(\beta z)$, which stay algebraically independent because β/m is also not in $\mathfrak{K}$. Together with their derivatives they take algebraic values at $z = 2^r u$. Otherwise just $f_1 = \wp(mz)$, $f_2 = \wp(\beta z)$ at $z = 2^r u$. □

Just as for exponential Gelfond–Schneider we have some easy consequences, such as the transcendence of the ratio of any non-zero periods provided this is not in $\mathfrak{K}$, or the transcendence of the ratio of any two u not in Ω such that $\wp(u)$ is algebraic also provided this ratio is not in $\mathfrak{K}$.

The first kind of ratio doesn't turn up when $\mathfrak{K} \neq \mathbb{Q}$. For example with $\Omega = \omega_0(\mathbb{Z} + i\mathbb{Z})$ as above the ratio is always in $\mathbb{Q}(i) = \mathfrak{K}$. But the second kind can again be expressed in terms of integrals. We deduce the transcendence for example of

$$\frac{\int_2^3 \frac{dX}{\sqrt{X^3-X}}}{\int_0^1 \frac{dX}{\sqrt{X^3-X}}},\quad \frac{\int_4^5 \frac{dX}{\sqrt{X^3-X}}}{\int_2^3 \frac{dX}{\sqrt{X^3-X}}},$$

provided they are not in $\mathbb{Q}(i)$. But how to check that? It looks a bit more difficult than the problem around (19.3)!

For the other example $g_2 = 28, g_3 = -24$ we have a new problem: what is $\mathfrak{K}$?

The famous conjecture of Gauss about class numbers of complex quadratic fields, proved independently by Baker and Stark after an earlier work of Heegner, implies that if g_2, g_3 are rational then $\mathfrak{K} \neq \mathbb{Q}$ if and only if the j-invariant

$$j = \frac{1728 g_2^3}{g_2^3 - 27 g_3^2}$$

(see Chapter 21) is one of

$$1728, 8000, 0, -3375, -32768, -884736, -884736000, -147197952000 \tag{20.11}$$

as well as -262537412640768000 (compare with (19.2) above) and

$$287496, 54000, 16581375, -12288000.$$

Ours is $148176/25$, so here $\mathfrak{K} = \mathbb{Q}$ (there are similar lists for each value of $d = [\mathbb{Q}(j) : \mathbb{Q}]$ up to 100, but one cannot do this so easily for general d).

Thus we can deduce the transcendence of say

$$\frac{\int_1^2 \frac{dX}{\sqrt{X^3-7X+6}}}{\int_1^{-3} \frac{dX}{\sqrt{X^3-7X+6}}}, \quad \frac{\int_2^3 \frac{dX}{\sqrt{X^3-7X+6}}}{\int_1^2 \frac{dX}{\sqrt{X^3-7X+6}}}, \quad \frac{\int_2^5 \frac{dX}{\sqrt{X^3-7X+6}}}{\int_2^3 \frac{dX}{\sqrt{X^3-7X+6}}},$$

provided they are not in $\mathbb{Q}$. But again how to check that? At least the first two are not in $\mathbb{R}$. But the third is, being $1.4353747\ldots$. Furthermore it is a fact, easily verified from the duplication formula and the addition formula, that

$$\frac{\int_2^{17} \frac{dX}{\sqrt{X^3-7X+6}}}{\int_2^3 \frac{dX}{\sqrt{X^3-7X+6}}} = 2$$

(see Exercise 20.55).

All these problems can be solved in a manner analogous to that discussed at the end of the previous chapter; we are now dealing with a quotient of elliptic logarithms.

So much for elliptic Gelfond–Schneider.

One can mix elliptic and exponential with the following result corresponding to the commutative group surface $\mathbb{G}_m \times E$, with $(e^w, \wp(z))$ the essential part of the exponential map.

Theorem 20.6 *Suppose that g_2, g_3 are algebraic, and that $\alpha \neq 0, \beta \neq 0$ are algebraic in $\mathbb{C}$. Then, for any non-zero determination of the logarithm, $\beta \log \alpha$ is not in Ω and $\wp(\beta \log \alpha)$ is transcendental.*

Proof We can now be brief. We leave it to the reader to check as in Lemma 20.3 the algebraic independence of $e^{\mu z}$ and $\wp(z)$ for any complex $\mu \neq 0$. In fact in the proof there it is enough to pick ω in Ω such that $\mu\omega$ is not in $2\pi i\mathbb{Q}$.

To prove that $\beta \log \alpha$ is not in Ω we use $f_1 = e^z$, $f_2 = 1/\wp(\beta z)$ at $z = n \log \alpha$.

Then to prove that $\wp(\beta \log \alpha)$ is transcendental we use $f_1 = e^z$, $f_2 = \wp(\beta z)$ at $z = 2^r \log \alpha$. □

As examples we get the transcendence of exotic animals like

$$\exp\left(2\int_0^1 \frac{dX}{\sqrt{X - X^3}}\right) = \exp\left(\frac{\Gamma(\frac{1}{4})^2}{\sqrt{2\pi}}\right), \quad \exp\left(\int_4^5 \frac{dX}{\sqrt{X^3 - 7X + 6}}\right)$$

or

$$\frac{1}{\pi}\int_0^1 \frac{\mathrm{d}X}{\sqrt{X-X^3}}, \quad \frac{1}{\log 2}\int_4^5 \frac{\mathrm{d}X}{\sqrt{X^3-7X+6}},$$

and so we get the transcendence of

$$\frac{\Gamma(\frac{1}{4})}{\pi^{3/4}}. \tag{20.12}$$

Next we move beyond Hermite–Lindemann and Gelfond–Schneider, to results that have no exponential analogue.

First with the zeta function. This is not doubly periodic, as we shall shortly see. But because $\zeta' = -\wp$ we see that for each period ω there is a so-called quasi-period $\eta = \eta(\omega)$ such that

$$\zeta(z+\omega) = \zeta(z) + \eta, \tag{20.13}$$

the so-called quasi-periodicity. If ω is not in 2Ω then putting $z = -\omega/2$ and using the fact that ζ is odd gives $\zeta(\omega/2) = \eta/2$.

We shall also need the famous Legendre relation. This says that if ω_1, ω_2 are any basis elements of Ω, then $\eta_1 = \eta(\omega_1)$, $\eta_2 = \eta(\omega_2)$ satisfy

$$\omega_2\eta_1 - \omega_1\eta_2 = \pm 2\pi i \tag{20.14}$$

(WW 20.411), where the sign (ignored by WW and others) depends on the relative positions of ω_1, ω_2 (see Exercise 20.33).

With the powerful method of auxiliary polynomials we have now developed so much machinery that we can dispose of a well-known conjecture in the negative.

Lemma 20.7 *The function $\zeta(z)$ has infinitely many non-real zeroes with real part different from $\frac{1}{2}$.*

Proof The function $f(z) = 1/\zeta(z)$ is analytic on a neighbourhood of the origin, with $f(0) = 0, f'(0) = 1$. So by the Open Mapping Theorem there is a neighbourhood N of the origin on which f takes all sufficiently small values w; and we can assume that the closure of N contains no zeroes of f except the origin.

Now the Legendre relation shows that η_1 and η_2 cannot both vanish. We choose $w = (n_1\eta_1 + n_2\eta_2)^{-1}$ for integers n_1 and n_2 such that $n_1\eta_1 + n_2\eta_2$ is sufficiently large. So there exists z_0 in N with $\zeta(z_0) = n_1\eta_1 + n_2\eta_2$. It follows that $\zeta(z) = 0$ for $z = z_0 - n_1\omega_1 - n_2\omega_2$.

It is now a simple matter to show that infinitely many of these zeroes z are not on the real line $\mathbb{R}$ or the critical line with imaginary part 1/2. In fact we can avoid any given finite set $\mathcal{S}$ of lines, as follows.

Without loss of generality we may assume that $\eta_1 \neq 0$. Also because ω_2 is not in $\mathbb{R}\omega_1$ we can fix n_2 such that the line $L = \mathbb{R}\omega_1 - n_2\omega_2$ is not in $\mathcal{S}$. Now as $n_1 \to \infty$ we have $n_1\eta_1 + n_2\eta_2 \to \infty$ as required above, and so $w \to 0$. This implies $z_0 \to 0$, and so z approaches the line L. But also $z \to \infty$; so for n_1 sufficiently large z cannot be on any line of $\mathcal{S}$. □

With slightly more effort it may be shown that there is a constant Z such that $\zeta(z)$ has a zero within every disc of radius Z.

We also need the appropriate analogue of Lemma 20.1.

Lemma 20.8 *For any complex μ the functions $\zeta(z) + \mu z, \wp(z)$ are algebraically independent over $\mathbb{C}$.*

Proof There is a period ω with $\eta + \mu\omega \neq 0$, where $\eta = \eta(\omega)$. In fact with ω_1, ω_2 as above, the equations $\eta_1 + \mu\omega_1 = \eta_2 + \mu\omega_2 = 0$ would contradict (20.14). We can now imitate the proof of Lemma 20.3 using just this single non-period ω of $f(z) = \zeta(z) + \mu z$, getting $m' \neq m''$ with $f(z + m'\omega) = f(z + m''\omega)$; but this comes down to $(m' - m'')(\eta + \mu\omega) = 0$. □

Finally we will use a quasi-duplication formula

$$\zeta(2z) = 2\zeta(z) + \wp'(z)\frac{12\wp(z)^2 - g_2}{4(4\wp(z)^3 - g_2\wp(z) - g_3)} \tag{20.15}$$

(WW ex. 20.23).

Here is our next transcendence result. This corresponds to a more subtle kind of commutative group surface E_{a}, with $(\wp(z), w + \zeta(z))$ the essential part of the exponential map (see Exercise 20.102). There is an exact sequence $0 \to \mathbb{G}_{\mathrm{a}} \to E_{\mathrm{a}} \to E \to 0$ (see Exercise 20.103).

Theorem 20.9 *Suppose that g_2, g_3 are algebraic, and that γ is algebraic in $\mathbb{C}$. If u is a complex number not in Ω such that $\wp(u)$ is algebraic, then $\zeta(u) + \gamma u$ is transcendental.*

Proof We start by proving that $x = \wp(\omega/m)$ is algebraic for any integer $m \geq 2$ and any ω in Ω not in $m\Omega$. For example if m is odd then there is $r \geq 1$ with $2^r \equiv 1$ modulo m, and we get the equation $R^{[r]}(x) = x$, where $R^{[r]}$ is the rational function R in (20.4) iterated r times. This equation is non-trivial because $\wp(2^r z) \neq \wp(z)$. If m is even with $m = 2^s m_1$ for odd $m_1 \geq 3$ then we get $R^{[s]}(x) = \wp(\omega/m_1)$ just shown to be algebraic. Finally if $m = 2^s$ there is maximal $r \leq s - 1$ with ω in $2^r\Omega$, and we use $R^{[s-r-1]}(x) = \wp(\omega'/2)$ with $\omega' = \omega/2^r$ in Ω but not in 2Ω.

Then also $\wp'(\omega/m)$ is algebraic by (20.2). All this can be done more efficiently with suitable multiplication formulae for $\wp(mz)$ generalizing (20.4). See for example Exercise 20.25.

In a similar way we next show that $t = \zeta(\omega/m) - \eta/m$ is algebraic for any $m \geq 2$ and any period ω not in $m\Omega$, with $\eta = \eta(\omega)$.

If m is odd, then there is $r \geq 1$ with $2^r \equiv 1$ modulo m, and repeated use of (20.15) shows that $\zeta(2^r z) - 2^r\zeta(z)$ is a rational function of $\wp(z), \wp'(z), \ldots,$ $\wp(2^{r-1}z), \wp'(2^{r-1}z)$ with algebraic coefficients. We can substitute $z = \omega/m$, and we get $(1 - 2^r)t$. If $m = 2^s m_1$ for odd $m_1 \geq 3$ then making the same substitution in $\zeta(2^s z) - 2^s\zeta(z)$ yields $-2^s t + t_1$ with $t_1 = \zeta(\omega/m_1) - \eta/m_1$ just shown to be algebraic, with a similar modification for $m_1 = 1$. As above, all this can be done more efficiently with suitable formulae for $\zeta(mz)$ generalizing (20.15).

Now suppose $\zeta(u) + \gamma u = \alpha$ is algebraic. We use

$$f_1(z) = \wp(z), \;\; f_2(z) = \zeta(z) + \gamma z, \;\; f_3(z) = \wp'(z) \tag{20.16}$$

with

$$f_1' = f_3, \;\; f_2' = -f_1 + \gamma, \;\; f_3' = 6f_1^2 - \frac{1}{2}g_2.$$

First suppose that $u = \omega/2$ for some period ω not in 2Ω. Then for every odd n we have

$$f_1(nu) = \wp(\omega/2), \;\; f_2(nu) = \zeta(n\omega/2) + \gamma n\omega/2 = n\alpha, \;\; f_3(z) = 0.$$

We get a contradiction to Theorem 19.1. It follows that

$$\zeta(u) + \gamma u = \frac{1}{2}(\eta + \gamma\omega)$$

is transcendental.

Now if u is any rational multiple of a period, it is ω/m for a period ω not in $m\Omega$, and then

$$\zeta(u) + \gamma u \;\; = \;\; \zeta(\omega/m) - \eta/m + \frac{1}{m}(\eta + \gamma\omega)$$

is also transcendental.

Finally if u is not a rational multiple of a period, then we just use $z = nu$ for powers n of 2 in (20.16), noting that $f_2(2z) - 2f_2(z) = \zeta(2z) - 2\zeta(z)$ ready for use in (20.15). □

We note that this result combined with Theorem 20.2 about the transcendence of u implies the linear independence of $1, u, \zeta(u)$ over the field of all algebraic numbers. In particular of $1, \omega, \eta$. So the transcendence of $\zeta(\omega/2)$.

Formally the differential equation gives

$$\int_{\wp(u_*)}^{\wp(u)} \frac{(-X+\mu)\mathrm{d}X}{\sqrt{4X^3-g_2X-g_3}} = \zeta(u)+\mu u-\zeta(u_*)-\mu u_* \tag{20.17}$$

after the substitution $X = \wp(z)$. So for example we find the transcendence of

$$\int_0^1 \frac{X\mathrm{d}X}{\sqrt{X-X^3}} = \frac{1}{2}\int_0^1 \frac{\mathrm{d}t}{t^{1/4}(1-t)^{1/2}} = \frac{1}{2}B\left(\frac{3}{4},\frac{1}{2}\right) = \frac{2\pi\sqrt{2\pi}}{\Gamma(\frac{1}{4})^2}$$

(which happens to be $2\zeta(\omega_0/2)$ in the (20.7) situation – see Exercise 20.58), and even the linear independence of

$$1,\ \frac{\Gamma(\frac{1}{4})^2}{2\sqrt{2\pi}},\ \frac{2\pi\sqrt{2\pi}}{\Gamma(\frac{1}{4})^2}. \tag{20.18}$$

Thus in particular the transcendence of

$$\frac{\Gamma(\frac{1}{4})}{\pi^{2/4}}. \tag{20.19}$$

And thanks to the quasi-addition formula

$$\zeta(z_1+z_2) = \zeta(z_1)+\zeta(z_2)+\frac{1}{2}\frac{\wp'(z_1)-\wp'(z_2)}{\wp(z_1)-\wp(z_2)} \tag{20.20}$$

(WW 20.53) also animals like

$$\int_4^5 \frac{(X-8)\mathrm{d}X}{\sqrt{X^3-7X+6}}.$$

Wonderful things! But careful: despite Apéry we still don't know even the irrationality (let alone the transcendence) of $\zeta(3)$, here

$$\int_0^1 \frac{X\mathrm{d}X}{\sqrt{X-X^3}} - \frac{1}{2}\int_{-1}^0 \frac{X\mathrm{d}X}{\sqrt{X^3-X}} + \frac{1}{2}\int_1^{\wp(3)} \frac{X\mathrm{d}X}{\sqrt{X^3-X}} = 3.840444614\ldots$$

for $\Omega = \omega_0(\mathbb{Z}+i\mathbb{Z})$, even though (or rather, precisely because) we do know the transcendence of $\wp(3) = 7.02942\ldots$.

We now go back a step to the root of all this business, the sigma function. This is also not periodic and not even quasi-periodic but shall we say quasi-quasi-periodic in the sense that

$$\sigma(z+\omega) = \chi(\omega)\sigma(z)e^{(z+\omega/2)\eta} \tag{20.21}$$

for $\eta = \eta(\omega)$, where $\chi(\omega) = -1$ if $\omega/2$ is not a period and otherwise $\chi(\omega) = 1$ (see WW 20.421 for the first case and Lang (1973) p. 241 in general).

From this we can see that the growth order $\rho = 2$ for σ. In fact (20.13) yields $\eta(\omega+\omega') = \eta(\omega)+\eta(\omega')$. It follows easily that $|\eta(\omega)| \le c|\omega|$ for some

c independent of ω. Thus taking z bounded in (20.21) we get $|(z+\omega/2)\eta| \leq c(1+|\omega|^2)$, which implies what we want (see also Exercises 20.29 and 20.30).

From the Schneider–Lang point of view σ has the nice differential equation $\sigma'/\sigma = \zeta$, but then ζ' involves $\wp$ and all we get is a weaker form of Theorem 20.9. For $\varphi(z) = \sigma(z+v)/\sigma(z)$ we have $\varphi'(z)/\varphi(z) = \zeta(z+v) - \zeta(z)$ which reminds us of (20.20); so with the adjustment $\tilde{\varphi}(z) = \varphi(z)e^{-z\zeta(v)}$ we see that $\tilde{\varphi}'(z)/\tilde{\varphi}(z) = \zeta(z+v) - \zeta(z) - \zeta(v)$ looks better, provided v is not in Ω. But we will also need analogues of the duplication formulae (20.4) and (20.15). We do have

$$\sigma(2z) = -\sigma(z)^4\wp'(z), \tag{20.22}$$

(WW ex. 20.24) but to get at $\sigma(2z+v)$ in $\varphi(2z)$ we will also need an addition formula. In fact a sort of master formula comes from the so-called Theorem of the Cube, which gives

$$\frac{\sigma(z_1+z_2+z_3)\sigma(z_1)\sigma(z_2)\sigma(z_3)}{\sigma(z_1+z_2)\sigma(z_2+z_3)\sigma(z_3+z_1)} = -\frac{\Delta_A}{\Delta_V},$$

where

$$\Delta_A = \Delta_A(z_1,z_2,z_3) = \frac{1}{2}\begin{vmatrix} 1 & \wp(z_1) & \wp'(z_1) \\ 1 & \wp(z_2) & \wp'(z_2) \\ 1 & \wp(z_3) & \wp'(z_3) \end{vmatrix},$$

$$\Delta_V = \Delta_V(z_1,z_2,z_3) = \begin{vmatrix} 1 & \wp(z_1) & \wp(z_1)^2 \\ 1 & \wp(z_2) & \wp(z_2)^2 \\ 1 & \wp(z_3) & \wp(z_3)^2 \end{vmatrix}$$

are determinants of type "Area" and "Vandermonde" (WW 20.53 ex. 1 and ex. 20.20). Putting $z_3 = v$ and making a final adjustment

$$\varphi_v(z) = \frac{\tilde{\varphi}(z)}{\sigma(v)} = \frac{\sigma(z+v)}{\sigma(z)\sigma(v)}e^{-z\zeta(v)},$$

we get a quasi-quasi-addition formula

$$\varphi_v(z_1+z_2) = -\varphi_v(z_1)\varphi_v(z_2)\frac{\Delta_A(z_1,z_2,v)}{\Delta_V(z_1,z_2,v)}. \tag{20.23}$$

This degenerates to a quasi-quasi-duplication formula

$$\varphi_v(2z) = -\frac{1}{2}\varphi_v(z)^2\frac{\wp(z)\wp''(z) - \wp'(z)^2 - \wp(v)\wp''(z) + \wp'(v)\wp'(z)}{\wp'(z)(\wp(z)-\wp(v))^2}. \tag{20.24}$$

It is valid at all z not in $\Omega/2$ except possibly z with $\wp(z) = \wp(v)$. However we see easily that the latter implies $z \pm v$ is in Ω for some choice of sign. For example with z not in $\Omega/2$ and $\wp(z) = \wp(v)$ we deduce that v is not in $\Omega/2$;

thus $\wp'(v) \neq 0$, so by changing the sign of z if necessary we can suppose $\wp'(z) \neq \wp'(v)$. Now from (20.10) we deduce that $z + v$ is in Ω.

Hence indeed (20.24) is valid for z not in $\Omega/2$ with $z \pm v$ both not in Ω.

By (20.21) we also have quasi-quasi-periodicity

$$\varphi_v(z + \omega) = e^{\lambda_v} \varphi_v(z) \tag{20.25}$$

for the quasi-quasi-periods

$$\lambda_v = \lambda_v(\omega) = \eta(\omega)v - \omega\zeta(v).$$

Finally using (20.20) we get

$$\varphi_v'(z) = \frac{1}{2}\varphi_v(z)\frac{\wp'(z) - \wp'(v)}{\wp(z) - \wp(v)}. \tag{20.26}$$

Here is the necessary independence result.

Lemma 20.10 *Suppose that v is not a rational multiple of a period. Then for any complex μ the functions $\varphi_v(z)e^{\mu z}, \wp(z)$ are algebraically independent over $\mathbb{C}$.*

Proof There is a period ω with $\kappa = \exp(\lambda_v(\omega) + \mu\omega)$ not a root of unity. In fact with basis elements ω_1, ω_2 as above, we could eliminate $\zeta(v)$ from the equations

$$\eta_1 v - \omega_1\zeta(v) + \mu\omega_1 = 2\pi i q_1, \quad \eta_2 v - \omega_2\zeta(v) + \mu\omega_2 = 2\pi i q_2$$

using the Legendre relation to get $v = \pm(q_1\omega_2 - q_2\omega_1)$, a contradiction if q_1, q_2 are both rational. We can now imitate the proof of Lemma 20.8 with $f(z) = \varphi_v(z)e^{\mu z}$, getting $m' \neq m''$ with $f(z + m'\omega) = f(z + m''\omega)$; but this comes down to $\kappa^{m'-m''} = 1$. □

Here we exclude v as a rational multiple of a period for the following reason. Say for example $v = \omega/2$ for a period ω not in 2Ω. Then (WW 20.53 ex. 1)

$$\varphi_{\omega/2}(z)^2 = \wp(z) - \wp(\omega/2) \tag{20.27}$$

(showing by the way that $\sqrt{\wp(z) - \wp(\omega/2)}$ is meromorphic). Thus the lemma would be false for $\mu = 0$. A similar thing happens for ω/m.

In fact we can regard v as being taken modulo Ω; this is because

$$\varphi_{v+\omega}(z) = \varphi_v(z) \tag{20.28}$$

for any period ω.

We are ready for our next transcendence result. This too corresponds to a more subtle kind of commutative group surface E_v, with $(\wp(z), e^w\varphi_v(z))$ the essential part of the exponential map (see Exercise 20.104). Now there is an exact sequence $\{1\} \to \mathbb{G}_{\mathrm{m}} \to E_v \to E \to 0$ (see Exercise 20.105).

Theorem 20.11 *Suppose that g_2, g_3 are algebraic, that δ is algebraic in $\mathbb{C}$, and that v is a complex number, not a rational multiple of a period, with $\wp(v)$ algebraic. If u is a complex number with $u, u+v$ not in Ω such that $\wp(u)$ is algebraic, then*

$$\varphi_v(u)e^{\delta u} = \frac{\sigma(u+v)}{\sigma(u)\sigma(v)}e^{-u(\zeta(v)-\delta)}$$

is transcendental.

In fact if v is a rational multiple of a period, this result reduces to Theorem 20.6, as one might guess from (20.27).

Proof We start as in the proof of Theorem 20.9 by showing that

$$t = \varphi_v(\omega/m)e^{-\lambda_v/m}$$

is algebraic for any $m \geq 2$ and any period ω not in $m\Omega$, where $\lambda_v = \lambda_v(\omega)$.

If m is odd, then there is $r \geq 1$ with $2^r \equiv 1$ modulo m, and repeated use of (20.24) shows that $\varphi_v(2^r z)/\varphi_v(z)^{2^r}$ is a rational function of $\wp(z), \wp'(z), \ldots, \wp(2^{r-1}z), \wp'(2^{r-1}z)$ with algebraic coefficients. From the discussion above we can substitute $z = \omega/m$. Using (20.25) we get t^{1-2^r}. If $m = 2^s m_1$ for odd $m_1 \geq 3$ then making the same substitution in $\varphi_v(2^s z)/\varphi_v(z)^{2^s}$ yields $t^{-2^s}t_1$ with $t_1 = \varphi_v(\omega/m_1)e^{-\lambda_v/m_1}$ just shown to be algebraic. For $m_1 = 1$ we modify as before to reduce down to $\varphi_v(\omega'/2)e^{-\lambda'_v/2}$ with $\omega' = \omega/2^r$ and $\lambda'_v = \lambda_v(\omega')$, which turns out to be nothing other than $\varphi_{\omega'/2}(v)$, and we conclude using (20.27) that

$$\varphi_v(\omega'/2)e^{-\lambda'_v/2} = \pm\sqrt{\wp(v) - \wp(\omega'/2)}. \tag{20.29}$$

As above, all this can be done more efficiently with suitable formulae for $\sigma(mz)$ generalizing (20.24).

We write for brevity $f(z) = \varphi_v(z)e^{\delta z}$. Now suppose that $f(u) = \alpha$ is algebraic.

With algebraic $\beta = \wp(v)$, $\gamma = \wp'(v)$ we use for example

$$f_1(z) = \wp(z),\ f_2(z) = f(z),\ f_3(z) = \wp'(z),\ f_4(z) = \frac{1}{\wp(z) - \beta} \tag{20.30}$$

with differential equations

$$f_1' = f_3,\ f_2' = \frac{1}{2}f_2(f_3f_4 - \gamma f_4 + 2\delta),\ f_3' = 6f_1^2 - \frac{1}{2}g_2,\ f_4' = -f_3f_4^2.$$

First assume that $u = \omega/2$ for a period ω not in 2Ω. Then for every odd n, we get $f_1(nu) = \wp(\omega/2), f_2(nu) = \alpha^n/\alpha_2^{n-1}$ for algebraic $\alpha_2 = \varphi_v(\omega/2)e^{-\lambda_v/2}$. This leads to a contradiction with Theorem 19.1.

Thus indeed $x_0 = f(\omega/2)$ is transcendental. We find that $x_0^2 e^{-\lambda_v(\omega)-\delta\omega} = \alpha_2^2$ and so the transcendence of

$$e^{\lambda_v(\omega)+\delta\omega} = \exp(\eta(\omega)v - \omega\zeta(v) + \delta\omega).$$

Next with ω not in $m\Omega$ we find $f(\omega/m)^m = \alpha_m^m(x_0/\alpha_2)^2$ with algebraic $\alpha_m = \varphi_v(\omega/m)e^{-\lambda_v/m}$; so also transcendental.

We may therefore suppose that u is not a rational multiple of a period.

Now $f(-v) = 0$; and indeed $u = -v$ is excluded from Theorem 20.11. We next take u as certain rational multiples of v, but we have to tiptoe around this exception $-v$.

Consider first $u = -2v$. We consider the functions (20.30) at $z = 2^r u$ ($r = 0, 1, 2, \ldots$), using (20.24) repeatedly and noting that the denominators don't vanish. We get another contradiction with Theorem 19.1. We also find using (20.22) that $f(-2v) = -x^{-4}/\wp'(v)$ for

$$x = \sigma(v)e^{-v\zeta(v)/2}e^{\delta v/2}$$

and so we deduce the transcendence of x. So in particular of $\sigma(v)e^{-v\zeta(v)/2}$; but just as for $\zeta(3)$, we don't know even the irrationality of $\sigma(3)$ when $\Omega = \omega_0(\mathbb{Z} + i\mathbb{Z})$ in (20.7). Or come to that, of $\sigma(\omega/2)$ for general Ω (with g_2, g_3 algebraic), even though we do know the transcendence of $\zeta(\omega/2)$. And at least the number

$$\sigma(\omega_0/2) = \frac{e^{\pi/8}}{2^{1/4}}$$

(compare Exercises 20.22 and 20.63) is transcendental, thanks to Gelfond–Schneider (and complex multiplication). But we digress.

Next we have $f(v) = -\wp'(v)x^2$, and so this too is transcendental.

And $f(-v/2) = -1/x$ is transcendental.

Now we can keep halving u. For example using (20.24) we find that $f(v) = f(v/2)^2 R$ for a rational function R of $\wp(v), \wp(v/2), \wp'(v), \wp'(v/2)$ with algebraic coefficients whose denominator is non-zero because $\wp(v) \neq \wp(v/2)$. As $f(v) \neq 0$ the numerator is also non-zero. By (20.4) the values $\wp(v/2), \wp'(v/2)$ are algebraic, so we see that $f(v/2)/x$ is algebraic and consequently $f(v/2)$ is transcendental. Similarly with $f(v/4)/\sqrt{x}$ and $f(v/4)$ involving $\wp(v/4)$; and so on.

And similarly with $f(-v/4)\sqrt{x}$ and $f(-v/4)$ and so on. Here we start to suffer a bit from not appealing to any theory of commutative group varieties, and I hope the reader will not think the above too much of a *tour de farce*. But this theory is not too elementary (see for example Exercise 20.104).

All this covers the cases when one of $u, \pm 2u, \pm 4u, \ldots$ is v. By (20.28) we can henceforth assume that none of $u - v, \pm 2u - v, \pm 4u - v, \ldots$ are in Ω.

Now the transcendence of $f(u)$ follows on considering (20.30) at $z = 2^r u$ ($r = 0, 1, 2, \ldots$) and Theorem 19.1. This completes the proof of Theorem 20.11. □

Formally the differential equation (20.26) gives

$$\frac{1}{2}\int_{\wp(u_*)}^{\wp(u)} \left(\frac{Y - \wp'(v)}{X - \wp(v)} + 2\mu\right)\frac{\mathrm{d}X}{Y} = \log(\varphi_v(u)e^{\mu u}) - \log(\varphi_v(u_*)e^{\mu u_*}) \tag{20.31}$$

after the substitution $X = \wp(z)$, where now we have abbreviated the expression $\sqrt{4X^3 - g_2X - g_3}$ to Y. We are still very far from the transcendence of general elliptic integrals

$$\int_\alpha^\beta R(X, Y)\mathrm{d}X, \quad \exp\left(\int_\alpha^\beta R(X, Y)\mathrm{d}X\right) \tag{20.32}$$

for rational R. But (20.31) does lead to the transcendence of unabbreviated monstrosities like

$$\exp\left(\int_4^5 \left(\frac{\sqrt{X^3 - 7X + 6} - 2\sqrt{3}}{X - 3} + 8\right)\frac{\mathrm{d}X}{\sqrt{X^3 - 7X + 6}}\right)$$

(see Exercise 20.64). With only slightly more complicated R some strange things can happen. For example

$$\exp\left(\int_0^1 \frac{62X - 46399}{2X - 1009}\frac{2\sqrt{6}\mathrm{d}X}{\sqrt{4X^3 - 3704403X + 1364775049}}\right) \tag{20.33}$$

has the value

$$\frac{A(B + C\sqrt{81664239})}{D(E + F\sqrt{8188650294})}$$

with

$$A = 111350967495666898903790720561084448 1$$

$$B = 10999636958786196285796828105081015440 1$$

$$C = 1217141786763081312264606680224796 0$$

$$D = 10873106619155067845289556481962032 01$$

$$E = 11196126954560674906091243923263431448 1$$

$$F = 1237200743807859503759266025749100$$

(see Exercise 20.65) because the indefinite expression is an algebraic function of X (whose writing out would take up a whole page). Thus taken between algebraic limits it will yield an algebraic number. (Underlying this example is a rational elliptic curve with rational 12-torsion.) Such a phenomenon almost certainly disappears if we change some coefficients in (20.33) – see Exercise 20.66. (That's enough integrals. Ed.)

Our final transcendence example involves a second elliptic function $\wp^*$ coming from a second lattice Ω^* with invariants g_2^*, g_3^* (corresponding to a second elliptic curve E^*). We then have to investigate the algebraic dependence of $\wp(z), \wp^*(\beta z)$. We saw in Lemma 20.4 that when $\wp^* = \wp$ there can be dependence for $\beta \neq 1$ or even irrational β.

In general, dependence can happen if there is non-zero μ with $\mu\Omega \subseteq \Omega^*/m$ for some positive integer m. It can be shown that if the set of all such multipliers μ together with 0 is not just the set $\{0\}$ itself (the so-called non-isogenous situation), then there is non-zero μ_0 such that it has the form $\mathbb{Q}\mu_0$ or $\mathfrak{K}\mu_0$ for a complex quadratic extension $\mathfrak{K}$ of $\mathbb{Q}$. In general we write $\mathfrak{M} = \mathfrak{M}_{\Omega,\Omega^*}$ for this set.

An example with $\mathfrak{M} = \{0\}$ involves $\Omega = \omega_0(\mathbb{Z}+i\mathbb{Z})$ and $\Omega^* = \omega_0^*(\mathbb{Z}+i^*\mathbb{Z})$ with $i^* = (-1+\sqrt{-3})/2$, even though $\mathfrak{K}_\Omega = \mathbb{Q}(i)$ and $\mathfrak{K}_{\Omega^*} = \mathbb{Q}(i^*)$ are both complex quadratic fields. Namely, it is geometrically clear that a square lattice cannot be rotated and stretched into a hexagonal one. Incidentally we now have $g_2^* = 0$, and with the choice

$$\omega_0^* = \frac{1}{3}B\left(\frac{1}{6}, \frac{1}{2}\right) = \frac{\Gamma(\frac{1}{3})^3}{2^{4/3}\pi},$$

we get $g_3^* = 4$.

Lemma 20.12 *For any β not in $\mathfrak{M}$ the functions $\wp(z), \wp^*(\beta z)$ are algebraically independent over $\mathbb{C}$.*

Proof As in the proof of Lemma 20.3 we find for each ω in Ω different integers m', m'' such that $\beta(m'-m'')\omega$ lies in Ω^*. Now taking ω as basis elements of Ω we conclude that β is in $\mathfrak{M}$, a contradiction. □

Here is our final result, a generalization of elliptic Gelfond–Schneider. It corresponds to the commutative group surface $E \times E^*$, with $(\wp(z), \wp^*(w))$ the essential part of the exponential map.

Theorem 20.13 *Suppose that g_2, g_3, g_2^*, g_3^* are algebraic, and that β in $\mathbb{C}$ is algebraic but not in $\mathfrak{M}$. If u is a complex number not in Ω such that $\wp(u)$ is algebraic, then βu is not in Ω^* and $\wp^*(\beta u)$ is transcendental.*

We can safely leave the proof to the reader; she must only modify the proof of Theorem 20.5 by inserting stars.

We can deduce the transcendence of ω_0^*/ω_0 above and so of

$$\frac{\Gamma(\frac{1}{3})^3}{\Gamma(\frac{1}{4})^2\sqrt{\pi}},$$

as well as quotients like

$$\frac{\int_8^9 \frac{\mathrm{d}X}{\sqrt{X^3-7X-6}}}{\int_4^5 \frac{\mathrm{d}X}{\sqrt{X^3-7X+6}}},$$

(I told you that's enough integrals. Ed.) provided they are not in $\mathfrak{M}$. But again how to check that, in view of things like Exercise 20.70? Now $g_2 = 28$, $g_3 = -24$ in the denominator but $g_2^* = 28, g_3^* = 24$ in the numerator, telling us not too much about Ω, Ω^* (see however Exercise 20.69). In fact it is a question of isogenies between elliptic curves, already mentioned in this connexion at the end of the previous chapter (see also Exercise 20.71).

It will be seen from (20.8), (20.12), (20.19) that we narrowly miss the transcendence of $\Gamma(1/4)$. That was first proved by Chudnovsky, who deduced it from the algebraic independence of the two numbers π and $\Gamma(1/4)$ (using elliptic functions – see Chapter 22). The same arguments on $Y^2 = 4X^3-4$ yield the transcendence of $\Gamma(1/3)$ (see also Exercise 20.61). Of course $\Gamma(1/2) = \sqrt{\pi}$ is also transcendental. This is often said to be essentially (that is, modulo fiddling with functional equations) all we know about the transcendence or even irrationality of values of Γ. However the transcendence of $\Gamma(i)$ can be deduced from more recent results of Nesterenko on the algebraic independence of the three numbers $\pi, e^\pi, \Gamma(1/4)$ (proved now using modular functions – see Chapter 21). Namely using the functional equations we can calculate $|\Gamma(i)|$ as the square root of

$$\Gamma(i)\overline{\Gamma(i)} = \Gamma(i)\Gamma(-i) = \frac{\Gamma(i)\Gamma(1-i)}{-i} = \frac{\pi}{-i\sin(i\pi)} = \frac{2\pi}{e^\pi - e^{-\pi}} \tag{20.34}$$

which by Nesterenko is transcendental; thus so is $\Gamma(i)$. This idea came to me in a dream; on waking up I saw that it was wrong but could be corrected. A similar proof works at any purely imaginary quadratic number and even at i^*.

In particular we don't yet know the irrationality of $\Gamma(1/5)$. As this occurs in the evaluation of

$$\int_0^1 \frac{\mathrm{d}X}{\sqrt{1-X^5}} = \frac{1}{5}B\left(\frac{1}{5}, \frac{1}{2}\right)$$

(This is your last warning. Ed.), one may guess that the transcendence can be approached using now the curve $Y^2 = 1 - X^5$. As this has genus 2, its Jacobian is yet another commutative group surface, an abelian variety A. In fact the Schneider–Lang result can be applied to the corresponding abelian functions and we get a whole host of transcendence results for quotients of such abelian integrals like

$$\frac{\int_3^4 \frac{X\mathrm{d}X}{\sqrt{X^5-X-2}}}{\int_3^4 \frac{\mathrm{d}X}{\sqrt{X^5-X-2}}}.$$

(You're fired. Ed.) Here there is no big problem to construct the abelian functions (in Baker (1907) there is even a $\wp$ function – see Grant (1990) also) and even their differential equations (we have integrals after all), but proving $\rho \le 2$ needs an analogue of σ. This is a theta function now of two complex variables. So we go no further in that direction, except to mention that Schneider, using a version of Theorem 19.1 in several complex variables (see Exercise 19.9) together with a more general concept of complex multiplication, was able to prove that $B(a, b) = \Gamma(a)\Gamma(b)/\Gamma(a + b)$ is transcendental for any rational non-integral a, b (see Schneider (1957) for references).

On the other hand, no-one has any idea at all how to prove even the irrationality of $\Gamma'(1)$, which is $-\gamma$ for Euler's constant.

We have now accumulated quite a list

$$\mathbb{G}_{\mathrm{a}} \times \mathbb{G}_{\mathrm{m}},\ \mathbb{G}_{\mathrm{m}} \times \mathbb{G}_{\mathrm{m}},\ \mathbb{G}_{\mathrm{a}} \times E,\ E \times E,\ \mathbb{G}_{\mathrm{m}} \times E,\ E_{\mathrm{a}},\ E_{\mathrm{v}},\ E \times E^*,\ A \qquad (20.35)$$

of commutative group surfaces over $\mathbb{C}$. If we throw in $\mathbb{G}_{\mathrm{a}} \times \mathbb{G}_{\mathrm{a}}$ (for which there is no transcendence result), then in fact the list becomes complete (modulo overlappings, for example E^* might be E) up to isogeny.

Schneider's results refer to all except the first two in (20.35) and E_{v}, although in his Third Problem (Schneider, 1957) he seemed to hint at the latter (though as mentioned he did not use these concepts). Serre during lectures of Waldschmidt pointed out the connexion between E_{v} and integrals like (20.31), and Waldschmidt worked out the transcendence consequences in his book (Waldschmidt, 1987). One can see this as a triumph for the commutative group varieties viewpoint (or one can consider that these things originally arose via integrals). Incidentally, this E_{v} has recently attracted some interest as Bertrand found a counterexample inside it for some extensions of Manin–Mumford to semiabelian schemes.

As for higher-dimensional group varieties, we have already met $\mathbb{G}_{\mathrm{m}}^n$ and $\mathbb{G}_{\mathrm{a}} \times \mathbb{G}_{\mathrm{m}}^n$ in Chapter 19. The list for general dimension is not so explicit as for surfaces. The corresponding transcendence (or linear independence) results were all proved by Wüstholz (1989a,b). They finally suffice to decide about for example the transcendence of general integrals like (20.32) where now X, Y are related by a general polynomial equation with algebraic coefficients.

For more about commutative group varieties see Waldschmidt (1987) and Baker and Wüstholz (2007) in connexion with transcendence, and Serre (1988) in connexion with algebraic geometry.

Exercises

20.1 Let k be in $\mathbb{N}$, and let $E_k(w) = (1-w)\exp(w + \frac{1}{2}w^2 + \cdots + \frac{1}{k}w^k)$. Show that $|E_k(w) - 1| \leq c_k|w|^{k+1}$ for all w in $\mathbb{C}$ with $|w| \leq 1$, where $c_k = 2\exp(1 + \frac{1}{2} + \cdots + \frac{1}{k}) + 1$ [*Hint*: $\log E_k$].

20.2 Let Ω be a lattice in $\mathbb{C}$, and let $s > 2$ in $\mathbb{R}$. Show that

$$\zeta_\Omega(s) = \lim_{W\to\infty} \sum_{\omega\in\Omega,\ 0<|\omega|\leq W} \frac{1}{|\omega|^s}$$

exists (compare Exercise 14.90).

20.3 Let $\Omega = \mathbb{Z}\omega_1 + \mathbb{Z}\omega_2$. Show that $\Delta = \begin{vmatrix} \omega_1 & \omega_2 \\ \overline{\omega_1} & \overline{\omega_2} \end{vmatrix}$ is in $i\mathbb{R}$; and that $i\Delta$ has the same sign as $\mathrm{Im}(\omega_2/\omega_1)$.

20.4 Let Ω be a lattice and let $d(z) = \min_{\omega\in\Omega}|z-\omega|$. Show that there is $\theta = \theta(\Omega)$ with $0 < \theta \leq 1$ and the following property. Let S be a finite subset of Ω with $m \geq 1$ elements. Then $\prod_{\omega\in S}|z-\omega| \geq \theta^{m-1}d(z)$ for all z in $\mathbb{C}$ (as a consequence one can show using $S = \{\omega \in \Omega;\ 0 \leq |\omega| \leq 3r\}$ that $|\sigma(z)| \geq r^{-cr^2}d(z)$ for all z in $\mathbb{C}$ with $|z| \leq r$ and $r \geq 2$).

20.5 For a lattice Ω let $\sigma(z) = z\lim_{W\to\infty}\prod_{\omega\in\Omega,\ 0<|\omega|\leq W} E_2(\frac{z}{\omega})$ be the associated Weierstrass function.

(a) Let $\Omega = \mathbb{Z} + \mathbb{Z}i$. Show that $\sigma(iz) = i\sigma(z)$ and that $\eta(\omega) = \pi\overline{\omega}$ [*Hint*: $i\Omega$].

(b) Let $\Omega = \mathbb{Z} + \mathbb{Z}i^*$ with $i^* = e^{2\pi i/3}$. Show that $\sigma(i^*z) = i^*\sigma(z)$ and that $\eta(\omega) = \frac{2}{\sqrt{3}}\pi\overline{\omega}$.

20.6 Deduce (20.4) from (20.10).

20.7 Show that $z, \wp(z)$ are algebraically independent over $\mathbb{C}$ [*Hint*: If $L = [\mathbb{C}(z,\wp(z)) : \mathbb{C}(\wp(z))]$ consider $z \to 2z$ and the duplication formula].

20.8 Let Ω be a lattice, and let $d(z) = \min_{\omega\in\Omega}|z-\omega|$ as in Exercise 20.4.

(a) Show that $\lim_{z\to 0, z\neq 0} d(z)^2|\wp(z)| = 1$.

(b) Show that there is $c = c(\Omega)$ such that $|\wp(z)| \leq cd(z)^{-2}$ on $\mathbb{C} \setminus \Omega$.

20.9 Let Ω be a lattice. Show that

$$\sum_{0\neq\omega\in\Omega}\frac{1}{\omega^8}=\frac{3}{7}\left(\sum_{0\neq\omega\in\Omega}\frac{1}{\omega^4}\right)^2$$

[*Hint*: Laurent coefficient of z^6 in $\wp(z)$ and the differential equation].

20.10 Let $\tau=\sqrt{-2}$ and $\Omega=\mathbb{Z}+\mathbb{Z}\tau$.

(a) Show that $\frac{1}{\tau}\Omega=\Omega\cup(\frac{\tau}{2}+\Omega)$.

(b) With corresponding $\wp(z)$ and $e=\wp(\frac{\tau}{2})$ show that

$$f(z)\;=\;(\wp(z)-e)\left(\wp(\tau z)+\frac{1}{2}\wp(z)\right)$$

is a constant [*Hint*: no pole].

(c) Show that

$$8\left(\sum_{0\neq\omega\in\Omega}\frac{1}{\omega^4}\right)^3=25\left(\sum_{0\neq\omega\in\Omega}\frac{1}{\omega^6}\right)^2$$

[*Hint*: Exercise 20.9; and eliminate e].

(d) Show that there is ω_0^{**} such that $g_2=6750, g_3=94500$ for $\omega_0^{**}\Omega$.

* * * * * * * * * * * * * * * * * * * *

20.11 Let S be discrete in $\mathbb{C}$ (i.e. $S\cap B$ is finite for every bounded subset B of $\mathbb{C}$). Show that there is an entire function f with simple zeroes on S and no other zeroes (the case $S=\{-1,-2,\ldots\}$ is dealt with in Exercise 19.17 – see also Exercise 19.18).

20.12 In Exercise 20.2 show that $\lim_{W\to\infty}W^{-2}\sum_{\omega\in\Omega,|\omega|\leq W}1\;=\;\pi/\det\Omega$ (compare with Corollary 10.5).

20.13 In Exercise 20.2 show that $\lim_{W\to\infty}(\log W)^{-1}\sum_{\omega\in\Omega,0<|\omega|\leq W}1/|\omega|^2=2\pi/\det\Omega$ (so the limit in Exercise 20.2 for $s=2$ does not exist).

20.14 In Exercise 20.3 show that $\Delta=\pm 2i\det\Omega$.

20.15 In Exercise 20.2 show that $\zeta_{\mathbb{Z}+\mathbb{Z}i}(s)\;=\;4\zeta(\frac{s}{2})L(\frac{s}{2};\chi)$, where L is the Dirichlet function associated to the unique character $\chi\neq\chi_0\pmod 4$.

20.16 In Exercise 20.4 show that there is $c=c(\Omega)$ such that $|\sigma(z)|\leq r^{cr^2}d(z)$ for all z with $|z|\leq r$ and $r\geq 2$.

20.17 Let Ω be a lattice and let $\sigma(z)$ be the associated Weierstrass function. If there are constants $k \neq 0, \pm 1$ and l with $\sigma(kz) = l\sigma(z)$, show that Ω is a complex multiple of the lattices in Exercise 20.5.

20.18 Let $\Omega = \mathbb{Z}\omega_1 + \mathbb{Z}\omega_2$, and let n be in $\mathbb{N}$. Show that

$$\frac{\sigma(nz)}{\prod_{0 \leq n_1 < n, 0 \leq n_2 < n} \sigma(z + \frac{n_1\omega_1 + n_2\omega_2}{n})}$$

is an entire function without zeroes.

20.19 Show that $\sigma'(\omega) = \chi(\omega)e^{\omega\eta/2}$ in (20.21).

20.20 (a) Show that $\wp''' = 12\wp\wp'$.
(b) Show that $\zeta'''' = -12\zeta'\zeta''$.
(c) Show that $\sigma^2\sigma''''' - 5\sigma\sigma'\sigma'''' + 2\sigma\sigma''\sigma''' + 8\sigma'^2\sigma''' - 6\sigma'\sigma''^2 = 0$.

20.21 Let $\Omega = \mathbb{Z} + \mathbb{Z}\tau$ with $\Im\tau > 0$. Show that

$$\sigma(z) = \frac{1}{2\pi i} e^{\eta z^2/2}(u - u^{-1}) \prod_{n=1}^{\infty} \frac{(1 - q^n u^2)(1 - q^n u^{-2})}{(1 - q^n)^2},$$

where $q = e^{2\pi i\tau}$, $u = e^{\pi iz}$ and $\eta = \frac{\pi^2}{3}(1 - 24\sum_{n=1}^{\infty} \frac{nq^n}{1-q^n})$.

20.22 Let $\Omega = \mathbb{Z} + \mathbb{Z}i$. Show that $\sigma(\frac{1}{2}) = \frac{e^{\pi/8}}{2^{5/4}I}$, where $I = \int_1^{\infty} \frac{dX}{\sqrt{4X^3 - 4X}}$.

20.23 In Exercise 20.8 show that there is $c = c(\Omega)$ such that $|\wp'(z)| \leq cd(z)^{-3}$ on $\mathbb{C} \setminus \Omega$.

20.24 In Exercise 20.9 show that

$$\sum_{0 \neq \omega \in \Omega} \frac{1}{\omega^{10}} = \frac{5}{11} \left(\sum_{0 \neq \omega \in \Omega} \frac{1}{\omega^4} \right) \left(\sum_{0 \neq \omega \in \Omega} \frac{1}{\omega^6} \right).$$

20.25 Show that $\wp(3z) = A_3(\wp(z))/B_3(\wp(z))$, where A_3 in $\mathbb{C}[X]$ has degree 9 and B_3 in $\mathbb{C}[X]$ has degree 8.

20.26 Let $2u$ be in Ω. Show that there are a, b, c, d with c, d not both zero and $\wp(z + u) = \frac{a\wp(z) + b}{c\wp(z) + d}$.

20.27 Let $3u$ be in Ω. Show that there are $a, b, c, d, e, f, g, h, k$ with g, h, k not all zero and

$$\wp(z + u) = \frac{a\wp'(z) + b\wp(z) + c}{g\wp'(z) + h\wp(z) + k}, \quad \wp'(z + u) = \frac{d\wp'(z) + e\wp(z) + f}{g\wp'(z) + h\wp(z) + k},$$

where e need not be the base of the natural logarithms.

20.28 Let $\Omega = \omega_0(\mathbb{Z} + \mathbb{Z}i)$ as in (20.7). Show that

$$\wp\left(z + \frac{\omega_0}{2} + \frac{i\omega_0}{2}\right) = -\frac{1}{\wp(z)}$$

thus providing the explanation of Exercise 19.21.

20.29 Let $\Omega = \mathbb{Z}\omega_1 + \mathbb{Z}\omega_2$, and let $\xi_1, \xi_2, \eta_1, \eta_2$ be in $\mathbb{C}$ with $\sigma(z + \omega_i) = \sigma(z)\exp(\xi_i + \eta_i z)$ $(i = 1, 2)$. Show that there is a quadratic polynomial Ξ such that $\sigma(z + n_1\omega_1 + n_2\omega_2) = \sigma(z)\exp(\Xi(n_1, n_2) + (n_1\eta_1 + n_2\eta_2)z)$ for all n_1, n_2 in $\mathbb{Z}$.

20.30 Show that there is $c = c(\Omega)$ with $|\sigma|_r \leq e^{cr^2}$ $(r \geq 2)$.

20.31 In Exercise 20.29 show that $\omega_2\eta_1 - \omega_1\eta_2$ lies in $2\pi i\mathbb{Z}$.

20.32 Let $\Omega = \mathbb{Z} + \mathbb{Z}\tau$ with $\Im\tau > 0$, and let $q = e^{2\pi i\tau}, u = e^{\pi iz}$. Show that $(2\pi i)^2\wp(z)$ is

$$\frac{1}{12} - 2\sum_{n=1}^{\infty}\frac{q^n}{1-q^n} + \frac{u^2}{(1-u^2)^2} + \sum_{n=1}^{\infty}q^n u^2\left(\frac{1}{(1-q^n u^2)^2} + \frac{1}{(u^2-q^n)^2}\right).$$

20.33 If $\Im\tau > 0$ for $\tau = \omega_2/\omega_1$, show in Exercise 20.31 that $\omega_2\eta_1 - \omega_1\eta_2 = 2\pi i$.

20.34 In Exercise 20.23 show that there is c, independent of Ω, such that

$$|\wp'(z)| \leq \frac{c}{d(z)^3} \quad (*)$$

on $\mathbb{C} \setminus \Omega$ (Masser and Wüstholz, 1990).

20.35 In Exercise 20.8(b) show that there is no c, independent of Ω, such that $|\wp(z)| \leq cd(z)^{-2}$ on $\mathbb{C} \setminus \Omega$ [*Hint*: $\Omega = \mathbb{Z} + \mathbb{Z}\tau, z = \tau/2, \tau = iy, y \to \infty$].

20.36 Let

$$\tilde{M} < M = \frac{\Gamma(1/3)^9}{(2\pi)^3 3^{3/2}} = 5.5137015767\ldots$$

Show that $(*)$ of Exercise 20.34 holds with $c = M$ but not with $c = \tilde{M}$ (Masser, 2003).

20.37 Let u, v be in $\mathbb{C}$. Solve the differential equation $f'^2 = (f-u)^2(f-v)$ and deduce that $g_2^3 \neq 27g_3^2$ for every lattice Ω [*Hint*: trigonometry].

20.38 Let $\mathcal{W}$ be a subgroup of $\mathbb{C}$ with $\mathbb{Z}$-rank 3, and let f be a quotient of entire functions with $f(z + w) = f(z)$ for all w in $\mathcal{W}$. Show that f is constant (thus there are no interesting triply periodic functions).

20.39 One defines

$$j(\Omega) = \frac{1728 g_2^3}{g_2^3 - 27 g_3^2}$$

for $g_2 = g_2(\Omega)$, $g_3 = g_3(\Omega)$, and then the elliptic modular function $j(\tau) = j(\mathbb{Z} + \mathbb{Z}\tau)$, usually only for $\Im\tau > 0$. Show that $j(i) = 1728$, $j(i^*) = 0$, and, from Exercise 20.10(d), that $j(\sqrt{-2}) = 8000$.

20.40 Let $\beta \neq 0$ be such that $\wp(z)$ and $\wp(\beta z)$ are algebraically dependent over $\mathbb{C}$. Show that β is algebraic of degree at most 2.

20.41 Let β be algebraic of degree at least 3 in $\mathbb{C}$ with $\Im\beta > 0$. From Theorem 20.5 and Exercise 20.40 show that $j(\beta)$ is transcendental.

20.42 Show that $j(\frac{-1+\sqrt{-163}}{2}) = -262537412640768000$.

20.43 Show that $j(\frac{-1+\sqrt{-427}}{2})$ is

$$-7805727756261891959906304000 - 999421027517377348595712000\sqrt{61}$$

(Kessler Master Thesis 1995).

20.44 Let $j = j(\frac{-1+\sqrt{-907}}{2})$. Show that $j^3 + Aj^2 + Bj + C = 0$ for

$$A = 12307208072119840239447759050683868774400 0,$$

$$B = 391815942080148196175658115753763143680000 00,$$

$$C = 14916127474652484132854589496927400755200000000 0.$$

20.45 Let β be algebraic of degree 2 in $\mathbb{C}$ with $\Im\beta > 0$. Show that $j(\beta)$ is algebraic and even integral.

20.46 Let K be imaginary quadratic with $\mathbb{Z}_K = \mathbb{Z} + \mathbb{Z}\beta$. Show that $[\mathbb{Q}(j(\beta)) : \mathbb{Q}]$ is the class number h_K.

20.47 Let $\Omega = \mathbb{Z}\omega_1 + \mathbb{Z}\omega_2$ be a lattice with fundamental domain $\mathcal{F} = [0,1)\omega_1 + [0,1)\omega_2$ having boundary $\partial\mathcal{F}$, and let $f \neq 0$ be a quotient of entire functions such that $f(z+\omega) = f(z)$ for all ω in Ω. Show that f has at most finitely many zeroes and poles on $\mathcal{F}$, and using $\int_{\partial\mathcal{F}} (f'(z)/f(z))dz$, that $\sum_{w\in\mathcal{F}} \mathrm{ord}_w f = 0$.

20.48 In Exercise 20.47 show that there are integers n_1, n_2 such that $\sum_{w\in\mathcal{F}} (\mathrm{ord}_w f) w = n_1\omega_1 + n_2\omega_2$.

20.49 If f is analytic at 0 in Exercises 20.47 and 20.48 with $f(0) \neq 0, f'(0) = 0$ show that $\sum_{w\in\mathcal{F}} (\mathrm{ord}_w f)\zeta(w) = n_1\eta_1 + n_2\eta_2$. This is related to the group variety E_{a}.

20.50 If further $f''(0) = 0$ in Exercise 20.49 show that $\sum_{w \in \mathcal{F}} (\mathrm{ord}_w f)\wp(w) = 0$. This is related to a group variety E_{aa} with $0 \to \mathbb{G}_{\mathrm{a}}^2 \to E_{\mathrm{aa}} \to E \to 0$.

20.51 Show that

$$\int_1^\infty \frac{\mathrm{d}X}{\sqrt{X^3 - X}}, \int_1^\infty \frac{\mathrm{d}X}{\sqrt{X^3 - 1}}, \int_{-60}^\infty \frac{\mathrm{d}X}{\sqrt{X^3 - 6750X - 189000}}$$

are transcendental.

20.52 Let $\Omega = \omega_0(\mathbb{Z} + \mathbb{Z}i)$ as in (20.7). Show that $\wp(\omega_0/2) = 1$ [Hint: why not 0 or -1?].

20.53 For $g_2 = 4, g_3 = 0$ show that

$$\wp\left(\frac{1}{2}\int_2^3 \frac{\mathrm{d}X}{\sqrt{X^3 - X}}\right) = 49.$$

20.54 Show that $\int_1^2 \frac{\mathrm{d}X}{\sqrt{X^3 - X}}$ and

$$\int_1^2 \frac{\mathrm{d}X}{\sqrt{X^3 - 7X + 6}}, \int_1^{-3} \frac{\mathrm{d}X}{\sqrt{X^3 - 7X + 6}}, \int_4^5 \frac{\mathrm{d}X}{\sqrt{X^3 - 7X + 6}}$$

are transcendental.

20.55 Show that

$$\frac{\int_2^{17} \frac{\mathrm{d}X}{\sqrt{X^3 - 7X + 6}}}{\int_2^3 \frac{\mathrm{d}X}{\sqrt{X^3 - 7X + 6}}} = 2$$

[*Hint*: $X = 2\frac{T^4 + 2T^3 + 14T^2 - 62T + 61}{T^4 - 8T^3 + 14T^2 + 8T + 1}$ in the numerator].

20.56 Show that

$$\exp\left(2\int_0^1 \frac{\mathrm{d}X}{\sqrt{X - X^3}}\right), \quad \exp\left(\int_4^5 \frac{\mathrm{d}X}{\sqrt{X^3 - 7X + 6}}\right)$$

and

$$\frac{1}{\pi}\int_0^1 \frac{\mathrm{d}X}{\sqrt{X - X^3}}, \quad \frac{1}{\log 2}\int_4^5 \frac{\mathrm{d}X}{\sqrt{X^3 - 7X + 6}}$$

are transcendental.

20.57 Show that

$$\frac{\int_2^{\alpha} \frac{\mathrm{d}X}{\sqrt{X^3 - 7X + 6}}}{\int_2^3 \frac{\mathrm{d}X}{\sqrt{X^3 - 7X + 6}}} = \frac{1}{2}$$

for

$$\alpha = 7 - 5\sqrt{2} + 2\sqrt{15} - \sqrt{30} = 2.1976733\ldots.$$

20.58 Let $\Omega = \omega_0(\mathbb{Z} + \mathbb{Z}i)$ as in (20.7). Show that $\zeta(\omega_0/2) = \pi/(2\omega_0)$.

20.59 Let $\Omega = \omega_0(\mathbb{Z} + \mathbb{Z}i)$ as in (20.7), with $\eta_0 = \eta(\omega_0)$.
(a) Show that $\wp(\omega_0/3) = \frac{1}{3}\sqrt{9 + 6\sqrt{3}}$.
(b) Show that $\wp(\omega_0/4) = 1 + \sqrt{2}$.
(c) Show that $\wp(\omega_0/5) = \sqrt{3 + 2\sqrt{5} + \sqrt{10 - 2\sqrt{5}} + \sqrt{10 + 2\sqrt{5}}}$.
(d) Show that $\zeta(\omega_0/3) - \eta_0/3 = \frac{1}{3}\sqrt[4]{9 + 6\sqrt{3}}$.
(e) Show that $\zeta(\omega_0/4) - \eta_0/4 = \frac{1}{2}(1 + \sqrt{2})$.

20.60 Show that

$$\int_0^1 \frac{X\mathrm{d}X}{\sqrt{X - X^3}}, \quad \int_4^5 \frac{(X - 8)\mathrm{d}X}{\sqrt{X^3 - 7X + 6}}$$

are transcendental.

20.61 Show that $\Gamma(1/3)/\pi^{1/3}$ and $\Gamma(1/3)/\pi^{2/3}$ are transcendental.

20.62 Show that

$$\zeta(3) = \int_0^1 \frac{X\mathrm{d}X}{\sqrt{X - X^3}} - \frac{1}{2}\int_{-1}^0 \frac{X\mathrm{d}X}{\sqrt{X^3 - X}} + \frac{1}{2}\int_1^{\wp(3)} \frac{X\mathrm{d}X}{\sqrt{X^3 - X}}$$

for $g_2 = 4, g_3 = 0$.

20.63 Let $\Omega = \omega_0(\mathbb{Z} + \mathbb{Z}i)$ as in (20.7).
(a) Show that $\sigma(\omega_0/2) = e^{\pi/8}/\sqrt[4]{2}$.
(b) Show that $\sigma(\omega_0/3) = \frac{1}{2}e^{\pi/18}\sqrt[12]{-576 + 384\sqrt{3}}$.

20.64 Show that

$$\exp\left(\int_4^5 \left(\frac{\sqrt{X^3 - 7X + 6} - 2\sqrt{3}}{X - 3} + 8\right)\frac{\mathrm{d}X}{\sqrt{X^3 - 7X + 6}}\right)$$

is transcendental [*Hint*: the v with $\wp(v) = 3, \wp'(v) = 2\sqrt{3}$ is not a rational multiple of a period].

20.65 Show that

$$\exp\left(\int_0^1 \frac{62X - 46399}{2X - 1009}\frac{2\sqrt{6}\mathrm{d}X}{\sqrt{4X^3 - 3704403X + 1364775049}}\right)$$

is algebraic [*Hint*: there are A, B in $\mathbb{Q}[U]$ with $A^2 - DB^2 = 1, B \neq 0$ – see also Exercise 18.12 – for $D = U(U^3 - 258U^2 + 9081U + 22680)$].

20.66 Show that

$$\exp\left(\int_0^1 \frac{Y-1}{2X-1177}\frac{2\sqrt{6}\mathrm{d}X}{\sqrt{4X^3-3704403X+1364775050}}\right)$$

is transcendental [*Hint*: non-torsion also here].

20.67 Show that for any $\beta \neq 0$ in $\mathfrak{M} = \mathfrak{M}_{\Omega,\Omega^*}$, the functions $\wp(z), \wp^*(\beta z)$ are algebraically dependent over $\mathbb{C}$.

20.68 Show that

$$\frac{\int_1^\infty \frac{\mathrm{d}X}{\sqrt{X^3-X}}}{\int_{-60}^\infty \frac{\mathrm{d}X}{\sqrt{X^3-6750X-189000}}}$$

is transcendental.

20.69 Show that

$$\frac{\int_{-5}^{-4} \frac{\mathrm{d}X}{\sqrt{6+7X-X^3}}}{\int_4^5 \frac{\mathrm{d}X}{\sqrt{X^3-7X+6}}} = 1.$$

20.70 Show that

$$\frac{\int_\tau^1 \frac{\mathrm{d}X}{\sqrt{-1+2X-X^3}}}{\int_{-3}^1 \frac{\mathrm{d}X}{\sqrt{X^3-7X+6}}} = 1$$

for $\tau = \frac{\sqrt{5}-1}{2}$ [*Hint*: $X = -\frac{T^2-T+1}{T-1}$ in the denominator]. This is related to isogenies of degree two.

20.71 (a) With

$$X = -\frac{T^2-T+1}{T-1}, \quad Y = i\frac{TU(T-2)}{(T-1)^2},$$

show that for any (T, U) with

$$U^2 = T^3 - 2T + 1, \quad T \neq 1$$

we have

$$Y^2 = X^3 - 7X + 6.$$

This defines an isogeny of degree two from the first elliptic curve above to the second.

(b) Find two rational functions in $\mathbb{C}(X, Y)$ similarly defining a map from the second elliptic curve above to the first. These will probably give the dual isogeny.

20.72 With $g_2 = 8, g_3 = -4$ and $g_2^* = 28, g_3^* = -24$ show that

$$\wp^*(iz) = -\frac{\wp(z)^2 - \wp(z) + 1}{\wp(z) - 1}.$$

20.73 Let Ω, Ω^* be lattices with Ω contained in Ω^*.

(a) Show that the corresponding Weierstrass functions satisfy

$$\wp^*(z) = \wp(z) + \sum\{\wp(z+\omega^*) - \wp(\omega^*)\},$$

where the sum is over all representatives ω^* of non-zero elements of Ω^*/Ω.

(b) Show that the corresponding invariants satisfy

$$g_2^* = g_2 + 10\sum \wp''(\omega^*), \quad g_3^* = g_3 + \frac{7}{6}\sum \wp''''(\omega^*).$$

20.74 Let

$$F(t) \;=\; F\left(\frac{1}{2}, \frac{1}{2}, 1; t\right) \;=\; \sum_{m=0}^{\infty} \frac{(2m)!^2}{2^{4m} m!^4} t^m$$

be a hypergeometric function, and let

$$K(t) \;=\; \frac{1}{3}(1-2t)F(t) + 2t(1-t)F'(t).$$

For complex λ with $|\lambda| < 1, |1-\lambda| < 1$ let Ω be a lattice with

$$g_2 \;=\; \frac{4}{3}(1-\lambda+\lambda^2), \quad g_3 \;=\; \frac{4}{27}(2-3\lambda-3\lambda^2+2\lambda^3).$$

Show that

$$\wp\left(\frac{\pi}{2}F(\lambda)\right) = \frac{1}{3}(2-\lambda), \quad \wp\left(\frac{\pi}{2}iF(1-\lambda)\right) = -\frac{1}{3}(1+\lambda),$$

$$\zeta\left(\frac{\pi}{2}F(\lambda)\right) = \frac{\pi}{2}K(\lambda), \quad \zeta\left(\frac{\pi}{2}iF(1-\lambda)\right) = -\frac{\pi}{2}iK(1-\lambda)$$

(Schmidt Doctoral Thesis 2015). This shows explicitly that the periods and quasi-periods are essentially hypergeometric.

20.75 With $\wp$ as in Exercise 20.74 show that $L + R = 0$, where

$$L = -36\wp^2\wp''^2 - 108\wp\wp'^2\wp'' - 432\wp^2\wp'' + 8\wp''^3 + 36\wp''^2 + 48\wp''$$

$$R = 432\wp^3\wp'^2 + 1296\wp^4 + 27\wp'^4 - 288\wp^2 + 16$$

(compare Exercise 20.20, where also the coefficients do not depend on the Weierstrass functions).

20.76 With F as in Exercise 20.74 show that ω_0 in (20.7) is

$$\int_0^1 \frac{\mathrm{d}X}{\sqrt{X-X^3}} = \frac{\pi\sqrt{2}}{2} F\left(\frac{1}{2}\right).$$

20.77 With F as in Exercise 20.74 show that a possible value of ω_0^{**} in Exercise 20.10 is

$$2\int_{15+\frac{45}{2}\sqrt{2}}^{\infty} \frac{\mathrm{d}X}{\sqrt{4X^3-6750X-94500}} = \frac{\pi}{30}(2-\sqrt{2})\sqrt{20+10\sqrt{2}}F(3-2\sqrt{2}).$$

20.78 Show that

$$2\int_{15+\frac{45}{2}\sqrt{2}}^{\infty} \frac{\mathrm{d}X}{\sqrt{4X^3-6750X-94500}} = \frac{\Gamma(\frac{1}{8})\Gamma(\frac{3}{8})}{12\sqrt{5\pi}}$$

(compare Chowla and Selberg (1967) – but this has to do with complex multiplication).

20.79 Let $g_2 = 6750, g_3 = 94500$.

(a) Show that

$$\wp(\sqrt{-2}z) = -\frac{1}{4}\frac{2\wp(z)^2+60\wp(z)+2025}{\wp(z)+30}$$

[*Hint*: Exercise 20.10].

(b) Show that

$$\zeta(\sqrt{-2}z)+\sqrt{-2}\zeta(z)-15\sqrt{-2}z = -\frac{\sqrt{-2}}{4}\frac{\wp'(z)}{\wp(z)+30}.$$

(c) Show that

$$\frac{\sigma(\sqrt{-2}z)^2}{\sigma(z)^4}e^{30z^2} = -2\frac{\wp(z)^2+60\wp(z)+900}{\wp(z)+30}.$$

(d) Show that $\frac{\sigma(\sqrt{-2}z)}{\sigma(z)^2}e^{15z^2}$ is not in $\mathbb{C}(\wp(z),\wp'(z))$.

20.80 (a) Find $R(X)$ in $\mathbb{C}(X)$ with

$$R(R(X)) = \frac{4X^4+13500X^2+756000X+11390625}{16X^3-27000X-378000}$$

[*Hint*: (20.4) above].

(b) Show that there is no $R(X)$ in $\mathbb{C}(X)$ with

$$R(R(X)) = \frac{4X^4+13500X^2+756000X+11390625}{16X^3-27000X-378001}.$$

20.81 Show that

$$\frac{\int_{-\frac{1967}{4}}^{-\frac{621}{4}} \frac{\mathrm{d}X}{\sqrt{94500+6750X-4X^3}}}{\int_{-29}^{-27} \frac{\mathrm{d}X}{\sqrt{4X^3-6750X-94500}}}$$

is algebraic [*Hint*: Exercise 20.79(a)].

20.82 a) Show that

$$\int \frac{\mathrm{d}X}{\sqrt[6]{X^{11}-11X^6-X}} = \frac{6}{5}\int \frac{\mathrm{d}T}{\sqrt{4T^3+6912}}$$

(with T a suitable algebraic function of X) is elliptic, even though the genus is 25 [*Hint*: Exercise 1.15].

(b) Show that

$$\int_\alpha^\beta \frac{\mathrm{d}X}{\sqrt[15]{X^{30}+522X^{25}-10005X^{20}-10005X^{10}-522X^5+1}}$$

is

$$\frac{B(\frac{1}{5},\frac{1}{3})+B(\frac{1}{3},\frac{7}{15})+B(\frac{7}{15},\frac{1}{5})}{5\sqrt[5]{1728}},$$

where

$$\alpha = \frac{\sqrt{5}-1-\sqrt{10-2\sqrt{5}}}{2}, \quad \beta = \frac{-\sqrt{5}-1+\sqrt{10-2\sqrt{5}}}{2}.$$

This is not elliptic as claimed in Greenhill (1892) (and the genus is now 196!).

(c) Do it all yourself for $\int \frac{\mathrm{d}X}{\sqrt[10]{X^{20}-228X^{15}+494X^{10}+228X^5+1}}$.

20.83 Let $\Omega = \mathbb{Z}\omega_1 + \mathbb{Z}\omega_2$ be such that $\mathfrak{K}_\Omega \neq \mathbb{Q}$ and g_2, g_3 are algebraic. With $\eta_1 = \eta(\omega_1), \eta_2 = \eta(\omega_2)$ show that there are algebraic σ, τ such that

$$\omega_2 = \tau\omega_1, \quad \eta_2 = \overline{\tau}\eta_1 + \sigma\omega_1$$

(Masser, 1975).

20.84 Let $\Omega = \mathbb{Z}+\mathbb{Z}\tau$ with $\tau = \sqrt{-2}$. Show that $\eta(\omega) = \frac{1}{\sqrt{2}}\pi\overline{\omega}+\frac{15}{2}(\omega_0^{**})^2\omega$ with ω_0^{**} as in Exercise 20.10 [*Hint*: Exercise 20.79(b)].

20.85 Show that $z, \wp(z), \zeta(z)$ are algebraically independent over $\mathbb{C}$.

20.86 Show that $z, \wp(z), \zeta(z), \sigma(z)$ are algebraically independent over $\mathbb{C}$.

20.87 Show that the Riemann zeta function and the Weierstrass zeta function are algebraically independent over $\mathbb{C}$ [*Hint*: the first is a function of s and the second is a function of z].

20.88 No, seriously, show that the Riemann zeta function $\zeta(z)$ and the Weierstrass zeta function $\zeta(z)$ are algebraically independent over $\mathbb{C}$. Well, you know what I mean...

20.89 Show that

$$G\frac{\partial \wp}{\partial g_2}(z) = \wp'(z)\left(-\frac{9}{2}g_3\zeta(z) + \frac{1}{4}g_2^2 z\right) - 9g_3\wp(z)^2 + \frac{1}{2}g_2^2\wp(z) + \frac{3}{2}g_2g_3,$$

$$G\frac{\partial \wp}{\partial g_3}(z) = \wp'(z)\left(3g_2\zeta(z) - \frac{9}{2}g_3 z\right) + 6g_2\wp(z)^2 - 9g_3\wp(z) - g_2^2,$$

$$G\frac{\partial \zeta}{\partial g_2}(z) = \frac{1}{2}\zeta(z)\left(9g_3\wp(z) + \frac{1}{2}g_2^2\right) - \frac{1}{2}g_2 z\left(\frac{1}{2}g_2\wp(z) + \frac{3}{4}g_3\right) + \frac{9}{4}g_3\wp'(z),$$

$$G\frac{\partial \zeta}{\partial g_3}(z) = -3\zeta(z)\left(g_2\wp(z) + \frac{3}{2}g_3\right) + \frac{1}{2}z\left(9g_3\wp(z) + \frac{1}{2}g_2^2\right) - \frac{3}{2}g_2\wp'(z),$$

for $G = g_2^3 - 27g_3^2$ (Halphen, 1886).

20.90 With g_2, g_3 as in Exercise 20.74 write $\wp_\lambda = \wp$. If $a \neq 0$ and b are constants, show that there is no non-empty open set of complex λ on which $\wp_\lambda(a\lambda) = b$ [*Hint*: first calculate $\zeta_\lambda(a\lambda)$ for the corresponding zeta function]. This is related to Ax–Schanuel for elliptic schemes.

20.91 Let $\Omega = \mathbb{Z} + \mathbb{Z}i$. Show that $\sum_{0\neq\omega\in\Omega} 1/\omega^k > 0$ $(k = 4, 8, 12, 16, \ldots)$.

20.92 For any a (not a pole of $\wp$) show that there is a non-zero polynomial P in $\mathbb{C}[X, Y]$, of degree at most 1 in X and of degree at most 1 in Y, such that $\operatorname{ord}_{z=a} P(z, \wp(z)) \geq 3$.

20.93 Show that there is a (not a pole of $\wp$) and a non-zero polynomial P in $\mathbb{C}[X, Y]$, of degree at most 1 in X and of degree at most 1 in Y, such that $\operatorname{ord}_{z=a} P(z, \wp(z)) \geq 4$.

20.94 Show that there is $\wp$ and a (not a pole of $\wp$) and a non-zero polynomial P in $\mathbb{C}[X, Y]$, of degree at most 1 in X and of degree at most 1 in Y, such that $\operatorname{ord}_{z=a} P(z, \wp(z)) \geq 5$.

20.95 Show that there is no $\wp$ and no a (not a pole of $\wp$) for which there exists a non-zero polynomial P in $\mathbb{C}[X, Y]$, of degree at most 1 in X and of degree at most 1 in Y, such that $\operatorname{ord}_{z=a} P(z, \wp(z)) \geq 6$.

20.96 Let Ω be a lattice, having basis elements ω_1, ω_2 and associated η_1, η_2, with determinant A.

(a) We already know that Legendre's $\Pi = \omega_2\eta_1 - \omega_1\eta_2$ is up to sign independent of the basis. Show that the same is true of $\Lambda = \overline{\omega_2}\eta_1 - \overline{\omega_1}\eta_2$ and $\Theta = \omega_2\overline{\omega_1} - \omega_1\overline{\omega_2}$ (compare Exercise 20.3); and that $\mu = \Lambda/\Theta$ and $\Pi/\Theta = \pi/A$ are literally independent of the basis.

(b) Show that

$$\zeta(z) + \mu z - \frac{\pi}{A}\bar{z}$$

is doubly periodic on Ω.

20.97 Show that there is no polynomial P in $\mathbb{C}[X,Y]$ with the property that $\sigma(z)\exp(P(z,\bar{z}))$ is doubly periodic on Ω.

20.98 (a) With μ as in Exercise 20.96(a) show that

$$\left|\sigma(z)\exp\left(\frac{\mu}{2}z^2\right)\right|\exp\left(-\frac{\pi}{2A}|z|^2\right)$$

is doubly periodic on Ω.

(b) For any κ and $\sigma_\kappa(z) = \sigma(z)\exp(\kappa z^2/2)$ show that

$$\limsup_{R\to\infty}\frac{\log|\sigma_\kappa|_R}{R^2} = \frac{\pi}{2A} + \frac{|\kappa-\mu|}{2}.$$

Thus it might have been slightly more efficient to use σ_μ instead of σ in the "analytic denominator" of $\wp$.

20.99 (a) For the function (2.27) show that

$$\sigma(z) = -\frac{i\omega_1}{2\pi}\exp\left(\frac{\eta_1 z^2}{2\omega_1} + \frac{\pi i z}{\omega_1}\right)\theta\left(\frac{z}{\omega_1} + \frac{\tau}{2} - \frac{1}{2}, \tau\right)\prod_{n=1}^{\infty}(1-q^n)^{-3}$$

with $q = \exp(2\pi i\tau)$, provided $\tau = \omega_2/\omega_1$ has positive imaginary part.

(b) Deduce that

$$e^{-\pi} + e^{-9\pi} + e^{-25\pi} + e^{-49\pi} + e^{-81\pi} + \cdots = \frac{(2^{1/4}-1)\Gamma(1/4)}{2^{11/4}\pi^{3/4}}.$$

20.100 (a) Show that the map from $\mathbb{C}\setminus\Omega$ to $\mathbb{C}^3$ defined by

$$\exp(z) = \sigma(z)^3(\wp(z), \wp'(z), 1)$$

with coordinates (X, Y, Z) defines also an isomorphism from $\mathbb{C}/\Omega$ to the variety E given in $\mathbb{P}_2$ by

$$Y^2Z = 4X^3 - g_2XZ^2 - g_3Z^3. \quad (*)$$

Thus E acquires from $\mathbb{C}/\Omega$ a commutative group law $\oplus$ (with $(0,1,0)$ as "origine du groupe").

(b) Show that for each π_1, π_2 in E there are polynomials $X_{\mathrm{add}}, Y_{\mathrm{add}}, Z_{\mathrm{add}}$ in $\mathbb{C}[X_1, Y_1, Z_1, X_2, Y_2, Z_2]$, bihomogeneous of the same bidegree in X_1, Y_1, Z_1 and in X_2, Y_2, Z_2, such that for all (ξ_1, η_1, ζ_1) in E near π_1 and all (ξ_2, η_2, ζ_2) in E near π_2 the point

$$(X_{\mathrm{add}}(\xi_1, \eta_1, \zeta_1, \xi_2, \eta_2, \zeta_2), Y_{\mathrm{add}}(\xi_1, \eta_1, \zeta_1, \xi_2, \eta_2, \zeta_2), Z_{\mathrm{add}}(\xi_1, \eta_1, \zeta_1, \xi_2, \eta_2, \zeta_2))$$

is $(\xi_1, \eta_1, \zeta_1) \oplus (\xi_2, \eta_2, \zeta_2)$ [*Hint*: (20.10) above]. Thus the group law is algebraically defined, making E into a group variety.

20.101 Define the group varieties

$$\mathbb{G}_{\mathrm{a}} \times E, \ E \times E, \ \mathbb{G}_{\mathrm{m}} \times E$$

as well as $E \times E^*$ for invariants g_2^*, g_3^*.

20.102 (a) Define Ω_{a} in $\mathbb{C}^2$ as the group of all $(\omega, -\eta(\omega))$ for ω in Ω. Show that the map from $(\mathbb{C} \setminus \Omega) \times \mathbb{C}$ to $\mathbb{C}^5$ defined by

$$\exp(z, w) = \sigma(z)^3(\wp(z), \wp'(z), 1, w + \zeta(z), (w + \zeta(z))\wp'(z) + 2\wp(z)^2)$$

with coordinates (X, Y, Z, W, V) defines also an isomorphism from $\mathbb{C}^2/\Omega_{\mathrm{a}}$ to the variety E_{a} given in $\mathbb{P}_4$ by $(*)$ of Exercise 20.100 and $VZ - YW = 2X^2$ with the line $X = Y = Z = 0$ removed. Thus E_{a} also acquires a commutative group law.

(b) As in Exercise 20.100(b) with π_1, π_2 in E_{a} and the polynomial ring $\mathbb{C}[X_1, Y_1, Z_1, W_1, V_1, X_2, Y_2, Z_2, W_2, V_2]$, making E_{a} also into a group variety [*Hint*: (20.20) above].

20.103 Show that the maps ι from $\mathbb{G}_{\mathrm{a}}$ to E_{a} defined by

$$\iota(W) = (0, 1, 0, 0, W)$$

and π from E_{a} to E defined by

$$\pi(X, Y, Z, W, V) = (X, Y, Z)$$

give rise to an exact sequence $0 \to \mathbb{G}_{\mathrm{a}} \to E_{\mathrm{a}} \to E \to 0$.

20.104 (a) For v in $\mathbb{C}$, not a rational multiple of a period, define Ω_v in $\mathbb{C}^2$ as the group of all $(\omega, -\lambda_v(\omega) + 2\pi ik)$ for ω in Ω and k in $\mathbb{Z}$. Show that the map from $(\mathbb{C} \setminus (\Omega \cup (\Omega + v) \cup (\Omega - v))) \times \mathbb{C}$ to $\mathbb{C}^5$ defined by

$$\exp(z, w) = \sigma(z)^3 \left(\wp(z), \wp'(z), 1, e^w \varphi_v(z), e^w \varphi_v(z)(\wp(z) + \frac{\wp'(z) - \wp'(v)}{\wp(z) - \wp(v)}) \right)$$

with coordinates (X, Y, Z, T, U) defines also an isomorphism from $\mathbb{C}^2/\Omega_v$ to a set E_v in $\mathbb{P}_4$ defined as follows. Write Σ for the projective surface given by $(*)$ of Exercise 20.100 and

$$XT(X-\wp(v)Z)+ZT(Y-\wp'(v)Z) \;=\; ZU(X-\wp(v)Z).$$

The plane Π_Z defined by $X=Z=0$ and the plane Π_v defined by $X=\wp(v)Z$, $Y=\wp'(v)Z$ lie in Σ (and in fact are components). The "elliptic curve" Ξ defined by (*) of Exercise 20.100 and $T=U=0$ also lies in Σ. Then E_v consists of the union of

(i) Σ deprived of Π_Z,Π_v,Ξ and the point $(\wp(v),-\wp'(v),1,0,0)$,

(ii) The line defined in Π_Z by $T=0$ but itself deprived of the points $(0,0,0,0,1)$ and $(0,1,0,0,0)$,

(iii) the line defined in Π_v by $U=\varrho_v T$ with $\varrho_v=\wp(v)+\wp''(v)/\wp'(v)$ but itself deprived of the points $(0,0,0,1,\varrho_v)$ and $(\wp(v),\wp'(v),1,0,0)$.

(It can be shown that E_v arises like E_{a} in Exercise 20.102(a) by removing some projective varieties from another projective variety, so is "quasi-projective" – see Exercise 20.106.) Thus E_v also acquires a commutative group law. The complications here suggest that our troubles in proving Theorem 20.11 may be difficult to avoid.

(b) As in Exercise 20.100(b) with π_1,π_2 in E_v and the polynomial ring $\mathbb{C}[X_1,Y_1,Z_1,T_1,U_1,X_2,Y_2,Z_2,T_2,U_2]$, making E_v also into a group variety [*Hint*: (20.23) – well, better not].

20.105 Show that the maps ι from $\mathbb{G}_{\mathrm{m}}$ to E_v defined by

$$\iota(T)=(0,-2,0,0,T)$$

and π from E_v to E defined by

$$\pi(X,Y,Z,T,U)=(X,Y,Z)$$

give rise to an exact sequence $1\to\mathbb{G}_{\mathrm{m}}\to E_v\to E\to 0$.

20.106 In the special case $g_2=0, g_3=8, \wp(v)=3, \wp'(v)=10$, show that E_v is defined by equations

$$Y^2Z=4X^3-8Z^3$$

$$XT(X-3Z)+ZT(Y-10Z) \;=\; ZU(X-3Z)$$

and

$$Y^2T-212XZT+28YZT-416Z^2T-4X^2U-16XZU+4YZU+124Z^2U=0$$

$$XYT+34XZT-4YZT+76Z^2T+4XZU-YZU-22Z^2U=0$$

(Kraft, personal communication, 2015), with Ξ and $\Lambda=\Pi_Z\cap\Pi_v$ from Exercise 20.104(a) removed (here Λ is defined by $X=Y=Z=0$). Thus indeed E_v is quasi-projective. As far as I know this is the first explicit example of such a multiplicative extension. I would not like to have to write down the addition law.

21

Modular functions

The star of this chapter is the function J, although a strong supporting role is played by j (which already had minor parts in some exercises of the previous chapter). The former can be defined by

$$J(q) = \frac{\left(1 + 240 \sum_{m=1}^{\infty} \frac{m^3 q^m}{1-q^m}\right)^3}{q \prod_{m=1}^{\infty}(1 - q^m)^{24}},$$

which makes sense for all complex q with $0 < |q| < 1$. For its properties we shall make constant reference to Lang (1973), here abbreviated to L, as well as Rankin (1977), abbreviated to R. It is analytic there (L pp. 39, 44, 249; or R p. 199), but its Laurent expansion

$$q^{-1} + 744 + 196884q + 21493760q^2 + 864299970q^3 + 20245856256q^4 + \cdots \tag{21.1}$$

(L p. 45 or R p. 199) is somewhat less illuminating than moonshine.

These q are exactly the values of $e^{2\pi i\tau}$ for $\tau = x + iy$ in the upper half-plane $\mathbb{H}$, that is, with x, y real and $y > 0$. Accordingly one defines

$$j(\tau) = J(e^{2\pi i\tau});$$

one of the most amazing functions in all of mathematics. For this we can prove a transcendence result relatively easily (that is, without a direct use of auxiliary polynomials).

Theorem 21.1 *Suppose α in $\mathbb{H}$ is algebraic but is not a quadratic irrational. Then $j(\alpha)$ is transcendental.*

The proof goes via Schneider–Lang, but not applied to j itself. In fact there is a differential equation

$$2j^2(j-1728)^2 j' j''' - 3j^2(j-1728)^2 j''^2 + (j^2 - 1968j + 2654208)j'^4 = 0$$

(not so easy to find in the literature – see Exercise 21.15) of order 3, but this is about as unhelpful as that for the sigma function. And anyway j is not meromorphic on $\mathbb{C}$ (or even defined there) as required by Theorem 19.1.

For the proof we have to know two things about j.

First, any τ in $\mathbb{H}$ determines a lattice $\mathbb{Z}+\mathbb{Z}\tau$, so invariants g_2, g_3 as in Chapter 20; and it is a fact that

$$j(\tau) = \frac{1728 g_2^3}{g_2^3 - 27 g_3^2} \tag{21.2}$$

(see L p.44 or R p.213). Second, it is a fact that $j(\tau_1) = j(\tau_2)$ if and only if the lattices $\mathbb{Z}+\mathbb{Z}\tau_1, \mathbb{Z}+\mathbb{Z}\tau_2$ are proportional (see L p.39 or R p.199). In particular

$$\mathbb{Z}+\mathbb{Z}\tau = \mathbb{Z}+\mathbb{Z}(\tau+1), \quad \mathbb{Z}+\mathbb{Z}\tau = \tau(\mathbb{Z}+\mathbb{Z}(-1/\tau))$$

and so

$$j(\tau) = j(\tau+1), \quad j(\tau) = j\left(-\frac{1}{\tau}\right) \tag{21.3}$$

(together with the generalization to $SL_2(\mathbb{Z})$ – see Exercise 21.3). Third, that j takes – no, three, that's three things we have to know (among the things we have to know are…). But that j takes all complex values is equivalent to the fact remarked in Chapter 20 that there is a Weierstrass function with any given invariants (see L p.39 or R p.199).

Now we can start the proof of Theorem 21.1. We assume that $j(\alpha)$ is algebraic and we will deduce that α is quadratic.

We can find algebraic g_2, g_3 such that $g_2^3 \neq 27 g_3^2$ and

$$j(\alpha) = \frac{1728 g_2^3}{g_2^3 - 27 g_3^2}. \tag{21.4}$$

For example we can take

$$g_2 = g_3 = \frac{27 j(\alpha)}{j(\alpha) - 1728}, \tag{21.5}$$

provided $j(\alpha) \neq 0, 1728$. If $j(\alpha) = 1728$ we take $g_2 = 4, g_3 = 0$, and if $j(\alpha) = 0$ we take $g_2 = 0, g_3 = 4$, as in Chapter 20.

Let $\wp$ be the Weierstrass function whose invariants are these g_2, g_3. It has a period lattice $\Omega = \mathbb{Z}\omega_1 + \mathbb{Z}\omega_2$. Now for any complex z not in $\mathbb{R}$, precisely one of $z = x+iy$ and $1/z = (x-iy)/(x^2+y^2)$ is in $\mathbb{H}$. Thus by a possible interchange of ω_1, ω_2 we can assume that $\tau = \omega_2/\omega_1$ is in $\mathbb{H}$. Now $\Omega = \omega_1(\mathbb{Z}+\mathbb{Z}\tau)$ and we deduce easily from the behaviour of g_2, g_3 under scalar multiplication of the lattice that (21.2) holds. Comparing with (21.4), we see that $\Omega = \tilde{\omega}_1(\mathbb{Z}+\mathbb{Z}\alpha)$.

We now apply elliptic Gelfond–Schneider, Theorem 20.5, with our $\wp$ and $u = \tilde{\omega}_1/2, \beta = \alpha$. Then u and $\beta u = \tilde{\omega}_1\alpha/2$ are both half-periods not in Ω, and so $\wp(u), \wp(\beta u)$ are both algebraic. The only way out is that $\beta = \alpha$ is in the set $\mathfrak{K} = \mathfrak{K}_\Omega$. We already pointed out in Chapter 20 that this set is either $\mathbb{Q}$ or a complex quadratic extension. As α is not real it must therefore be quadratic. This completes the proof of Theorem 21.1. The reader may observe that, in contrast to earlier situations, we did not need a preliminary result on algebraic independence of functions, here τ and $j(\tau)$. She may consult Exercise 21.13 to fill the gap.

The exceptions in Theorem 21.1 are genuine: if α in the upper half plane is quadratic irrational, then $j(\alpha)$ is algebraic. We sketch a proof using well-known properties of Wronskians (compare the proof of Proposition 12.2).

The formula (21.5) can be used with any complex $\xi \neq 0, 1728$ in place of $j(\alpha)$ to construct $\wp_\xi$ and Ω_ξ. For each d, L in $\mathbb{N}$ we can then define a set $S(d, L)$ consisting of all the Laurent coefficients at $z = 0$ of the Wronskian of the $\wp_\xi(z)^i\wp_\xi(\sqrt{-d}z)^j$ $(i, j = 0, 1, \ldots, L)$. By the remarks around (20.3) we see that the elements of $S(d, L)$ lie in $\mathbb{Q}[\xi, 1/(\xi - 1728)]$.

Now if α as above is quadratic irrational, there are $a \neq 0, b, c$ in $\mathbb{Z}$ with $a\alpha^2 + b\alpha + c = 0$. We check easily that $a\alpha(\mathbb{Z} + \mathbb{Z}\alpha) \subseteq \mathbb{Z} + \mathbb{Z}\alpha$. So α is in $\mathfrak{K} = \mathfrak{K}_{\mathbb{Z}+\mathbb{Z}\alpha}$.

Thus also $\sqrt{-d}$ is in $\mathfrak{K}$ for any d in $\mathbb{N}$ with $\mathbb{Q}(\alpha) = \mathbb{Q}(\sqrt{-d})$ (for example $d = 4ac - b^2 < 0$).

If $j(\alpha) = 0, 1728$ we are done; and otherwise the above construction for $\xi = j(\alpha)$ gives $\wp, \Omega$. As in the proof of Theorem 21.1 we see that $\Omega = \tilde{\omega}_1$ $(\mathbb{Z} + \mathbb{Z}\alpha)$. Now $\mathfrak{K} = \mathfrak{K}_\Omega$ and so Lemma 20.4 implies the algebraic dependence of $\wp(z), \wp(\sqrt{-d}z)$. Thus there is L such that all elements of $S(d, L)$ vanish at $\xi = j(\alpha)$.

On the other hand, not all the elements of this $S(d, L)$ vanish at a generic ξ and indeed at any $\eta = j(\beta)$ with $\mathfrak{K}_{\mathbb{Z}+\mathbb{Z}\beta} = \mathbb{Q}$ (for example non-real $\beta = \sqrt[3]{2}$), else $\wp_\eta(z), \wp_\eta(\sqrt{-d}z)$ would be algebraically dependent, contradicting Lemma 20.3. Thus some element of $S(d, L)$ leads to a non-trivial algebraic equation for $j(\alpha)$.

Better proofs come from modular function theory (see L p. 57 or p.123), and using this we can give some examples. Thus

$$j(i) = 1728, \quad j(2i) = 287496, \quad j(3i) = 76771008 + 44330496\sqrt{3} \tag{21.6}$$

(of which the first comes from $g_3 = 0$) and

$$\begin{aligned} j(4i) &= 41113158120 + 29071392966\sqrt{2}, \\ j(5i) &= 22015749613248 + 9845745509376\sqrt{5}, \end{aligned}$$

but $j(6i) = \beta_1 + \beta_2$ where β_1 is

$$5894625992142600 + 3403263903336192\sqrt{3},$$

and β_2 is

$$7056\sqrt{139580696456091316002410 4 + 80586952672599764258761 4}.$$

And with $i^* = e^{2\pi i/3}$

$$j(i^*) = 0, \ \ j(2i^*) = 54000, \ \ j(3i^*) = -12288000$$

and

$$\begin{aligned} j(4i^*) &= 1417905000 + 818626500\sqrt{3}, \\ j(5i^*) &= -327201914880 - 146329141248\sqrt{5}, \end{aligned}$$

but

$$j(6i^*) = 50337742902000 + 39953093016000\sqrt[3]{2} + 31710790944000\sqrt[3]{4}.$$

And of course

$$j\left(\frac{-1+\sqrt{-163}}{2}\right) = -262537412640768000$$

(as in Exercise 20.42) explaining (10.2) and (19.2) with the help of (21.1). For a couple more see Exercises 20.43 and 20.44.

An analogue of Theorem 21.1 holds for J but the proof involves new considerations as the natural domain, the unit disc $\mathbb{D}$ ($\neq \mathbb{D}_1$) defined by $|q| < 1$, is now yet smaller than the upper half plane.

Theorem 21.2 *Suppose β is algebraic in $\mathbb{C}$ with $0 < |\beta| < 1$. Then $J(\beta)$ is transcendental.*

This in turn has an elliptic formulation. We already saw in the previous chapter that the irrational ratio τ of two periods of a Weierstrass function with algebraic invariants is transcendental except in the CM case. Now we see that $e^{2\pi i\tau}$ is transcendental without exception.

We start the proof again on traditional lines. By now we well recognize the necessity of the following result.

Lemma 21.3 *The functions q and $J(q)$ are algebraically independent over $\mathbb{C}$.*

Proof This has to be done in the upper half plane. It then takes a familiar pattern. If the lemma is false, there is non-zero P in $\mathbb{C}[X,Y]$ with $P(e^{2\pi i\tau}, j(\tau)) = 0$. Fix any non-quadratic τ_0 with $P_0(X) = P(X, j(\tau_0)) \neq 0$ (again we are using j-surjectivity). Then for any integer d and $\sigma_d = -1/(\tau_0 + d)$ we have $j(\sigma_d) = j(\tau_0)$ by (21.3). We deduce $P_0(e^{2\pi i \sigma_d}) = 0$. But it is easy to check that all the σ_d are different modulo $\mathbb{Z}$. □

Now for the auxiliary polynomial.

Lemma 21.4 *For any $L \geq 1, T \geq 1$ in $\mathbb{Z}$ with*

$$(L+1)^2 \geq 2(L+T) \tag{21.7}$$

there is P in $\mathbb{Z}[X, Y]$, of degree at most L in each variable and with coefficients of absolute values at most $2^{96L}L^{24L}$, such that

$$\phi(q) = P(q, J(q)) \tag{21.8}$$

is not identically zero and

$$\operatorname{ord}_{q=0} \phi(q) \geq T. \tag{21.9}$$

Proof If we use the Laurent expansion (21.1) directly, then we get a slightly worse estimate for the coefficients (but still adequate for our needs). It is easier to use the original definition. This involves

$$Q = Q(q) = 1 + 240 \sum_{m=1}^{\infty} \frac{m^3 q^m}{1 - q^m}$$

in the numerator, now analytic on $\mathbb{D}$. To simplify the denominator

$$\Delta = \Delta(q) = q \prod_{m=1}^{\infty} (1 - q^m)^{24}$$

(there are already too many meanings of Δ around, but each is more traditional than the others) we need also

$$R = R(q) = 1 - 504 \sum_{m=1}^{\infty} \frac{m^5 q^m}{1 - q^m},$$

also analytic on $\mathbb{D}$. Here Δ, Q, R are all modular forms, and it is classical that $1728\Delta = Q^3 - R^2$ (see L p.249 again or R p.197).

We now find rational integers p_{ij}, not all zero, such that in

$$\sum_{i=0}^{L} \sum_{j=0}^{L} p_{ij} q^i (1728 Q^3)^j (Q^3 - R^2)^{L-j} \tag{21.10}$$

the Taylor coefficients of $1, q, \ldots, q^{T+L-1}$ are zero. This gives $M = T + L$ equations in $N = (L+1)^2$ unknowns, and the Siegel exponent $M/(N-M) \leq 1$ in Proposition 8.3.

It is handy to use the standard majorizing notation $A \ll \overline{A}$ for $A = \sum a_n q^n$ in $\mathbb{C}[[q]]$ and $\overline{A} = \sum \overline{a_n} q^n$ in $\mathbb{R}[[q]]$ to mean $|a_n| \le \overline{a_n}$ for all n (compare Exercise 4.39). Thus for

$$Q = 1 + 240 \sum_{n=1}^{\infty} q^n \sum_{d|n} d^3 = 1 + 240 \sum_{n=1}^{\infty} n^3 q^n \sum_{d|n} d^{-3},$$

we get $Q \ll 1440\zeta(3)(1-q)^{-4}$. Similarly $R \ll 60480\zeta(5)(1-q)^{-6}$.

It follows that

$$q^i (1728 Q^3)^j (Q^3 - R^2)^{L-j} \ll 2^{48L} q^i (1-q)^{-12L} \ll 2^{48L}(1-q)^{-12L-1}.$$

Thus the coefficient of q^t in our linear equations is of absolute value at most

$$2^{48L}\binom{12L+t}{t} = 2^{48L}\binom{12L+t}{12L}, \tag{21.11}$$

which for $t \le T + L - 1$ is at most $2^{48L}(13L+T)^{12L}$. As $T \le (L^2+1)/2$ we end up with the required bound for the $|p_{ij}|$. The lemma follows after dividing (21.10) by $(Q^3 - R^2)^L$, which has order exactly L at $q = 0$. Finally note that $\phi \ne 0$ by Lemma 21.3. □

Write $\tilde{T} = \mathrm{ord}_{q=0}\, \phi(q)$, so that $\tilde{T} \ge T$ and $\psi(q) = \phi(q)/q^{\tilde{T}}$ is analytic on $\mathbb{D}$. The next result is a concealed application of the Schwarz Lemma.

Lemma 21.5 *For $\Psi = (Q^3 - R^2)^L \psi / q^L$ and any r with $0 \le r < 1$ we have*

$$|\Psi|_r \le 2^{168L} L^{36L} \tilde{T}^{12L} (1-r)^{-12L-1}.$$

Proof By (21.11) we have $(Q^3 - R^2)^L \phi = \sum_{t=0}^{\infty} c_t q^t$ with c_t in $\mathbb{Z}$ satisfying

$$|c_t| \le 2^{48L}\binom{12L+t}{12L} \sum_{i=0}^{L}\sum_{j=0}^{L} |p_{ij}|$$

and $c_0 = c_1 = \cdots = c_{\tilde{T}+L-1} = 0$. Thus

$$\Psi(q) = \sum_{t=\tilde{T}+L}^{\infty} c_t q^{t-\tilde{T}-L} = \sum_{n=0}^{\infty} c_{\tilde{T}+L+n} q^n \tag{21.12}$$

is majorized by

$$2^{48L}(L+1)^2 2^{96L} L^{24L} \sum_{n=0}^{\infty} \binom{13L+\tilde{T}+n}{12L} q^n.$$

Here the binomial coefficient is (since $x+y+z \leq 3xyz$ for $x \geq 1, y \geq 1, z \geq 1$)

$$\frac{1}{(12L)!}\prod_{l=1}^{12L}(l+L+\tilde{T}+n) \leq \frac{1}{(12L)!}\prod_{l=1}^{12L} 3L\tilde{T}(l+n) = (3L\tilde{T})^{12L}\binom{12L+n}{12L}.$$

This leads easily to the required result. It was important here to avoid any exponential dependence on $\tilde{T}$; for one thing we don't yet have upper bounds for this quantity. □

Lemma 21.6 *Given any $q_0 \neq 0$ in $\mathbb{D}$, there is C depending only on q_0, and a positive integer*

$$s \leq C\sqrt{L\log(L\tilde{T})}$$

such that $\phi(q_0^s) \neq 0$.

Proof Let S be a positive integer with $\phi(q_0^s) = 0$ $(s = 1, \ldots, S)$. Then

$$\Psi(q)/\prod_{s=1}^{S}(q - q_0^s)$$

is analytic on $\mathbb{D}$, and the usual maximum modulus principle gives

$$\frac{|\Psi(0)|}{\prod_{s=1}^{S}|q_0|^s} \leq \frac{|\Psi|_r}{\prod_{s=1}^{S}(r - |q_0|^s)}$$

for any fixed r with $|q_0| < r < 1$.

On the other hand in the notation (21.12) we have $\Psi(0) = c_{\tilde{T}+L} \neq 0$, so recalling from the misty past the legendary Fundamental Theorem we get $|\Psi(0)| \geq 1$. Using Lemma 21.5 now gives

$$\frac{S(S+1)}{2}\log|q_0|^{-1} \leq C_0(L\log L + L\log\tilde{T} + S)$$

with C_0 depending only on q_0. This leads quickly to the desired result, and at the same time lifts the burden of calculating the constants. □

The above result is of course a zero estimate. In fact it is unusually (and at first sight implausibly) strong: we expect (see Exercise 21.7) that $\tilde{T}$ is of order at most L^2, so we get s of order at most $\sqrt{L\log L}$, much smaller than the usual bound of order L^2 for the number of zeroes. An explanation is the clustering effect of the zeroes $q_0, q_0^2, \ldots, q_0^S$, somewhat as in the proof of the Bombieri–Pila Theorem 18.2. We also exploited crucially the fact that P is in $\mathbb{Z}[X,Y]$, even with smallish coefficients, which is not usually assumed in zero estimates such as Proposition 8.1 in Mahler's Method (but see the remarks about impurity near the end of Chapter 12).

In fact there are closer analogues with this method, because of a functional equation relating $J(q)$ and $J(q^2)$. It is

$$F_2(J(q), J(q^2)) = 0 \tag{21.13}$$

with the far-from-obvious

$$F_2(X,Y) = X^3 + Y^3 - X^2Y^2 + 1488(X^2Y + XY^2) - 162000(X^2 + Y^2)$$

$$+ 40773375XY + 8748000000(X + Y) - 157464000000000.$$

(see R p.230). But up to now no-one has succeeded in applying Mahler's Method to prove Theorem 21.2.

Actually we will need a whole family

$$F_s(J(q), J(q^s)) = 0 \quad (s = 1, 2, 3, \ldots)$$

of such equations, with $F_s(X, Y)$ in $\mathbb{Z}[X, Y]$, the modular transformation polynomials, defined by

$$F_s(j(\tau), Y) = \prod_{C_s} \left(Y - j\left(\frac{a\tau + b}{d}\right)\right). \tag{21.14}$$

Here C_s is the set of (a, b, d) in $\mathbb{Z}^3$ with

$$a \geq 1,\ d \geq 1,\ 0 \leq b < d,\ ad = s$$

and a, b, d coprime. The cardinality of C_s is

$$\lambda(s) = s \prod_{p|s} \left(1 + \frac{1}{p}\right),$$

so this is also the degree of F_s in Y. Surprisingly $F_s(X, Y) = F_s(Y, X)$ for $s > 1$, so it is also the degree in X (for all this see L pp.51–55 or R pp.209–210).

The reader may imagine that the coefficients here get rather large. According to Oliver Atkin (personal communication, 1985) the constant term of F_{17} has 213 decimal digits. See also Exercises 21.20, 21.21 and 21.22.

By the way, these F_s give a better proof that if α in $\mathbb{H}$ is quadratic, then $j(\alpha)$ is algebraic (and even an algebraic integer). For example inserting $q = e^{-2\pi i/\sqrt{-2}}$ into (21.13) we get $F_2(j, j) = 0$ for $j = j(\sqrt{-2})$. This has three solutions $1728, -3375, 8000$, all in the list (20.11); and a rough numerical test proves $j(\sqrt{-2}) = 8000$. Similarly $j(2\sqrt{-2}) = 26125000 + 18473000\sqrt{2}$.

Here we shall use the F_s to estimate the (logarithmic) height of some J-values. But first a simple analytic result.

Lemma 21.7 *For x, y real with $y > 0$ we have*

$$\max\{1, |j(x+iy)|\} \le e^{2\pi(1+y+1/y)}.$$

Proof It is easy to see that all the coefficients in (21.1) are non-negative. Denoting them by j_n we deduce

$$|j(x+iy)| \le \sum_{n=-1}^{\infty} j_n e^{-2\pi ny} = j(iy).$$

If $y \ge 1$ this is at most

$$e^{2\pi y} + \sum_{n=0}^{\infty} j_n e^{-2\pi n} \;=\; e^{2\pi y} + j(i) - e^{2\pi} \;\le\; e^{2\pi y}(1 + (j(i) - e^{2\pi})e^{-2\pi})$$

which is

$$e^{2\pi y} j(i) e^{-2\pi} \le e^{2\pi(1+y)}$$

by (21.6). If $y \le 1$ we use the same argument noting that $j(iy) = j(i/y)$. □

By the way, these j_n can get quite large, much larger than the corresponding Taylor coefficients of Q, R (see Exercise 21.4).

Lemma 21.8 *Given any $q_0 \ne 0$ in $\mathbb{D}$ such that $J(q_0)$ is algebraic, there is C depending only on q_0 such that for any s in $\mathbb{N}$ the number $\alpha_s = J(q_0^s)$ is algebraic with $[\mathbb{Q}(\alpha_s) : \mathbb{Q}] \le Cs^{3/2}$ and*

$$[\mathbb{Q}(\alpha_s) : \mathbb{Q}] h(\alpha_s) \le Cs^2.$$

Proof Write $q_0 = e^{2\pi i\tau}$. We have $F_s(\alpha_1, \alpha_s) = 0$ for $\alpha_1 = J(q_0) = j(\tau)$, so certainly α_s is algebraic, with degree $d_s \le d_1\lambda(s)$ with d_1 the degree of α_1. Here

$$\lambda(s) \;\le\; \frac{3}{2}s \prod_{2<p|s} \sqrt{p} \;\le\; \frac{3}{2}s^{3/2},$$

giving the required bound for d_s.

To get at the height we have to investigate the conjugates of α_s. Those just over $\mathbb{Q}(\alpha_1)$ are among the $j((a\tau + b)/d)$ in (21.14). We can get all of them by writing each conjugate $\sigma(\alpha_1)$ of α_1 over $\mathbb{Q}$ as $j(\tau_\sigma)$. They are then among the $j((a\tau_\sigma + b)/d)$.

Now consider the resultant $A_s(Y)$ with respect to X of the minimal polynomial $A_1(X)$ of α_1 over $\mathbb{Z}$ and $F_s(X, Y)$. Then $A_s(\alpha_s) = 0$. It is easily checked with the Sylvester determinant (5.3) that the leading coefficient of A_s is $a_1^{\lambda(s)}$, where a_1 in $\mathbb{N}$ is the leading coefficient of A_1. As A_s is divisible by the minimal polynomial of α_s over $\mathbb{Z}$, we get using (14.3)

$$d_s h(\alpha_s) \le \lambda(s) \log a_1 + \sum_{C_{s,\sigma}} \log \max \left\{1, \left| j\left(\frac{a\tau_\sigma + b}{d}\right)\right|\right\}.$$

By Lemma 21.7 the sum here is at most

$$2\pi \sum_{C_{s,\sigma}} \left(1 + \frac{ay_\sigma}{d} + \frac{d}{ay_\sigma}\right) \le 3\pi \sum_{C_{s,\sigma}} \left(\frac{ay_\sigma}{d} + \frac{d}{ay_\sigma}\right)$$

with $y_\sigma = \Im\tau_\sigma$ for the imaginary part. Thus up to constants depending only on q_0 we get

$$\sum_{C_s} \left(\frac{s}{d^2} + \frac{d^2}{s}\right) \le \sum_{d|s} \left(\frac{s}{d} + \frac{d^3}{s}\right) \le s^2 + s^2 \sum_{d|s} \frac{1}{d^3} \le 3s^2. \quad (21.15)$$

This completes the proof. □

In fact the s^2 in (21.15) can be replaced by $s \log s \log\log s$ for $s \ge 3$; but the proof needs upper bounds for the coefficients of F_s, not a classical matter. The asymptotics for these were proved only in 1984 by Paula Cohen (Cohen, 1984), who showed that the bad terms d^3/s in (21.15) get averaged away. Our weaker Lemma 21.8 is compensated by the strength of Lemma 21.6. See also Exercise 21.1(b).

The proof of Theorem 21.2 is now quite fast.

First we choose L large in Lemma 21.4, and then fix $T = [(L^2+1)/2]$ to secure (21.7). Thus $\tilde{T} \ge T \ge L^2/2$. From now on we use constants C depending only on β.

By Lemma 21.6 there is $s \le C\sqrt{L \log \tilde{T}}$ with $\phi(\beta^s) \ne 0$.

Clearly $|\Delta(\beta^s)/\beta^s| \ge C^{-1}$ and now by Lemma 21.5 with $r = |\beta^s|$ we get

$$\log |\phi(\beta^s)| \le CL \log \tilde{T} - s\tilde{T} \log |\beta|^{-1}. \quad (21.16)$$

On the other hand $\phi(\beta^s) = P(\beta^s, J(\beta^s))$ giving via Proposition 14.7 and Lemma 21.8

$$h(\phi(\beta^s)) \le CL(\log L + s + s^2/d_s)$$

for $d_s = [\mathbb{Q}(J(\beta^s)) : \mathbb{Q}]$.

By the Liouville inequality of Proposition 14.13 we get

$$\log |\phi(\beta^s)| \ge -[\mathbb{Q}(\beta) : \mathbb{Q}] d_s h(\phi(\beta^s)) \ge -CL(d_s \log L + d_s s + s^2).$$

Therefore the $s\tilde{T}$ in (21.16) has to beat large multiples of

$$L \log \tilde{T},\ L d_s \log L,\ L d_s s,\ L s^2$$

and by now the checking of this sort of thing (for L large enough) should be a piece of cake.

This completes the proof of "Le Théorème Stéphanois", having been found in 1996 by Barré, Diaz, Gramain, and Philibert all at Saint-Étienne. Their proof (Barré et al., 1996) is in the spirit of Schneider's Second Problem (Schneider, 1957). The result had been conjectured by Mahler back in 1969 and was also mentioned by him in the first of two papers dealing also with the coefficients of F_s. It was independently mentioned by Manin in 1971. The proof led soon afterwards to Nesterenko's work on modular forms (see the next chapter).

I know of no higher-dimensional extensions, for example to Siegel modular functions. Recently the j-function has attracted much interest due to conjectures of André–Oort (see for example André (1989) and chapter 4 of Zannier (2012) for more). In their very simplest forms, these involve solving polynomial equations in values $j(\tau)$ with τ algebraic of degree 2 (such as those in (21.6) *et seq.*). These values are considered to be analogues of roots of unity in the conjectures of Manin–Mumford mentioned in Chapter 14. The less simple forms do involve Siegel modular functions.

Exercises

21.1 (a) For $\tau = iy$ and $1 \leq y \leq 2$ show that the length $\mathcal{L}(F_s(j(\tau), Y)) \leq e^{10\pi s^2}$.

(b) Deduce that there is C with $\mathcal{L}(F_s(X, Y)) \leq C^{s^2}$ for every $s \geq 1$ [*Hint*: Lagrange or Newton].

21.2 Show that $j(\sqrt{-2}/2) = 8000$.

21.3 Show that $j(\tau) = j(\frac{a\tau+b}{c\tau+d})$ for any integers a, b, c, d with $ad - bc = 1$.

21.4 Let $J(q) = q^{-1} + \sum_{n=0}^{\infty} j_n q^n$ be the Laurent expansion.
(a) Show that $0 \leq j_n \leq \exp(8\pi\sqrt{n})$ ($n = 1, 2, \ldots$).
(b) Show that $\limsup_{n\to\infty}(\log j_n)/\sqrt{n} > 0$.

* * * * * * * * * * * * * * * * * * *

21.5 Show that

$$j(2\tau) + j\left(\frac{\tau}{2}\right) + j\left(\frac{\tau+1}{2}\right) = j(\tau)^2 - 1488j(\tau) + 162000.$$

21.6 Let $\alpha, \beta, \gamma, \delta$ be real with $\alpha\delta - \beta\gamma > 0$ such that $j(\tau) = j(\frac{\alpha\tau+\beta}{\gamma\tau+\delta})$. Show that there is real θ and integers a, b, c, d with $\alpha = \theta a, \beta = \theta b, \gamma = \theta c, \delta = \theta d$ and $ad - bc = 1$.

21.7 Show that there is C such that $\mathrm{ord}_{q=0} P(q, J(q)) \leq CL^2$ for any non-zero polynomial P in $\mathbb{C}[X, Y]$ of degree at most L in each variable (Philibert, 1997).

21.8 Show that $F_s(j(\tau), j(\frac{a\tau+b}{c\tau+d})) = 0$ for any integers a, b, c, d with $ad - bc = n > 0$, where $s = n/k^2$ for the highest common factor k of a, b, c, d.

21.9 Let $\alpha, \beta, \gamma, \delta$ be real with $\alpha\delta - \beta\gamma > 0$ such that $j(\tau)$ and $j(\frac{\alpha\tau+\beta}{\gamma\tau+\delta})$ are algebraically dependent over $\mathbb{C}$. Show that there is real θ and integers a, b, c, d with $\alpha = \theta a, \beta = \theta b, \gamma = \theta c, \delta = \theta d$ [*Hint*: $\tau \to \tau + n$ and use Exercise 4.36 to show that α/γ is rational]. For vast generalizations see Pila (2011).

21.10 For $E_4(\tau) = Q(e^{2\pi i\tau})$ and $E_6(\tau) = R(e^{2\pi i\tau})$ show that

$$E_4\left(-\frac{1}{\tau}\right) = \tau^4 E_4(\tau), \quad E_6\left(-\frac{1}{\tau}\right) = \tau^6 E_6(\tau).$$

21.11 Show that $q, Q(q), R(q)$ are algebraically independent over $\mathbb{C}$.

21.12 Show that $j(\tau), j(\tau^2)$ are algebraically independent over $\mathbb{C}$.

21.13 Show that $\tau, e^{2\pi i\tau}, j(\tau)$ are algebraically independent over $\mathbb{C}$.

21.14 For τ in $\mathbb{H}$ show that

$$g_2(\mathbb{Z} + \mathbb{Z}\tau) = \frac{4\pi^4}{3} E_4(\tau), \quad g_3(\mathbb{Z} + \mathbb{Z}\tau) = \frac{8\pi^6}{27} E_6(\tau).$$

21.15 Show that

$$2j^2(j - 1728)^2 j' j''' - 3j^2(j - 1728)^2 j''^2 + (j^2 - 1968j + 2654208)j'^4 = 0.$$

[*Hint:* Exercises 1.16 and 19.15]

21.16 Show that $j'(i) = j'(i^*) = j''(i^*) = 0$ and

$$j''(i) = -162\frac{\Gamma(\frac{1}{4})^8}{\pi^4}, \quad j'''(i^*) = -162i\frac{\Gamma(\frac{1}{3})^{18}}{\pi^9}.$$

21.17 With ω_0^{**} as in Exercise 20.77 show that

$$j'(\sqrt{-2}) = -\frac{1008000i(\omega_0^{**})^2}{\pi}.$$

21.18 Show that there are no τ_1, τ_2 in $\mathbb{H}$, algebraic of degree 2, with $j(\tau_1) + j(\tau_2) = 1$ (Kühne, 2013; Bilu et al., 2013).

21.19 Show that there are no τ_1, τ_2 in $\mathbb{H}$, algebraic of degree 2, with $P(j(\tau_1), j(\tau_2)) = 0$ for the polynomial P of Exercise 1.14 (this implies that there are no $\lambda \neq 0, 1, -1$ such that the elliptic curves $Y^2 = X(X-1)(X-\lambda)$ and $Y^2 = X(X-1)(X+\lambda)$ both have complex multiplication).

21.20 Show that $F_3(J(q), J(q^3)) = 0$ for $F_3(X, Y)$ given by

$$\begin{aligned}&X^4 + Y^4 - X^3Y^3 + 2232X^2Y^2(X + Y) - 1069956XY(X^2 + Y^2)\\&+ 2587918086X^2Y^2 + 36864000(X^3 + Y^3) + 8900222976000XY(X + Y)\\&+ 452984832000000(X^2 + Y^2) - 770845966336000000XY\\&+ 1855425871872000000000(X + Y).\end{aligned}$$

Why is there no constant term?

21.21 Show that the constant term of $F_4(X, Y)$ is

$$2809493747221953721096406250000000000000 = 2^{12}3^{27}5^{18}11^9$$

[*Hint*: resultants].

21.22 Show that the constant term of $F_{23}(X, Y)$ is

$$2^{378}3^{72}5^{72}11^{27}17^{27}23^{3}41^{18}53^{9} > 10^{308}.$$

22
Algebraic independence

We have seen plenty of this already, but only for functions over $\mathbb{C}$. The property for numbers over $\mathbb{Q}$ is rather more difficult to prove. We quote some results relevant to the various chapters so far.

Regarding Chapter 2, we already mentioned that even the transcendence of the numbers in Theorem 2.1 is unknown.

Regarding Chapter 3, much is known. For example with Mahler's $f(z) = \sum_{k=0}^{\infty} z^{2^k}$ one can prove the algebraic independence of $f(\alpha_1), \ldots, f(\alpha_m)$ for algebraic $\alpha_1, \ldots, \alpha_m$ with $0 < |\alpha_1|, \ldots, |\alpha_m| < 1$, provided a necessary and sufficient condition arising from functional equations holds. This condition is not so simple; besides $f(z^2) = f(z) - z$ there are other relations such as $f(z) + f(i^*z) + f(i^{*2}z) = 0$ for $i^* = e^{2\pi i/3}$. The original proof (Loxton and van der Poorten, 1978) was also not so simple, involving Mahler's elaborate auxiliary polynomial construction so far applicable apparently only in his Method (see also Nishioka (1996) for more references).

Regarding Chapter 11, we know from Theorem 11.2 the transcendence of $\sum_{k=1}^{\infty} \alpha^{k!}$ for algebraic α with $0 < |\alpha| < 1$. Nishioka generalized this to $\alpha_1, \ldots, \alpha_m$ under the now simple necessary and sufficient condition that no α_i/α_j $(i < j)$ is a root of unity. Now no direct auxiliary polynomial is needed, but still we illustrate the simplest basic idea with $\alpha_1 = 1/2, \alpha_2 = 1/3$ by proving the algebraic independence of

$$\xi = \sum_{k=1}^{\infty} \frac{1}{2^{k!}}, \quad \eta = \sum_{k=1}^{\infty} \frac{1}{3^{k!}}$$

(see also Exercise 11.18).

If not, then there is non-zero P in $\mathbb{Z}[X, Y]$ with $P(\xi, \eta) = 0$. Here P must involve X, and we can choose it with minimal degree in X. It follows that for $P_X = \partial P/\partial X$ we have

$$P_X(\xi,\eta) \neq 0. \tag{22.1}$$

We consider the truncations

$$\xi_n = \sum_{k=1}^{n} \frac{1}{2^{k!}}, \quad \eta_n = \sum_{k=1}^{n} \frac{1}{3^{k!}}.$$

As in the proof of Theorem 11.2 we find that $|P(\xi_n,\eta_n)| \leq c2^{-(n+1)!}$ for some c independent of n. There is a denominator a fixed power of $6^{n!}$, and it follows that $P(\xi_n,\eta_n) = 0$ for all sufficiently large n.

Now a Mean Value Theorem in two variables gives

$$P(\xi,\eta) \;=\; P(\xi_n,\eta_n) + P_X(\tilde{\xi}_n,\tilde{\eta}_n)(\xi-\xi_n) + P_Y(\tilde{\xi}_n,\tilde{\eta}_n)(\eta-\eta_n)$$

for $\xi_n < \tilde{\xi}_n < \xi$, $\eta_n < \tilde{\eta}_n < \eta$ and $P_Y = \partial P/\partial Y$. As $\xi - \xi_n > 2^{-(n+1)!}$ it follows that

$$|P_X(\tilde{\xi}_n,\tilde{\eta}_n)| \leq c\left(\frac{2}{3}\right)^{(n+1)!}$$

for all sufficiently large n. Finally making n tend to infinity we get a contradiction with (22.1). This proof works because $|\alpha_1| \neq |\alpha_2|$. But in general if some of $|\alpha_1|,\ldots,|\alpha_m|$ are equal, then one needs Schlickewei's extension of the Subspace Theorem (see Chapter 12) to finish the argument; see Nishioka (1986).

Regarding Chapter 13, the algebraic independence of $e^{\alpha_1},\ldots,e^{\alpha_m}$ for algebraic $\alpha_1,\ldots,\alpha_m$ linearly independent over $\mathbb{Q}$ was achieved relatively early by Lindemann and Weierstrass. It can be proved in two pages (see Baker (1990) for example) but there too one sees no auxiliary polynomial. If the reader needs such a thing, she can find it in Exercise 22.5 used for a proof in Exercise 22.7.

However it is somewhat shocking that, even if we know everything about the transcendence of $\log\beta$ for algebraic β, we still cannot prove that two among the entire collection of these $\log\beta$ are algebraically independent (or solve Schneider's First Problem – see Schneider (1957) again).

Everything here and much much more is implied by Schanuel's Conjecture, which says that if $\sigma_1,\ldots,\sigma_m$ are complex numbers linearly independent over $\mathbb{Q}$, then at least m among the numbers

$$\sigma_1,\ldots,\sigma_m, e^{\sigma_1},\ldots,e^{\sigma_m}$$

are algebraically independent (see Lang (1966a) for example). In particular we could deduce the algebraic independence of π and e, also unknown, perhaps due to the dangerously close

$$\pi^4 + \pi^5 - e^6 \;=\; -.0000176\ldots.$$

But the Conjecture is proved only for $m = 1$ (when it is Hermite–Lindemann).

We now come to Chapter 19 and the Gelfond–Schneider Corollary 19.2 on the transcendence of α^β. Despite the announcement (Gelfond, 1934) (which remains premature), not much is known, even the algebraic independence of $2^{\sqrt{2}}$ and $2^{\sqrt{3}}$ (also covered by Schanuel, as well as Schneider's Seventh Problem (Schneider, 1957) on $\alpha^{\beta_1}, \ldots, \alpha^{\beta_m}$). But the situation is not hopeless, and in 1949 Gelfond succeeded in proving the algebraic independence of numbers like $2^{\sqrt[3]{2}}$ and $2^{\sqrt[3]{4}}$ (see Gelfond (1960) for references). He actually proved the following generalization with an extra condition; the latter was eliminated by Tijdeman (1971b).

Theorem 22.1 *Let ξ, η, ζ be complex numbers linearly independent over $\mathbb{Q}$, and let $\sigma_1, \sigma_2, \sigma_3$ be complex numbers also linearly independent over $\mathbb{Q}$. Then at least two among the 12 numbers*

$$\xi, \eta, \zeta, e^{\xi\sigma_1}, e^{\eta\sigma_1}, e^{\zeta\sigma_1}, e^{\xi\sigma_2}, e^{\eta\sigma_2}, e^{\zeta\sigma_2}, e^{\xi\sigma_3}, e^{\eta\sigma_3}, e^{\zeta\sigma_3} \tag{22.2}$$

are algebraically independent.

As with the Schneider–Lang Theorem on transcendence, this at first sight seems to say nothing about the algebraic independence of two specific numbers. But for an algebraic number $\alpha \neq 0$ let $\log\alpha$ be any non-zero determination of the logarithm, and let β be an algebraic number of degree three. We deduce that $\alpha^\beta = e^{\beta\log\alpha}$ and $\alpha^{\beta^2} = e^{\beta^2\log\alpha}$ are algebraically independent simply by taking

$$\xi = 1,\ \eta = \beta,\ \zeta = \beta^2,\ \sigma_1 = \log\alpha,\ \sigma_2 = \beta\log\alpha,\ \sigma_3 = \beta^2\log\alpha \tag{22.3}$$

and noting that β^3, β^4 lie in $\mathbb{Q} + \mathbb{Q}\beta + \mathbb{Q}\beta^2$.

For the proof we need several preliminaries.

The first concerns the Fundamental Theorem of Transcendence. In the proof of Gelfond–Schneider (see Lemma 19.4) we took a parameter N (for example T there), and a corresponding auxiliary polynomial P in $\mathbb{Z}[X, Y]$ and the function $P(e^z, e^{\beta z})$. From its many zeroes we extrapolated to find many small values, and then an inductive process enabled us to find a non-zero value among them. This is a contradiction by the Fundamental Theorem (in case the value is in $\mathbb{Z}$) if N is sufficiently large; or (in general) by its various refinements for a number field containing α^β (and α, β).

Such a procedure has to be modified here because we already know that α^β is transcendental. It turns out that we have to allow P in $\mathbb{Z}[\alpha^\beta][X, Y, Z]$ (and $P(e^z, e^{\beta z}, e^{\beta^2 z})$ in the above set-up). We end up with a non-zero value in $\mathbb{Z}[\alpha^\beta]$ (in the simplest case). But no sort of Fundamental Theorem is available for this.

Some lower bound would come from a transcendence measure asserting that α^β is not too near an algebraic number. But we have seen how technical mere irrationality measures can become. Gelfond found a way of avoiding such things by using all sufficiently large N above, not just a single one. Via the contrapositive this translates into the following.

Proposition 22.2 *Let ω in $\mathbb{C}$ be transcendental, and let $A_N \neq 0$ be a sequence of polynomials in $\mathbb{Z}[W]$ of degree at most N and of length $\mathcal{L}(A_N) \leq 2^N$ ($N = 1, 2, \ldots$). Then there are infinitely many N with*

$$|A_N(\omega)| \geq 2^{-19N^2}. \tag{22.4}$$

We prove this using two lemmas. The first observation shows that two coprime polynomials cannot be very small at the same point.

Lemma 22.3 *Suppose A, B in $\mathbb{Z}[W]$ are non-zero, without common factor in $\mathbb{Q}[W]$, of degrees at most $r \geq 1, s \geq 1$ respectively. Then for any complex number w we have*

$$\max\{|A(w)|, |B(w)|\} \geq \frac{1}{2}\mathcal{L}(A)^{-s}\mathcal{L}(B)^{-r}.$$

Proof It is not difficult to see that the sum of the absolute values of all the maximal minors of a complex square matrix is at most twice the product of the l^1 (sum or length) norms of its rows, provided each of these norms is at least 1.

The lemma is clear if at least one of A, B is constant. So by decreasing r, s if necessary we can assume that they are the exact degrees. Then by Proposition 5.3(ii) we see that $\rho = \text{res}(A, B) \neq 0$. Now by Proposition 5.3(i) there are P, Q in $\mathbb{Z}[W]$, whose coefficients are maximal minors (in different positions) of the Sylvester matrix (as in (5.3) for example), with $PA + QB = \rho$; and from the remark above we deduce $\mathcal{L}(P) + \mathcal{L}(Q) \leq 2\mathcal{L}(A)^s\mathcal{L}(B)^r$.

If $|w| \leq 1$ then the lemma follows from taking $W = w$ and using – yes, the Fundamental Theorem!

If $|w| > 1$ then we consider the corresponding reciprocal polynomials $V^rA(1/V)$,$V^sB(1/V)$, having the same lengths and also coprime (they have no common zero, even at $V = 0$). Applying the above to their values at the reciprocal $V = 1/w$ gives the lemma in even stronger form. See also Exercises 5.8 and 5.21. □

The second observation simplifies the problem of proving coprimality.

Lemma 22.4 *Suppose A in $\mathbb{Z}[W]$ is non-zero of degree at most $r \geq 1$, and w is a complex number with $|A(w)| < 1$. Then there is F in $\mathbb{Z}[W]$, irreducible in $\mathbb{Z}[W]$, and a positive integer e such that $G = F^e$ divides A in $\mathbb{Z}[W]$ with*

$$|G(w)| \leq 2^{8r^2}\mathcal{L}(A)^{2r}|A(w)|.$$

Proof The factorization of A in $\mathbb{Z}[W]$ gives $k \geq 1$ and $A = aG_1 \cdots G_k$ with coprime prime-powers $G_1, \ldots, G_k$ as above. Thus if $g_1, \ldots, g_k$ are their values at w we have $g_1 \cdots g_k \leq |A(w)| < 1$. So if $A(w)$ is small then certainly one of $g_1, \ldots, g_k$ is small. This may remind the reader of the situation in the proof of Theorem 12.9; however it is a bit more subtle here.

Indisputably the first element of the sequence

$$1,\ g_1,\ g_1g_2, \ldots,\ g_1 \cdots g_{l-1},\ g_1 \cdots g_l, \ldots,\ g_1 \cdots g_k$$

exceeds the first element of the sequence

$$g_1 \cdots g_k,\ g_2 \cdots g_k,\ g_3 \cdots g_k, \ldots,\ g_l \cdots g_k,\ g_{l+1} \cdots g_k, \ldots,\ g_k,\ 1.$$

With the last elements it is the other way round. So at the crossover point we get l (possibly 1 or k) such that

$$g_1 \cdots g_{l-1} \geq g_l \cdots g_k, \quad g_1 \cdots g_l \leq g_{l+1} \cdots g_k.$$

For the first of these we apply Lemma 22.3 to $P = G_1 \cdots G_{l-1}$ and $Q = G_l \cdots G_k$ to get

$$g_1 \cdots g_{l-1} \geq \frac{1}{2}\mathcal{L}(P)^{-r}\mathcal{L}(Q)^{-r}.$$

By Proposition 14.1(i) we have $\mathcal{L}(P)\mathcal{L}(Q) \leq 8^r\mathcal{L}(A)$. So

$$g_1 \cdots g_{l-1} \geq \frac{1}{2}8^{-r^2}\mathcal{L}(A)^{-r};$$

and a similar argument for the second gives the same lower bound for $g_{l+1} \cdots g_k$. Finally multiplying the two gives the required result with $G = G_l$. □

We now prove Proposition 22.2.

If it is false, then for all sufficiently large N there is such an A_N with $|A_N(\omega)| < 2^{-19N^2}$. By Lemma 22.4 there is F_N in $\mathbb{Z}[W]$, irreducible in $\mathbb{Z}[W]$, and a positive integer e_N such that $G_N = F_N^{e_N}$ divides A_N in $\mathbb{Z}[W]$ with

$$|G_N(w)| \leq 2^{8N^2}(2^N)^{2N}|A(w)| \leq 2^{-9N^2}. \tag{22.5}$$

Also by Proposition 14.1(i) we have $\mathcal{L}(G_N) \leq 2^{4N}$. Now we compare the situations at N and $N+1$. If G_N, G_{N+1} are coprime, then Lemma 22.3 would give

$$2^{-9N^2} \geq \frac{1}{2}(2^{4N})^{-(N+1)}(2^{4(N+1)})^{-N}$$

a contradiction for large N.

It follows that $F_N = F_{N+1}$ (up to sign) for all sufficiently large N, and so eventually we get the same polynomial, say F. But now (22.5) gives

$|F(\omega)| \le 2^{-9N}$. Making N tend to infinity yields $F(\omega) = 0$ contradicting the transcendence of ω. This establishes Proposition 22.2.

In the proof of Gelfond–Schneider, we could find a non-zero value for our auxiliary function by a trick involving an infinite induction process. Here we cannot afford such luxuries, and we have to generalize zero estimates like (13.1). This was Tijdeman's contribution (Tijdeman, 1971a). In the following we regard the zero polynomial as having degree -1.

Proposition 22.5 *Let $\lambda_1, \ldots, \lambda_n$ be distinct complex numbers with absolute values at most Λ. Then for any $R_1, \ldots, R_n$ not all zero in $\mathbb{C}[z]$ and any $Z \ge 0$ the exponential polynomial $\sum_{i=1}^n R_i(z)e^{\lambda_i z}$ has at most*

$$18\Lambda Z + 5\sum_{i=1}^{n}(1 + \deg R_i)$$

zeroes (counting multiplicity) in any disc of radius Z.

Proof Write $r_i = 1 + \deg R_i$ $(i = 1, \ldots, n)$ and $r = \sum_{i=1}^n r_i$. The function f in question satisfies a differential equation $P(\mathrm{d}/\mathrm{d}z)f = 0$, where

$$P(X) = \prod_{i=1}^{n}(X - \lambda_i)^{r_i} = X^r - \mu_1 X^{r-1} - \cdots - \mu_r \tag{22.6}$$

(here the coefficients are up to sign the elementary symmetric polynomials in the zeroes). By differentiating $k - r$ times we see that the $u_k = f^{(k)}(0)$ satisfy a recurrence relation

$$u_k = \mu_1 u_{k-1} + \cdots + \mu_r u_{k-r} \quad (k = r, r+1, \ldots).$$

Then for the generating function $F(X) = \sum_{k=0}^{\infty} u_k X^k$ and the reciprocal

$$G(X) = X^r P\left(\frac{1}{X}\right) = 1 - \mu_1 X - \cdots - \mu_r X^r,$$

we deduce in time-honoured fashion

$$F(X)G(X) = \sum_{s=0}^{r-1} v_s X^s \tag{22.7}$$

where (with $\mu_0 = -1$)

$$v_s = u_s - (\mu_1 u_{s-1} + \cdots + \mu_s u_0) = -\sum_{t=0}^{s} \mu_{s-t} u_t \quad (s = 0, 1, \ldots, r-1).$$

We solve (22.7) for F by multiplying it by

$$\frac{1}{G(X)} = \frac{1}{\prod_{i=1}^{n}(1-\lambda_i X)^{r_i}} = \sum_{p=0}^{\infty} \nu_p X^p \tag{22.8}$$

(now the symmetric polynomials are the so-called homogeneous product sums) in $\mathbb{C}[[X]]$. Equating coefficients gives (with $m = s - t$)

$$u_k = -\sum_{t=0}^{r-1} u_t \sum_{m=0}^{r-1-t} \mu_m \nu_{k-m-t} \quad (k = r, r+1, \ldots) \tag{22.9}$$

(expressing the later elements of a recurrence sequence in terms of the earliest elements).

So much for algebra; now comes analysis. Without loss of generality we may assume that the disc in question is centred at the origin. And also that $Z = 1$. And even that $|f|_1 = 1$ in our usual notation. Now dividing out by the N zeroes of f and using Maximum Modulus gives in the standard way $1 \le 2^{-N}|f|_5$ or

$$N \log 2 \le \log |f|_5. \tag{22.10}$$

We proceed to estimate the right-hand side.

Clearly

$$|f|_5 \le \sum_{k=0}^{\infty} \frac{|u_k|}{k!} 5^k \tag{22.11}$$

and $|u_k| \le k!$ by Cauchy. Thus for the earliest terms

$$\sum_{k=0}^{r-1} \frac{|u_k|}{k!} 5^k \le 5^r. \tag{22.12}$$

For the later terms we use (22.9). We have $|\mu_m| \le 2^r \Lambda^m$ $(m = 0, 1, \ldots, r-1)$ from (22.6). Now (22.8) is majorized by

$$\frac{1}{(1-\Lambda X)^r} = \sum_{p=0}^{\infty} (-1)^p \binom{-r}{p} \Lambda^p X^p = \sum_{p=0}^{\infty} \binom{r+p-1}{r-1} \Lambda^p X^p.$$

We deduce $|\nu_p| \le 2^{r+p-1}\Lambda^p$. Thus

$$|u_k| \le \sum_{t=0}^{r-1} |u_t| \sum_{m=0}^{r-1-t} 2^{2r-m-1}(2\Lambda)^{k-t} \le 4^r \sum_{t=0}^{r-1} t!(2\Lambda)^{k-t} \quad (k = r, r+1, \ldots).$$

The right-hand side here is

$$4^r(r-1)!(2\Lambda)^{k-r+1} \sum_{t=0}^{r-1} \frac{(2\Lambda)^{r-1-t}}{(r-1-t)!} \Big/ \binom{r-1}{t} \le 4^r(r-1)!(2\Lambda)^{k-r+1} e^{2\Lambda}.$$

We get

$$\sum_{k=r}^{\infty} \frac{|u_k|}{k!} 5^k \le 4^r (r-1)! e^{2\Lambda} \sum_{k=r}^{\infty} \frac{(2\Lambda)^{k-r+1}}{k!} 5^k.$$

By the same fiendish trick the right-hand side here is

$$5^{r-1} 4^r e^{2\Lambda} \sum_{k=r}^{\infty} \frac{(10\Lambda)^{k-r+1}}{(k-r+1)!} \Big/ \binom{k}{r-1} < \frac{1}{5} 20^r e^{12\Lambda}.$$

Combining this with (22.12) we get $|f|_5 \le 20^r e^{12\Lambda}$ in (22.11); and recalling (22.10) we complete the proof. See also Baker's 1990 book for another method using Newton Interpolation. □

The result is doubly sharp in the sense that neither of the constants 18 or 5 can be improved too much. We already know (see (13.1) and the remark following) that there can be a single zero with multiplicity at least $-1 + \sum_{i=1}^{n}(1 + \deg R_i)$. And $\phi(z) = e^z - 1$ with $\Lambda = 1$ has at least $Z/(2\pi)$ zeroes $z = 2\pi i k$ with $k \ge 0$ in $\mathbb{Z}$ and $|z| \le Z$. The result easily solves Exercise 18.9, for example.

We shall soon be dealing with a field K, finitely generated over $\mathbb{Q}$, of transcendence degree 1 over $\mathbb{Q}$. Any such K can be written as $\mathbb{Q}(\omega, \chi)$ with ω transcendental and χ algebraic over $\mathbb{Q}(\omega)$. As just before the statement of Proposition 11.1, we can modify χ so that it is integral over $\mathbb{Z}[\omega]$.

We can work with denominators in K; because it is also $\mathbb{Q}(\omega)[\chi]$ every element can be multiplied by some non-zero element of $\mathbb{Q}[\omega]$ and even $\mathbb{Z}[\omega]$ so it lands in $\mathbb{Z}[\omega, \chi]$.

But we have no adequate height machinery even on $\mathbb{Q}(\omega)$, and worse, we don't have a very good norm or size $||\cdot||$ even on $\mathbb{Z}[\omega]$ as in Chapter 11. This necessitates the following remark, where $d = [\mathbb{Q}(\omega, \chi) : \mathbb{Q}(\omega)]$. Note that if $d = 1$ then we don't need χ at all; but in this case we may pretend that $\chi = 1$.

Lemma 22.6 *Suppose $d > 1$. Then there is a constant c, depending only on ω, χ, such that for any integer $l \ge 0$ we have $\chi^l = \sum_{i=0}^{d-1} A_{li}\chi^i$ for A_{li} in $\mathbb{Z}[\omega]$ of degrees at most cl in ω whose coefficients in $\mathbb{Z}$ have absolute values at most c^l.*

Proof This is an easy induction using the minimal equation $\chi^d = \sum_{i=0}^{d-1} A_i\chi^i$ for χ over $\mathbb{Z}[\omega]$ (compare Lemma 12.3). We find the recurrence

$$A_{li} = A_{l-1,i-1} + A_{l-1,d-1}A_i \quad (i = 1, \dots, d-1)$$

with a term missing if $i = 0$. □

We now start the proof of Theorem 22.1, vaguely following the Schneider–Lang exposition but also that of the Hermite–Lindemann.

As in Chapter 13, we assume the result false and we aim for a contradiction. Then the numbers (22.2) lie in a field K, finitely generated over $\mathbb{Q}$, of transcendence degree $t \le 1$ over $\mathbb{Q}$. Here $t = 1$, otherwise for example $\alpha = e^{\xi\sigma_1}$ and $\alpha^\beta = e^{\eta\sigma_1}$ would be algebraic with $\beta = \eta/\xi$ also algebraic but irrational. Thus we have ω, d and (if $d > 1$) also χ as above.

The first lemma, constructing the auxiliary polynomial, is now able to imitate Lemma 13.4, but using unnumbered constants c in the standard lazy way.

Lemma 22.7 *For any $L \ge 1, S \ge 1, T \ge 1$ in $\mathbb{Z}$ with*

$$(L+1)^3 \ge 2dS^3T \tag{22.13}$$

there is P in $\mathbb{Z}[\omega][X,Y,Z]$, of degree at most L in each variable, and with coefficients in $\mathbb{Z}[\omega]$ of degrees at most $c(T + LS)$ in ω whose coefficients in $\mathbb{Z}$ have absolute values at most $L^{cT}c^{LS}$, such that

$$\phi(z) = P(e^{\xi z}, e^{\eta z}, e^{\zeta z}) \tag{22.14}$$

is not identically zero and

$$\mathrm{ord}_{z=\sigma}\, \phi(z) \ge T \tag{22.15}$$

for every

$$\sigma = s_1\sigma_1 + s_2\sigma_2 + s_3\sigma_3 \quad (s_1, s_2, s_3 = 1, \ldots, S). \tag{22.16}$$

Proof We introduce M, soon to be chosen. With

$$P(X,Y,Z) = \sum_{m=0}^{M}\sum_{i=0}^{L}\sum_{j=0}^{L}\sum_{k=0}^{L} p_{mijk}\omega^m X^i Y^j Z^k$$

we have

$$\phi(z) = \sum_{m=0}^{M}\sum_{i=0}^{L}\sum_{j=0}^{L}\sum_{k=0}^{L} p_{mijk}\omega^m e^{(i\xi+j\eta+k\zeta)z}$$

and we obtain the equations

$$\sum_{m=0}^{M}\sum_{i=0}^{L}\sum_{j=0}^{L}\sum_{k=0}^{L} E_{ijk}(s_1,s_2,s_3,t)p_{mijk}\omega^m = 0 \quad (t = 0, 1, \ldots, T-1) \tag{22.17}$$

for the ranges in (22.16) with

$$E_{ijk}(s_1,s_2,s_3,t) = (i\xi + j\eta + k\zeta)^t \prod_{h=1,2,3} (e^{\xi\sigma_h})^{is_h}(e^{\eta\sigma_h})^{js_h}(e^{\zeta\sigma_h})^{ks_h}. \tag{22.18}$$

If δ in $\mathbb{Z}[\omega]$ is a common denominator for the numbers in (22.2), then the $\delta^{T+9LS}E_{ijk}(s_1,s_2,s_3,t)$ are in $\mathbb{Z}[\omega,\chi]$ with degrees in ω,χ at most $c(T+LS)$ whose coefficients in $\mathbb{Z}$ have absolute values at most $L^{cT}c^{LS}$. But using Lemma 22.6 we can get the degree in χ down to $d-1$ without rocking the boat.

When these are inserted into (22.17), we get by equating coefficients of the various $\omega^I\chi^J$ a total of $M' \leq cd(T+LS)S^3T$ equations in the $N'=(M+1)(L+1)^3$ unknowns p_{mijk}. In Lemma 13.4 we had the assumption $(L+1)^2 \geq 2dST$ for a safe Siegel Lemma, and our analogous assumption $(L+1)^3 \geq 2dS^3T$ leads to $N' \geq 2M'$ just by choosing $M=[c(T+LS)]$. In fact the role of d here is not very important, and we could also build it into M. But we like to signal the analogy, already hinted above, between finite extensions of $\mathbb{Q}$ and finite extensions of $\mathbb{Q}(\omega)$.

Finally using the prehistoric Proposition 8.3 we get the required bound for the coefficients, and we conclude by noting that $e^{\xi z}, e^{\eta z}, e^{\zeta z}$ are algebraically independent, even over $\mathbb{C}$, because of the linear independence of ξ,η,ζ over $\mathbb{Q}$. □

We now consider increasing the order of vanishing as in Chapter 13, but thanks to our zero estimate we need no inductive scheme. Indeed we cannot afford such a scheme here because we have to consider all large values of the starting parameter T and not just a single one; further we have no algebraic numbers and so we cannot hope for higher order zeroes as in Lemma 13.6 (in fact it is a "non-zero" that we need here). The next result is a preliminary step as in Lemma 13.5. But it uses a new parameter $\mathcal{Y}>1$ as in Lemma 10.8. For brevity we write Σ for the set in (22.16), and C for a distinguished constant $|\sigma_1|+|\sigma_2|+|\sigma_3|$, so that $|\sigma| \leq CS$ on Σ.

Lemma 22.8 *For any integer $T_1 \geq T$ the derivatives up to order T_1-1 satisfy*

$$|\phi^{(t)}(\sigma)| \leq \mathcal{Y}^{-S^3T}T_1^{T_1}L^{cT}c^{L\mathcal{Y}S} \quad (t=0,1,\ldots,T_1-1)$$

for every σ in Σ.

Proof This is fairly routine by now, taking into account the "large radius". We divide out by the S^3T zeroes and compare on radii $2CS$ and $4C\mathcal{Y}S$. Then we use (12.29) with $m=t, r=CS, R=2CS$ to get the result. □

Next we find a non-zero value. We need another distinguished constant $C'=|\xi|+|\eta|+|\zeta|$, so that $|i\xi+j\eta+k\zeta| \leq C'L$ in ϕ above.

Lemma 22.9 *For any integer T_1 with*

$$S^3T_1 > 5(L+1)^3+18CC'LS \tag{22.19}$$

there is $t=0,1,\ldots,T_1-1$ and σ in Σ with $\phi^{(t)}(\sigma) \neq 0$.

Proof If not, then ϕ would have at least S^3T_1 zeroes with $|z| \leq Z = CS$. It has the shape in Proposition 22.5 with $n = (L+1)^3$, $\Lambda \leq C'L$ and constant $R_1, \ldots, R_n$, so this would lead to a contradiction. □

We are close to finishing. Calculations as in the proof of Lemma 22.7 show that on multiplying by a suitable denominator, we get from the $\phi^{(t)}(\sigma)$ of Lemma 22.9 a polynomial B in $\mathbb{Z}[W, U]$ of degree at most $c(T_1 + LS)$ in W and degree at most $d-1$ in U whose coefficients have absolute values at most $L^{cT_1}c^{LS}$, such that

$$0 < |\beta| \leq \mathcal{Y}^{-S^3T}T_1^{T_1}L^{cT_1}c^{L\mathcal{Y}S} \tag{22.20}$$

for $\beta = B(\omega, \chi)$. We would like to get a final contradiction using Proposition 22.2, but we have to eliminate χ first (if $d > 1$, that is, and also if B actually involves U). The expert would probably take norms from $\mathbb{Q}(\omega, \chi)$ to $\mathbb{Q}(\omega)$ but that requires a tad too much extra theory. One can often do norms (and a lot more besides – see Exercises 5.28, 5.29 and 5.30) using resultants, as follows.

There is an irreducible polynomial B_0 in $\mathbb{Z}[W, U]$ of degree d in U such that $B_0(\omega, \chi) = 0$. As $B(\omega, \chi) \neq 0$ the resultant $A(W)$ of $B(W, U)$ and $B_0(W, U)$ with respect to U is defined and non-zero. Thus $A(\omega) \neq 0$. From the Sylvester matrix (which has bounded size) we find the same bounds for degree and coefficients of A as for B. The usual equation $Q_0B + QB_0 = A$ gives $Q_0(\omega, \chi)\beta = A(\omega)$, and from the expression for Q_0 in terms of maximal minors we deduce the same bounds for $A(\omega)$ as in (22.20).

Thus in Proposition 22.2 we have found our $A = A_N$ satisfying the degree and length bounds with some integer N of order $T_1 \log L + LS$. So we will be home if the smallness in (22.20) beats the N^2 in (22.4).

This means that $S^3T \log \mathcal{Y}$ has to beat (in our usual sense of the living daylights out of)

$$T_1 \log T_1, \ T_1 \log L, \ L\mathcal{Y}S \tag{22.21}$$

as well as

$$(T_1 \log L)^2, \ (LS)^2; \tag{22.22}$$

not forgetting the conditions already imposed in (22.13), (22.19) (and $T_1 \geq T$).

How one solves these inequalities is a matter of personal taste; in more complicated situations it is practically a problem of linear programming. One can start with the assumption that all logarithms are essentially equal, say to $\log T$. So even if $\mathcal{Y}$ is a tiny power of T its logarithm will still pull the necessary weight. Then one can ignore (22.21) against (22.22), and systematically begin eliminating with the aim of expressing everything in terms of T. Here everything is fairly tight except for the choice of $\mathcal{Y}$ as a power of T. After an hour or so I ended up with L, S, T_1 of orders

$$T^{2/3}(\log T)^{7/9},\ T^{1/3}(\log T)^{6/9},\ T(\log T)^{4/9}$$

respectively, in which the powers of T are rigid but the powers of $\log T$ not quite. And $\mathcal{Y}$ could be as large as T.

That completes the proof of Theorem 22.1.

We wish to draw attention to another way, algebraic instead of analytic, of proving zero estimates like Proposition 22.5. It uses the fact that our function in Lemma 22.7 is a polynomial P in three functions; this results in the $\lambda_1, \ldots, \lambda_n$ having a special "additive" form $i\xi + j\eta + k\zeta$ $(i, j, k = 0, 1, \ldots, L)$. Not only that, the zeroes being considered have a similar special form $s_1\sigma_1 + s_2\sigma_2 + s_3\sigma_3$ $(s_1, s_2, s_3 = 1, \ldots, S)$.

By way of illustration, we now sketch an algebraic proof that for any complex ξ, η linearly independent over $\mathbb{Q}$, any $\phi(z) = P(e^{\xi z}, e^{\eta z})$ with $P \neq 0$, and any complex $\sigma_1, \sigma_2, \sigma_3$ also linearly independent over $\mathbb{Q}$, not too many of the $\phi(s_1\sigma_1 + s_2\sigma_2 + s_3\sigma_3)$ can vanish. So we have one exponential function fewer; and also no multiplicities.

We begin by noting that $\phi(z + \sigma_1) = (\mathcal{E}_1 P)(e^{\xi z}, e^{\eta z})$ for the map $\mathcal{E}_1$ from $\mathbb{C}[X, Y]$ to itself defined by $(\mathcal{E}_1 P)(X, Y) = P(e^{\xi\sigma_1}X, e^{\eta\sigma_1}Y)$. Compare with the endomorphism σ in Lemma 8.2. Similarly $\mathcal{E}_2, \mathcal{E}_3$ describing translations by σ_2, σ_3; these are all automorphisms of $\mathbb{C}[X, Y]$. More generally for any r_1, r_2, r_3 in $\mathbb{Z}$ and $\rho = r_1\sigma_1 + r_2\sigma_2 + r_3\sigma_3$, we have

$$\phi(z + \rho) \ = \ (\mathcal{E}(\rho)P)(e^{\xi z}, e^{\eta z})$$

for $\mathcal{E}(\rho) = \mathcal{E}_1^{r_1}\mathcal{E}_2^{r_2}\mathcal{E}_3^{r_3}$. Thus if P corresponds to $\phi(z)$ vanishing on a large set $\Sigma = \Sigma(S)$ as in (22.16), then $\mathcal{E}(\rho)P$ corresponds to $\phi(z + \rho)$ vanishing on $\Sigma(S) - \rho$, equally large.

We can even hope to handle many ρ at once; if for example

$$0 \le r_1, r_2, r_3 \le R \tag{22.23}$$

for some integer $R < S$, then every $\Sigma(S) - \rho$ contains $\Sigma(S - R)$. In this connexion the rule $\mathcal{E}(\rho)\mathcal{E}(\rho') = \mathcal{E}(\rho + \rho')$ will be useful, for any ρ, ρ' in the group $\overline{\Sigma} = \mathbb{Z}\sigma_1 + \mathbb{Z}\sigma_2 + \mathbb{Z}\sigma_3$.

From such $\mathcal{E}(\rho)P$ we are going to construct two coprime polynomials and then take their resultant. But first we must remove all powers of X and Y from P (as in Exercise 8.8); these do not affect the vanishing of ϕ on $\Sigma(S)$. Let k be the number of irreducible factors Q left. For each Q we will find ρ such that Q does not divide $\mathcal{E}(\rho)P$.

If some ρ fails, then $\mathcal{E}(-\rho)Q$ divides $\mathcal{E}(-\rho)\mathcal{E}(\rho)P = P$. It is easy to convince oneself that $\mathcal{E}(-\rho)Q$ is also irreducible (see Exercise 22.4(a) below); thus there are at most k possibilities for it up to constants.

So if more than k different ρ fail, the Box Principle provides $\rho' \neq \rho''$ with $\mathcal{E}(-\rho')Q, \mathcal{E}(-\rho'')Q$ equal up to constants. This holds also for $\mathcal{E}(\rho')\mathcal{E}(-\rho')Q = Q$ and $\mathcal{E}(\rho')\mathcal{E}(-\rho'')Q = \mathcal{E}(\rho' - \rho'')Q$.

We can also verify that this is impossible provided $\xi, \eta, \sigma_1, \sigma_2, \sigma_3$ are "sufficiently general"; more precisely, if $e^{\xi\sigma}, e^{\eta\sigma}$ are multiplicatively independent for each $\sigma \neq 0$ in $\overline{\Sigma}$ (see Exercise 22.4(b) below). For the moment we assume this is so.

Now we choose ρ from the set $\Sigma_0(R)$ defined by (22.23), with $R = [k^{1/3}]$ so containing $(R+1)^3 > k$ elements. Thus for each of the irreducible factors, say $Q_1, \ldots, Q_k$, of P there is ρ_i in $\Sigma_0(R)$ with Q not dividing $\mathcal{E}(\rho_i)P$.

We next find a linear combination $\tilde{P}$ of $\mathcal{E}(\rho_1)P, \ldots, \mathcal{E}(\rho_k)P$, with complex coefficients, not divisible by any of $Q_1, \ldots, Q_k$. In fact a "generic" such $\tilde{P}$ will do. We can find a more explicit one by considering for example $\tilde{P}_\lambda = \sum_{i=1}^k \lambda^{i-1}\mathcal{E}(\rho_i)P$. For each j there are at most k complex values of λ such that $\tilde{P}_\lambda$ is divisible by Q_j, otherwise the non-vanishing of the Vandermonde determinant would show that all $\mathcal{E}(\rho_i)P$ are divisible by Q_j. Thus we only have to exclude at most k^2 values of λ to arrive at $\tilde{P}$.

So P and $\tilde{P}$ are coprime. Their resultant with respect to Y (subject to the usual definability quibbles) is non-zero in $\mathbb{C}[X]$ and the corresponding element of $\mathbb{C}[e^{\xi z}]$ vanishes on $\Sigma(S-R)$. Now the $e^{\xi\sigma}$ are distinct for σ in $\overline{\Sigma}$; this is a special case of our "sufficiently general" assumption. As the resultant has degree at most $2L^2$, we conclude $(S-R)^3 \leq 2L^2$. Finally by considering total degree we get $k \leq 2L$ and so

$$S \leq (2L^2)^{1/3} + (2L)^{1/3} \tag{22.24}$$

(see also Exercise 22.8). In other words, for large L, if S^3 is asymptotically bigger than $2L^2$, there is σ in $\Sigma(S)$ with $\phi(\sigma) \neq 0$. This compares very well with (22.19) for $T_1 = 1$ with $(L+1)^2$ in place of $(L+1)^3$, and the term $18CC'LS$ there, coming from analysis, has now been eliminated. However determinantal considerations (as in Chapter 8) might lead one to expect L^2 in place of $2L^2$. It remains faintly irritating that we still cannot prove such things.

The assumption that we made above is usually harmless in applications. For example in (22.3) we have for integers r, s not both zero $(e^{\xi\sigma})^r(e^{\eta\sigma})^s = e^{\gamma \log\alpha}$ for algebraic $\gamma \neq 0$; and then if α is not a root of unity this cannot be 1 by Gelfond–Schneider (if α is a root of unity then the arguments have to be slightly modified leading to a second term on the right-hand side of (22.24) that is a bit bigger).

We can extend things to $\phi(z) = P(e^{\xi z}, e^{\eta z}, e^{\zeta z})$, and we get in a similar way $\tilde{P}$ prime to P. Now taking resultants with respect to Z gives $P^*(e^{\xi z}, e^{\eta z})$ having degree of order at most L^2 in each variable, so with this approach and induction we need S^3 of order at least L^4. This is much worse than (22.19) and indeed inadequate for our application. In fact one probably cannot reach (22.19), even for $T_1 = 1$, using only resultants (see however Exercise 22.3). More fundamentally, the resultant of Chapter 5 hardly generalizes at all; for example there is no reasonable resultant of three polynomials in two variables, despite Exercise 5.35. One problem is the lack of well-defined leading coefficients, not unrelated to trouble at infinity. Macaulay defined a resultant by working projectively, as in Exercises 5.34 and 5.36; but that is seldom useful for our purposes (Macaulay (1916) – see also Lang (1993) for a modern treatment).

What is needed is an extra step to reach $P^{**}(e^{\xi z})$ with P^{**} having degree of order at most L^3 not L^4, and this can be done using the prime ideals of the ideal in $\mathbb{C}[X, Y, Z]$ generated by $P, \tilde{P}$ together with degree theory from commutative algebra (or geometrically the components of a curve in three dimensions). Such concepts are crucial for modern developments in the theory of transcendence as well as algebraic independence.

We can extend to arbitrary multiplicities T_1 by representing differentiation in a similar algebraic way; for example

$$\frac{d}{dz}P(e^{\xi z}, e^{\eta z}, e^{\zeta z}) = (\mathcal{D}P)(e^{\xi z}, e^{\eta z}, e^{\zeta z}) \tag{22.25}$$

with

$$\mathcal{D}P = \xi X\frac{\partial P}{\partial X} + \eta Y\frac{\partial P}{\partial Y} + \zeta Z\frac{\partial P}{\partial Z}$$

defining a derivation on $\mathbb{C}[X, Y, Z]$ as in Chapter 19. We sketch the use of such derivations below in an elliptic situation.

Using such techniques, Diaz (see Nesterenko and Philippon (2001) for references) has proved that at least $[\frac{d+1}{2}]$ of the numbers $\alpha^\beta, \alpha^{\beta^2}, \ldots, \alpha^{\beta^{d-1}}$ are algebraically independent for any algebraic number $\alpha \neq 0$ (and any non-zero determination of the logarithm), and any algebraic number β of degree d. Schanuel's Conjecture predicts that they all are (as we just proved for $d = 3$).

That completes our discussion around Chapter 19.

Regarding Chapter 20 on elliptic functions, the situation for algebraic independence is not so good. For example, when g_2, g_3 are algebraic and there is no complex multiplication, we do not know that among all the periods ω there are two which are algebraically independent. This would presumably follow from some sort of elliptic Schanuel. But there is a nice result of Chudnovsky reminiscent of our Theorem 20.9. Namely, suppose that g_2, g_3 are algebraic, and that $\omega \neq 0$ is in Ω, with corresponding $\eta = \eta(\omega)$. If u is a complex number not in $\mathbb{Q}\omega$ such that $\wp(u)$ is algebraic, then $\zeta(u) - (\eta/\omega)u$ and η/ω are algebraically independent (see Chudnovsky (1984) for references). Taking u as another period and using the Legendre relation, we deduce the algebraic independence of π/ω and η/ω. Taking in turn $\Omega = \omega(\mathbb{Z} + i\mathbb{Z})$ and recalling (20.18), we get the algebraic independence of π and $\Gamma(1/4)$. And that is the simplest way to prove that $\Gamma(1/4)$ is transcendental. Similarly for $\Gamma(1/3)$.

We also know the elliptic analogue of Gelfond for $\wp(\beta u), \wp(\beta^2 u)$, but only in the case of complex multiplication (Masser and Wüstholz, 1986), and then with β cubic over the associated quadratic field $\mathfrak{K}$ (note that there is still no elliptic analogue of the analytic Proposition 22.5, so here we must use algebra). Also for complex multiplication we know the algebraic independence of $\wp(\alpha_1), \ldots, \wp(\alpha_m)$ for algebraic $\alpha_1, \ldots, \alpha_m$ linearly independent over $\mathfrak{K}$, thanks to Philippon (1983) and Wüstholz (1983) independently.

Here we interpolate a discussion of the use of derivations in zero estimates to get an upper bound for the order of vanishing of $P(z, \wp(z))$ at a single point. Unfortunately due to (20.2) the analogue of (22.25) gives a map $\mathcal{D}$ from $\mathbb{C}[X, Y]$ only to $\mathbb{C}[X, Y, Y']$ with $Y'^2 = 4Y^3 - g_2Y - g_3$. We can descend by using field norms and this leads to the awkward-looking

$$\tilde{P} = \left(\frac{\partial P}{\partial X}\right)^2 - (4Y^3 - g_2Y - g_3)\left(\frac{\partial P}{\partial Y}\right)^2.$$

As in the proof of Proposition 8.1 we start with irreducible P and take the resultant of $P, \tilde{P}$. A little effort is needed to check its non-vanishing, but a reasonable upper bound comes out. Then factorizing P gives the conclusion that if $P \neq 0$ in $\mathbb{C}[X, Y]$ has degree at most L in each variable, then $P(z, \wp(z))$ cannot have a zero of order more than $4(L+1)^2$ at any z_0 not in Ω (Brownawell and Masser, 1980). Feldman could improve this to asymptotically $5L^2/2$ (Feldman, 1951), but I am not aware of any subsequent improvement. Presumably it holds with asymptotic L^2. However the determinantal considerations of Chapter 8 leading to $L^2 + 2L$ break down, and it can be shown that order $L^2 + 2L + 1$ is possible by carefully choosing z_0 (for $L = 1$ see Exercise 20.93). There are grounds for believing that even $L^2 + 2L + 2$ is possible by carefully choosing $\wp$ as well (for $L = 1$ see Exercise 20.94).

Finally regarding Chapter 21 on modular functions, the appearance of Theorem 21.2 led very quickly, also in 1996, to the vast generalization by Nesterenko. Define also

$$P = P(q) = 1 - 24\sum_{m=1}^{\infty} \frac{mq^m}{1 - q^m}$$

(not a modular form) to go with Q and R in the proof of Lemma 21.4. He showed that if $q \neq 0$ is in $\mathbb{D}$, then at least three among $q, P(q), Q(q), R(q)$ are algebraically independent (Nesterenko, 1996). In fact at $q = e^{-2\pi}$ (corresponding to $\tau = i$) we have

$$P = \frac{3}{\pi}, \quad Q = 3\frac{\Gamma(\frac{1}{4})^8}{(2\pi)^6}, \quad R = 0;$$

the second is related to (20.7). We thus obtain the algebraic independence of the three numbers $\pi, e^\pi, \Gamma(1/4)$ used in (20.34). And that is the only way to prove that π and e^π are algebraically independent. Similarly at $q = e^{-\pi\sqrt{3}}$ we get the algebraic independence of the three numbers $\pi, e^{\pi\sqrt{3}}, \Gamma(1/3)$.

But we still don't know if $e^\pi, e^{\pi\sqrt{3}}$ are algebraically independent!

The Lindemann–Weierstrass Theorem has been extended to values of what are known as E-functions; these are power series whose coefficients satisfy certain analytic and arithmetic growth conditions, of which the archetypal example is e^z. An early result was that of Siegel (yet another Siegel (1929) breakthrough) on J and J' (see Exercise 19.2). Mahler (1968) was able to deduce the transcendence of

$$\frac{\pi}{2}\frac{Y(-1)}{J(-1)} - \gamma,$$

where Y is a normalized Bessel function of the second kind defined (say for real $z < 0$) by

$$Y(z) = \frac{2}{\pi}\left(\gamma + \frac{1}{2}\log(-z)\right)J(z) - \frac{2}{\pi}\sum_{k=1}^{\infty}\frac{h_k z^k}{k!^2}$$

with $h_k = \sum_{l=1}^{k} 1/l$; in fact for one week he thought he could prove the transcendence of Euler's constant itself.

Shidlovsky was the first to obtain results about general E-functions. See Shidlovsky (1989) and also Mahler (1976).

There is also a theory of G-functions, but as yet it yields only algebraic independence results restricted by degree. A notable example by Galochkin (1974) is that

$$\log\left(1 + \frac{1}{a}\right)\log\left(1 - \frac{1}{a}\right)$$

is irrational for all integers $a > e^{795}$. This has the same style as (12.48); and indeed $\sqrt[3]{1+z}$ is a G-function.

Much of the modern theory of algebraic independence is based on a generalization of Proposition 22.2 to several variables $W, W', \ldots$ due to Philippon. For a good account see Nesterenko and Philippon (2001).

And that is the last chapter of our book. But it ain't over till it's over (Yogi Berra too), and sometimes also not until the fat lady sings, which she will do in the appendix.

Exercises

22.1 Show that $\sum_{k=1}^{\infty} 1/2^{k!}$, $\sum_{k=1}^{\infty} 1/3^{k!}$, $\sum_{k=1}^{\infty} 1/4^{k!}$ are algebraically independent over $\mathbb{Q}$.

22.2 Let $\alpha \neq 0, \beta, \gamma$ be algebraic with any determination $\log \alpha$. Deduce from Schanuel's Conjecture that e^{γ} and $\alpha^{\beta} = \exp(\beta \log \alpha)$ are algebraically independent, provided α^{β} is transcendental and $\gamma \neq 0$.

22.3 Let β in $\mathbb{C}$ be irrational. Show that $\mathrm{ord}_{z=0} P(z, e^z, e^{\beta z}) \leq (L+1)^3 - 1$ for any non-zero P in $\mathbb{C}[X, Y, Z]$ with degree at most L in each variable.

22.4 For non-zero u, v in $\mathbb{C}$ and P in $\mathbb{C}[X, Y]$ define $\mathcal{E}P$ by $\mathcal{E}P(X, Y) = P(uX, vY)$.

(a) If P is irreducible show that $\mathcal{E}P$ is irreducible.

(b) If further P is not divisible by X or Y and $\mathcal{E}P = \lambda P$ for some λ in $\mathbb{C}$, show that u, v are multiplicatively dependent.

* * * * * * * * * * * * * * * * * * *

22.5 (a) Let $\beta_1, \ldots, \beta_n$ be in a number field K. Show that there is c with the following property. For any even $L \geq 0$ in $\mathbb{Z}$ there are $P_1, \ldots, P_n$ in $\mathbb{Z}_K[z]$, not all zero, of degrees at most L and with coefficients of sizes at most $c^L L^L$, such that $\mathrm{ord}_{z=0} \phi \geq (n - \frac{1}{2})L$ for

$$\phi(z) = P_1(z)e^{\beta_1 z} + \cdots + P_n(z)e^{\beta_n z}$$

(if non-zero) [*Hint*: be careful, be very very careful, about factorials; and use Proposition 13.2].

(b) From now on assume that $\beta_1, \ldots, \beta_n$ are linearly independent over $\mathbb{Q}$. Show that ϕ is indeed non-zero. From now on assume too that $L \geq 2n - 3$. Show that each of $P_1, \ldots, P_n$ is non-zero [*Hint*: Exercise 13.19].

(c) We can write the derivatives as

$$\phi^{(k)}(z) = P_{1k}(z)e^{\beta_1 z} + \cdots + P_{nk}(z)e^{\beta_n z} \quad (k = 0, 1, 2, \ldots)$$

for polynomials P_{jk}. In (b) show that the matrix with entries $P_{jk}(z)$ ($j = 1, \ldots, n$; $k = 0, \ldots, n-1$) has non-zero determinant $\Delta(z)$.

(d) Show that $\mathrm{ord}_{z=1} \Delta \leq \frac{1}{2}L + n - 1$.

(e) Show that the matrix with entries $P_{jk}(1)$ ($j = 1, \ldots, n; k = 0, \ldots, \frac{L}{2} + 2n - 2$) has rank n.

(f) Show that

$$|\phi^{(k)}(1)| \leq c^L L^{-(n-3)L} \quad (k = 0, \ldots, \frac{1}{2}L + 2n - 2)$$

for c depending only on $\beta_1, \ldots, \beta_n$ and K [*Hint*: large radius].

(g) Deduce that at least $\frac{n}{6[K:\mathbb{Q}]}$ among $e^{\beta_1}, \ldots, e^{\beta_n}$ are linearly independent over K [*Hint*: if not, then adjoin suitable $\phi^{(k)}(1)$ to the vanishing forms, find a small non-zero determinant, and count powers of L^L as in the proof of Theorem 19.1].

22.6 Show that $e, e^{\sqrt{2}}$ are algebraically independent [*Hint*: multiply a relation $P(e, e^{\sqrt{2}}) = 0$ by many $e^{r+s\sqrt{2}}$].

22.7 If $\alpha_1, \ldots, \alpha_m$ are algebraic numbers linearly independent over $\mathbb{Q}$, show that $e^{\alpha_1}, \ldots, e^{\alpha_m}$ are algebraically independent (Lindemann–Weierstrass).

22.8 Let ξ, η be complex numbers linearly independent over $\mathbb{Q}$, let $\sigma_1, \sigma_2, \sigma_3$ be complex numbers also linearly independent over $\mathbb{Q}$, and suppose $e^{\xi\sigma}, e^{\eta\sigma}$ are multiplicatively independent for each non-zero σ in $\mathbb{Z}\sigma_1 + \mathbb{Z}\sigma_2 + \mathbb{Z}\sigma_3$. Show that at least one among

$$P(e^{\xi\sigma}, e^{\eta\sigma}), \quad \sigma = s_1\sigma_1 + s_2\sigma_2 + s_3\sigma_3 \quad (0 \leq s_1, s_2, s_3 \leq 4L^{2/3})$$

is non-zero for any non-zero P in $\mathbb{C}[X, Y]$ of degree at most L in each variable.

22.9 With the aid of Schanuel's Conjecture answer the question in Exercise 11.23.

22.10 With the aid of Schanuel's Conjecture prove that there is only $z = 0$ in Exercise 18.2.

22.11 Let $\alpha \neq 0, \beta \neq 0, \gamma$ be algebraic with any determinations $\log\alpha, \log\beta$, and let $\beta^\gamma = \exp(\gamma \log\beta)$. Deduce from Schanuel's Conjecture that

$$\alpha^{\beta^\gamma}$$

(of course $\exp(\beta^\gamma \log\alpha)$ – what else?) is transcendental, provided $\log\alpha \neq 0$ and β^γ is irrational (see also Exercise 19.27).

22.12 For $E_2(\tau) = P(e^{2\pi i\tau})$ show that $E_2(-1/\tau) = \tau^2 E_2(\tau) - 6i\tau/\pi$.

22.13 Show that $E_2(\tau) = 3\eta(1)/\pi^2$ for the quasi-period associated with the period 1 of the lattice $\mathbb{Z} + \mathbb{Z}\tau$. This is an analogue of Exercise 21.14 – see also Exercise 20.21.

22.14 Show that the functions $q, P(q), Q(q), R(q)$ are algebraically independent over $\mathbb{C}$.

22.15 The error function $\mathrm{erf}(z)$ is defined by

$$\mathrm{erf}(z) = \frac{2}{\sqrt{\pi}} \int_0^z e^{-t^2} \mathrm{d}t = \frac{2}{\sqrt{\pi}} \sum_{n=0}^{\infty} \frac{(-1)^n z^{2n+1}}{n!(2n+1)}.$$

With $g(z) = e^{z^2} \mathrm{erf}(z)$ show that there is C such that $\mathrm{ord}_{z=0} P(z, g(z)) \leq CL^2$ for any non-zero P in $\mathbb{C}[X, Y]$ with degree at most L in each variable [*Hint*: for $\phi = P(z, g(z))$ write $\phi' = Q(z, g(z))$ and show that P cannot divide Q if $\phi(0) = 0$; then assume P irreducible].

22.16 The exponential-integral (non-modular) function $E_1(z)$ is defined by

$$E_1(z) = \int_z^{\infty} \frac{e^{-t}}{t} \mathrm{d}t = -\gamma - \log z - \sum_{n=1}^{\infty} \frac{(-1)^n z^n}{n!n}$$

where γ is Euler's constant and the logarithm is branched along the negative real axis in the standard way. With $g(z) = e^z E_1(z)$ show that there is C such that $\mathrm{ord}_{z=1} P(z, g(z)) \leq CL^2$ for any non-zero P in $\mathbb{C}[X, Y]$ with degree at most L in each variable.

22.17 Show that there is C such that

$$\mathrm{ord}_{z=0} P(z, e^z, e^{e^z}) \leq CL^4$$

for any non-zero P in $\mathbb{C}[X, Y, Z]$ with degree at most L in each variable.

22.18 Show that there is C such that

$$\mathrm{ord}_{z=0} P(z, e^z, e^{e^z}) \leq CL^3$$

for any non-zero P in $\mathbb{C}[X, Y, Z]$ with degree at most L in each variable (Brownawell (1983) – see also Nesterenko (1983) for generalizations to several functions).

22.19 Is there C such that

$$\mathrm{ord}_{z=0} P(z, e^{e^z}) \leq CL^2$$

for any non-zero P in $\mathbb{C}[X, Y]$ with degree at most L in each variable? I don't know.

22.20 Show that there is C such that $\mathrm{ord}_{z=a} P(z, \wp(z)) \leq CL^2$ for any $\wp$ and any a, not a pole of $\wp$, and any non-zero P in $\mathbb{C}[X, Y]$ with degree at most L in each variable (Brownawell and Masser, 1980).

22.21 Show that there is $C = C(g_2, g_3)$ such that for any $R \geq 3$ in $\mathbb{R}$ and any $L \geq 1$ in $\mathbb{Z}$ and any non-zero polynomial P in $\mathbb{C}[X, Y]$ of degree at most L in each variable, the function $P(z, \wp(z))$ has at most $C(L^2 + LR^2)$ zeroes in the disc $\mathbb{D}_R$ [*Hint*: Lagrange].

22.22 Show that there is $C = C(g_2, g_3)$ such that for any $R \geq 3$ in $\mathbb{R}$ and any $L \geq 1$ in $\mathbb{Z}$ and any non-zero polynomial P in $\mathbb{C}[X, Y, Z]$ of degree at most L in each variable, the function $P(z, \wp(z), \zeta(z))$ has at most $C(L^3 + LR^2)$ zeroes in $\mathbb{D}_R$.

22.23 Let $\zeta(z)$ still be the Weierstrass zeta function. Is there $C = C(g_2, g_3)$ such that for any $R \geq 3$ in $\mathbb{R}$ and any $L \geq 1$ in $\mathbb{Z}$ and any non-zero polynomial P in $\mathbb{C}[X, Y]$ of degree at most L in each variable, the function $P(z, \zeta(z))$ has at most $C(L^2 + LR^2)$ zeroes in $\mathbb{D}_R$? Compare with Exercise 18.18.

22.24 For real $\theta \geq 0$ show that there is c with the following property. Let ω in $\mathbb{C}$ be transcendental, and let $A_N \neq 0$ be a sequence of polynomials in $\mathbb{Z}[W]$ of degree at most N and of length at most $2^{N(\log N)^\theta}$ $(N = 1, 2, \ldots)$. Then there are infinitely many N with $|A_N(\omega)| \geq 2^{-cN^2(\log N)^\theta}$.

22.25 Let ξ, η be complex numbers linearly independent over $\mathbb{Q}$, and let σ_1, σ_2 be complex numbers also linearly independent over $\mathbb{Q}$ such that $e^{\xi\sigma_2}, e^{\eta\sigma_2}$ are both algebraic. Using P in $\mathbb{Z}[\omega][X, Y, Z]$ of degrees at most M in ω and Z and of degree at most L in X and Y for $L < M$, with $\phi(z) = P(e^{\xi z}, e^{\eta z}, z)$ at every $s_1\sigma_1 + s_2\sigma_2$ $(s_1 \leq S_1,\ s_2 \leq S_2)$ for $S_1 < S_2$, show that at least two among the six numbers

$$\xi, \eta, \sigma_1, \sigma_2, e^{\xi\sigma_1}, e^{\eta\sigma_1}$$

are algebraically independent. The case $\xi = \sigma_1 = 1,\ \eta = \sigma_2 = e$ implies that at least one of e^e, e^{e^2} is transcendental (Schneider's Eighth Problem as solved independently by Brownawell (1974) and Waldschmidt (1973) – the announcement (Hensel, 1905) was also premature).

22.26 Let $\sigma \neq 0$ be in $z\mathbb{C}[[z]]$, and define e^σ in $\mathbb{C}[[z]]$ as in Exercise 4.31. Show that σ, e^σ are not both in $\mathbb{C}(z)$ [*Hint*: differentiate $\tau = e^\sigma$]. This is an extremely special case of "functional Schanuel" or "Ax–Schanuel" for $m = 1$.

22.27 Let ω be transcendental over $\mathbb{Q}$ and let χ be algebraic of degree d over $\mathbb{Q}(\omega)$. Are there $\chi_1, \ldots, \chi_d$ such that the integral closure $\mathbb{Z}[\omega]_{\mathbb{Q}(\omega,\chi)}$ of $\mathbb{Z}[\omega]$ in $\mathbb{Q}(\omega, \chi)$ has the shape $\mathbb{Z}[\omega]\chi_1 + \cdots + \mathbb{Z}[\omega]\chi_d$?

Appendix: Néron's square root

Let ξ be an algebraic number. In Proposition 14.10 for one variable we estimated the height of a rational function of ξ from above in terms of the height $H(\xi)$ of ξ. In Exercise 14.77 we verified that the upper bound is usually asymptotically best possible as $H(\xi)$ gets large. Here we find an analogous upper bound on a fixed elliptic curve; that is, we consider algebraic numbers ξ, η related by $\eta^2 = 4\xi^3 - g_2\xi - g_3$ or for simplicity

$$\eta^2 = \xi^3 + a\xi + b \tag{A.1}$$

with fixed a, b such that

$$4a^3 + 27b^2 \neq 0. \tag{A.2}$$

If a, b are also algebraic, then we want to estimate from above the height of a rational function in ξ and η in terms of $H(\xi)$ or better its logarithm $h(\xi)$.

This is in itself not difficult; for example with $\eta^2 = \xi^3 + 3\xi + 4$ we can quickly see that $h(\eta) \leq (3/2)h(\xi) + c$ for an absolute constant c. And almost as easily that

$$h\left(\frac{\eta}{\xi}\right) \leq \frac{3}{2}h(\xi) + c \tag{A.3}$$

(but note that for example Proposition 14.10 for two variables gives only $5/2$ in place of $3/2$). We get with similar bounds for $h(\alpha)$, where

$$\alpha = \frac{\eta}{\xi + 1}. \tag{A.4}$$

Here (A.3) cannot be asymptotically improved for large $h(\xi)$ (see Exercise A.2). But in view of the identity

$$\alpha^2 = \frac{\xi^2 - \xi + 4}{\xi + 1}, \tag{A.5}$$

we can deduce at once from Proposition 14.10 the asymptotic improvement

$$h(\alpha) \leq h(\xi) + \frac{1}{2}\log 6. \tag{A.6}$$

Similarly

$$h\left(\frac{3\xi - 4\eta + 7}{\xi^2}\right) \leq 2h(\xi) + c, \tag{A.7}$$

with a similar bound for $h(\beta)$, where

$$\beta = \frac{3\xi - 4\eta + 8}{\xi^2}. \tag{A.8}$$

Again (A.7) cannot be improved (see Exercise A.3). But in fact we will soon see that $h(\beta)$ is asymptotically $h(\xi)$! Further this cannot be checked by means of a simple identity for β. Thus we see that for elliptic curves the matter is not so clear.

Things are most naturally expressed in terms of degree theory on algebraic curves. But for the time being we may give an elementary definition as follows, which makes sense even for complex a, b not necessarily algebraic. The equation $y^2 = x^3 + ax + b$ defines a quadratic extension $\mathbb{C}(x, y)$ of $\mathbb{C}(x)$. Suppose $R(X, Y)$ in $\mathbb{C}(X, Y)$ is such that $\rho = R(x, y)$ is not in $\mathbb{C}(x)$. Then we have an equation

$$A(x)\rho^2 + B(x)\rho + C(x) = 0 \tag{A.9}$$

for $A(X) \neq 0, B(X), C(X)$ in $\mathbb{C}[X]$; and we may suppose that they have no common factor. We define the degree of $R(x, y)$ as half the maximum of the degrees of A, B, C (the maximum itself is in fact the geometric degree). If $R(x, y)$ is in $\mathbb{C}(x)$ so $P(x)/Q(x)$, then we define its degree in the more obvious way, consistent with an equation $Q(x)^2\rho^2 - P(x)^2 = 0$.

For example y has degree 3/2 and x degree 1; and with $y^2 = x^3 + 3x + 4$ and $\rho = y/x$, as in (A.3), we have $x^2\rho^2 - (x^3 + 3x + 4) = 0$ so degree 3/2; but with $\rho = y/(x + 1)$, as in (A.4), now $(x + 1)\rho^2 - (x^2 - x + 4) = 0$ so degree suddenly 1. Similarly with $\rho = (3x - 4y + 7)/x^2$, as in (A.7), we have

$$x^4\rho^2 - (6x^3 + 14x^2)\rho - (16x^3 - 9x^2 + 6x + 15) = 0$$

so degree 2; but with $\rho = (3x - 4y + 8)/x^2$, as in (A.8), we have

$$x^2\rho^2 - (6x + 16)\rho - (16x - 9) = 0$$

so the degree slumps to 1.

Here is the main result of this appendix. When it is remembered that the constant appearing should be exponentiated, we have not so much another atrocity as more of a barbarity.

Theorem A.1 *Given algebraic numbers ξ, η with $\eta^2 = \xi^3 + 3\xi + 4$, and P, Q in $\mathbb{Z}[X, Y]$ of total degree at most $D \geq 0$ and lengths $\mathcal{L}(P), \mathcal{L}(Q)$, with $Q(\xi, \eta) \neq 0$ and $P(x, y)/Q(x, y)$ of degree d for $y^2 = x^3 + 3x + 4$, we have*

$$h\left(\frac{P(\xi,\eta)}{Q(\xi,\eta)}\right) \leq dh(\xi) + 2000(D + \log\max\{\mathcal{L}(P), \mathcal{L}(Q)\})(\sqrt{h(\xi)} + 1). \tag{A.10}$$

Thus the main term involves our degree d, consistent with the above remarks. It can be shown (for every P, Q) that d cannot be replaced by anything smaller (see Exercise A.4), so this gives the correct asymptotics, which had been found (in much more general situations) by Siegel in 1929.

A new feature is the $\sqrt{h(\xi)}$. We will see that it cannot be avoided in general, even though it is missing from the special cases (A.3), (A.6) and (A.7). It is associated with the name of Néron, who derived it from quadratic properties of the Néron–Tate height on Jacobians. The reader may therefore suspect some subtleties involved in the proof. Here it is just another job for auxiliary-polynomial-man. There are also ingredients based on a well-known theorem of Eisenstein on power series, but here we can avoid any explicit reference to that. Following on from Proposition 14.10 we have elected to calculate all the constants explicitly (more in the style of bicycle-repair-man); this makes the proof quite a bit longer, which we hope does not obscure the general strategy. The latter can be viewed without explicit constants in Exercise A.24.

An example is

$$h(\beta) \leq h(\xi) + 10000(\sqrt{h(\xi)} + 1) \tag{A.11}$$

in (A.8).

A result similar to Theorem A.1 holds for any algebraic a, b in (A.1), but for simplicity of exposition we have restricted to $y^2 = x^3 + 3x + 4$; this has the technical advantage that there is a rational point $(x, y) = (0, 2)$ (which we will also use in combination with β above to show that Néron's square root cannot be avoided). A second such point $(-1, 0)$ is responsible for the anomaly with α above.

For the proof we have to extend our notion of order of vanishing at a complex number ξ to a point $\pi = (\xi, \eta)$ with $\eta^2 = \xi^3 + a\xi + b$; here also a, b can be complex numbers with (A.2). This order will be attached to a non-zero $\rho = R(x, y)$ in the function field $\mathbb{C}(x, y)$.

If $\eta \neq 0$ we define S in $\mathbb{C}[z, w]$ by

$$S(z, w) = (\xi + z)^3 + a(\xi + z) + b - (\eta + w)^2.$$

Then the conditions of Lemma 4.3 are satisfied (with $\mathcal{K}' = \mathcal{K} = \mathbb{C}$). So there is a unique power series ψ in $z\mathbb{C}[[z]]$ with $S(z, \psi(z)) = 0$.

Now if Q is in $\mathbb{C}[X, Y]$ with $Q(x, y) \neq 0$, then $Q(\xi + z, \eta + \psi(z)) \neq 0$ in $\mathbb{C}[[z]]$. For S is clearly irreducible and so otherwise the minimal polynomial $S(z, Y - \eta)$ of $\eta + \psi$ would have to divide $Q(\xi + z, Y)$ in $\mathbb{C}[z, Y]$. But then substituting $z = x - \xi, Y = y$ in $\mathbb{C}[x, y]$ would give a contradiction, as $S(x - \xi, y - \eta) = 0$ there.

It follows that for our ρ in the quotient field of $\mathbb{C}[x, y]$, the function

$$\Psi(z) = R(\xi + z, \eta + \psi(z)) \tag{A.12}$$

is defined and non-zero in the quotient field of $\mathbb{C}[[z]]$. We define (in this $\eta \neq 0$ case)

$$\mathrm{ord}_\pi\, \rho = \mathrm{ord}_{z=0}\, \Psi(z). \tag{A.13}$$

With the example $\pi = (0, 2)$ on $y^2 = x^3 + 3x + 4$, we find

$$\psi(z) \;=\; -2 + \sqrt{4 + 3z + z^3} \;=\; 2\sum_{k=1}^{\infty} \binom{1/2}{k} \left(\frac{3z + z^3}{4}\right)^k \tag{A.14}$$

as in (4.28). This is

$$\frac{3}{4}z - \frac{9}{64}z^2 + \frac{155}{512}z^3 - \frac{1941}{16384}z^4 + \frac{8613}{131072}z^5 - \frac{117197}{2097152}z^6 + \cdots. \tag{A.15}$$

We find for $P(x, y) = 3x - 4y + 7$ that

$$P(z, 2 + \psi(z)) \;=\; -1 + \cdots$$

and so for $R(x, y) = (3x - 4y + 7)/x^2$ in (A.7) we get

$$R(z, 2 + \psi(z)) \;=\; -\frac{1}{z^2} + \cdots$$

so $\mathrm{ord}_\pi\, R = -2$. But for $P(x, y) = 3x - 4y + 8$ we get

$$P(z, 2 + \psi(z)) \;=\; \frac{9}{16}z^2 + \cdots$$

so $\mathrm{ord}_\pi\, P = 2$, and thus for $R(x, y) = (3x - 4y + 8)/x^2$ as in (A.8) we get $\mathrm{ord}_\pi\, R = 0$.

If $\eta = 0$ in π we have to modify thanks to "ramification". We note first that $c = 3\xi^2 + a \neq 0$ because of the identity

$$(-18aX + 27b)(X^3 + aX + b) + (6aX^2 - 9bX + 4a^2)(3X^2 + a) \;=\; 4a^3 + 27b^2$$

at $X = \xi$. Then we take

$$S(z, w) \;=\; \frac{(\xi + z^2)^3 + a(\xi + z^2) + b}{cz^2} - (1 + w)^2$$

(still a polynomial) to get ψ with $S(z,\psi)=0$. We choose a determination $\sqrt{c}$, and use

$$\Psi(z)=R(\xi+z^2,\sqrt{c}z(1+\psi(z))) \tag{A.16}$$

to define $\mathrm{ord}_\pi\,\rho$ as in (A.13).

By the way, it can be easily seen that ψ is a power series in z^2, and so changing the sign of z in (A.16) amounts to changing the sign of $\sqrt{c}$; thus the resulting definition of $\mathrm{ord}_\pi\,\rho$ is independent of the determination.

In all these cases for π, the z plays the role of a choice of "local parameter" at π; there are of course many other choices.

With the example $\pi=(-1,0)$ on $y^2=x^3+3x+4$, we find $c=6$ and

$$\psi(z) \;=\; \frac{\sqrt{6-3z^2+z^4}}{\sqrt{6}}-1 \;=\; \sum_{k=1}^{\infty}\binom{1/2}{k}\left(\frac{-3z^2+z^4}{6}\right)^k,$$

which is

$$-\frac{1}{4}z^2+\frac{5}{96}z^4+\frac{5}{384}z^6+\frac{35}{18432}z^8-\frac{5}{24576}z^{10}-\frac{415}{1769472}z^{12}+\cdots.$$

We find for $P(x,y)=y$ that

$$P(-1+z^2,\sqrt{6}z(1+\psi(z))) \;=\; \sqrt{6}z+\cdots$$

and so for $R(x,y)=y/x$ in (A.3) we get $\mathrm{ord}_\pi\,R=1$. But for $R(x,y)=y/(x+1)$ as in (A.4) we get $\mathrm{ord}_\pi\,R=-1$.

So far the π considered can be viewed as analogues of the $\mathfrak{p}$ in Chapter 14. But we need analogues also of the σ there. This will be just $\pi=\infty$. In that case we take

$$S(z,w) \;=\; z^6\left(\left(\frac{1}{z^2}\right)^3+a\left(\frac{1}{z^2}\right)+b\right)-(1+w)^2$$

to get ψ. We use

$$\Psi(z)=R\left(\frac{1}{z^2},\frac{1+\psi(z)}{z^3}\right) \tag{A.17}$$

to define $\mathrm{ord}_\pi\,\rho$ as in (A.13).

For example on $y^2=x^3+3x+4$ we find

$$\psi(z) \;=\; \sqrt{1+3z^4+4z^6}-1 \;=\; \sum_{k=1}^{\infty}\binom{1/2}{k}\left(3z^4+4z^6\right)^k,$$

which is

$$\frac{3}{2}z^4+2z^6-\frac{9}{8}z^8-3z^{10}-\frac{5}{16}z^{12}+\frac{27}{4}z^{14}+\frac{747}{128}z^{16}-\frac{103}{8}z^{18}+\cdots.$$

We find for $R(x, y) = x/y$ that

$$R\left(\frac{1}{z^2}, \frac{1+\psi(z)}{z^3}\right) = z + \cdots$$

so $\mathrm{ord}_\infty(x/y) = 1$.

Now we can prove something related to an analogue of Proposition 14.3.

Proposition A.2 *Given $\rho \neq 0$ in $\mathbb{C}(x, y)$ there are at most finitely many π with $\mathrm{ord}_\pi\, \rho \neq 0$, and ρ has degree*

$$-\frac{1}{2}\sum_\pi \min\{0, \mathrm{ord}_\pi\, \rho\} = \frac{1}{2}\sum_\pi \max\{0, \mathrm{ord}_\pi\, \rho\}. \tag{A.18}$$

Proof We note first that ord_π for $\pi = (\xi, \eta)$ is related to ord_ξ for $\rho = R(x)$ in $\mathbb{C}(x)$. More precisely they are equal except when $\eta = 0$, and then

$$\mathrm{ord}_\pi\, \rho = 2\, \mathrm{ord}_\xi\, R;$$

also

$$\mathrm{ord}_\infty\, \rho = -\,\mathrm{ord}_{z=0}\, R(1/z^2)$$

which is twice the degree of R when it is a polynomial (see Exercise A.8). From these it is relatively easy to check the Proposition for ρ in $\mathbb{C}(x)$ (see also Exercise 14.46).

Now $\mathbb{C}(X, Y)$ has an automorphism over $\mathbb{C}(X)$ taking Y to $-Y$, and it induces on the function field $\mathbb{C}(x, y)$ the unique non-trivial automorphism over $\mathbb{C}(x)$. We denote the images of R or ρ under this by $\overline{R}$ or $\overline{\rho}$; for example $\overline{3x - 4y + 8} = 3x + 4y + 8$. There is a corresponding definition of $\overline{\pi}$ for $\pi = (\xi, \eta)$ as $(\xi, -\eta)$, with $\overline{\infty} = \infty$. It is not difficult to verify that

$$\mathrm{ord}_\pi(\overline{\rho}) = \mathrm{ord}_{\overline{\pi}}(\rho) \tag{A.19}$$

(see Exercise A.9).

Next for ρ not in $\mathbb{C}(x)$ we have (A.9), and the same equation holds for the conjugate $\overline{\rho} \neq \rho$. It follows that

$$AT^2 + BT + C = A(T - \rho)(T - \overline{\rho})$$

in $\mathbb{C}(x, y)[T]$. We are going to apply the Gauss Lemma as in the proof of Proposition 14.4. The point is that each π gives rise to an ultrametric valuation defined by $|\rho|_\pi = \exp(-\,\mathrm{ord}_\pi(\rho))$. We deduce the very useful

$$\max\{|A|_\pi, |B|_\pi, |C|_\pi\} = |A|_\pi \max\{1, |\rho|_\pi\} \max\{1, |\overline{\rho}|_\pi\}. \tag{A.20}$$

Now there are at most finitely many π with $|A|_\pi \neq 1$; they correspond to the zeroes ξ of A and ∞. It follows that there are at most finitely many π with

$|\rho|_\pi > 1$. Applying the same argument to $1/\rho$, we get the first assertion of the present proposition.

But since the number of zeroes of A with multiplicity is its degree, we see even that

$$\prod_\pi |A|_\pi = 1 \tag{A.21}$$

taken over all π. Also the coprimality of A, B, C implies that the left-hand side of (A.20) is 1 for any $\pi \neq \infty$; and for $\pi = \infty$ it is $\exp(4d)$ for the degree d of ρ. Then taking the product over all π and using (A.19) yields

$$\exp(4d) = \prod_\pi \max\{1, |\rho|_\pi\}^2;$$

which on taking logarithms shows that d is the left-hand side of (A.18). Finally applying this to $1/\rho$, which also has degree d, leads to the right-hand side of (A.18). This completes the proof (and when ρ is in $\mathbb{C}(x)$ the same arguments can be applied to $Q^2\rho^2 - P^2$). □

The equation (A.18) says that ρ has the same number of poles as zeroes (with multiplicity). It is the Sum Formula

$$\sum_\pi \operatorname{ord}_\pi \rho = 0, \tag{A.22}$$

which is equivalent to

$$\prod_\pi |\rho|_\pi = 1$$

generalizing (A.21); and the analogue of (14.17) or (14.38).

We give an example for $\rho = (3x - 4y + 8)/x^2$ on $y^2 = x^3 + 3x + 4$ as in (A.8).

For $P(X, Y)$ in $\mathbb{C}[X, Y]$ with $P(x, y) \neq 0$, we have $\operatorname{ord}_\pi P(x, y) \geq 0$ for $\pi = (\xi, \eta)$ (see Exercise A.5(a) below) and $\operatorname{ord}_\pi P(x, y) \geq 1$ if and only if $P(\xi, \eta) = 0$ (see Exercise A.5(b) below). With the numerator of ρ this leads to just $\pi_1 = (0, 2)$ and $\pi_2 = (9/16, 155/64)$. We already calculated the order of P as 2 at π_1. We could calculate the order at π_2 directly as 1, or we could verify this the back way by calculating the order at ∞ more easily as -3 and using (A.22). Similarly for x in the denominator of ρ we get orders 1 at π_1 and $(0, -2) = \overline{\pi_1}$, and order -2 at ∞. Thus $\operatorname{ord}_\pi \rho \neq 0$ precisely at $\pi = \pi_2, \infty, \overline{\pi_1}$ with orders respectively $1, 1, -2$. This fits with degree 1 in (A.18).

We can now turn to the proof of Theorem A.1. Broadly speaking this goes along the lines laid down by Thue as in Chapter 12. But in detail the arguments resemble more closely those used in Chapter 17.

We start with some preliminary skirmishings.

It will be convenient to ensure that $\rho = P(x,y)/Q(x,y)$ is "regular" at $\infty, \pi_0 = (0,2)$ and $\overline{\pi_0}$, and even "non-zero" at ∞; that is,

$$|\rho|_{\pi_0} \leq 1, \quad |\rho|_{\overline{\pi_0}} \leq 1, \quad |\rho|_\infty = 1. \tag{A.23}$$

We can do this by simple tricks. For example, there is $a = 0,1,2,3$ such that ρ is either non-regular or not $-a$ at $\infty, \pi_0, \overline{\pi_0}$. Then $\rho' = 1/(\rho + a)$ is regular at these points. Next there is $b = 0,1$ such that $\rho'' = \rho' + b$ is non-zero at ∞. So ρ'' has the convenient properties above. Since $\rho = -a + 1/(\rho'' - b)$ and

$$h\left(-a + \frac{1}{\alpha - b}\right) \leq h(\alpha) + \log 12,$$

we may indeed assume (A.23).

Next we would like also to have $Q(0,2) \neq 0$. This is not so simple, as it might happen that both $P(x,y)$ and $Q(x,y)$ have high order zeroes at $(0,2)$ (compatible with the first of (A.23), that is). At a bit of expense we can do it as follows. Write $\mathcal{B} = \max\{\mathcal{L}(P), \mathcal{L}(Q)\}$.

Lemma A.3 *Given P, Q as in Theorem A.1 we have $d \leq 3D$. Further there are A, A_0, A_1 in $\mathbb{Z}[X]$, of degree at most $2d$ and lengths at most $2^{39D}\mathcal{B}^2$, such that for*

$$P_*(X,Y) = A_0(X) + A_1(X)Y, \quad Q_*(X,Y) = 2A(X)$$

we have

$$Q_*(0,2) \neq 0 \tag{A.24}$$

and

$$\frac{P_*(x,y)}{Q_*(x,y)} = \frac{P(x,y)}{Q(x,y)}.$$

Proof Here $2d$ by (A.18) is the number of poles of $P(x,y)/Q(x,y)$ with multiplicity. These can occur only at poles of $P(x,y)$ or zeroes of $Q(x,y)$. In $\mathbb{Q}[X,Y]$ they have total degrees at most D so in $\mathbb{Q}[x,y]$ degrees at most $3D/2$. So at most $6D$ poles and zeroes.

Next it is easily seen that A, B, C in (A.9) for $\rho = P(x,y)/Q(x,y)$ may be taken in $\mathbb{Q}[X]$, and we may even suppose them to be in the unique factorization domain $\mathbb{Z}[X]$ with no common factor. This determines A up to sign.

As $\mathbb{Q}(x,y) = \mathbb{Q}(x) + \mathbb{Q}(x)y$ we can write $2A(x)\rho = A_0(x) + A_1(x)y$ for A_0, A_1 in $\mathbb{Q}(X)$. In fact they are in $\mathbb{Q}[X]$. This is clear for

$$A_0(x) = \mathrm{Tr}(A(x)\rho) = -B(x)$$

with the trace from $\mathbb{Q}(x, y)$ to $\mathbb{Q}(x)$; and also from

$$A_0(x)^2 - A_1(x)^2(x^3 + 3x + 4) \;=\; \mathrm{Nm}(2A(x)\rho) \;=\; 4A(x)C(x)$$

with the norm, on recalling that $X^3 + 3X + 4$ is squarefree. We see that A_0, A_1 are even in $\mathbb{Z}[X]$ because this cubic is primitive.

To check $A(0) \neq 0$ in (A.24) we appeal to (A.20). At $\pi = \pi_0$ we get using (A.23) that $|A(x)|_\pi = 1$ indeed. And of course A, B, C have degrees at most $2d$ (as polynomials in X). Now the degree bounds for A, A_0, A_1 are clear from the trace and norm equations above.

Finally for the length bounds we note a second equation

$$\tilde{A}(x)\rho^2 + \tilde{B}(x)\rho + \tilde{C}(x) = 0,$$

with $\tilde{A}, \tilde{B}, \tilde{C}$ obtained from

$$Q\overline{Q}, \;\; -(P\overline{Q} + \overline{P}Q), \;\; P\overline{P} \tag{A.25}$$

after expressing these as polynomials in X and Y^2 and replacing Y^2 by $X^3 + 3X + 4$. It follows that $\tilde{A} = \Delta A, \tilde{B} = \Delta B, \tilde{C} = \Delta C$ for some Δ in $\mathbb{Z}[X]$. Also $\tilde{A}, \tilde{B}, \tilde{C}$ have degrees at most $3D$. The lengths of (A.25) are at most $2\max\{\mathcal{L}(P), \mathcal{L}(Q)\}^2 = 2\mathcal{B}^2$ and so those of $\tilde{A}, \tilde{B}, \tilde{C}$ at most this multiplied by $\mathcal{L}(X^3 + 3X + 4)^{3D} = 8^{3D}$. Thus by Proposition 14.1(i) the lengths of A, B, C are at most $L_* = 2.8^{6D}\mathcal{B}^2$. So also the length of A_0. For A_1 we first note that $A_0^2 - 4AC$ has length at most

$$\mathcal{L}(A_0)^2 + 4\mathcal{L}(A)\mathcal{L}(C) \;\leq\; 5L_*^2.$$

Its degree is at most $6D$, and it is also $A_1^2(X^3 + 3X + 4)$, so with length at least

$$8^{-6D}\mathcal{L}(A_1^2) \geq 8^{-12D}\mathcal{L}(A_1)^2.$$

This leads to what we want. □

Some of these arguments could be avoided if we had defined some sort of height or length of ρ directly through A, B, C as we defined its degree. This would have enabled us to eliminate D altogether in Theorem A.1. But we preferred to use P, Q in line with Proposition 14.10.

We now advance in traditional style with our auxiliary polynomial. As before we introduce δ with $0 < \delta \leq 1$.

Lemma A.4 *For any $L \geq 1, n \geq 1$ in $\mathbb{Z}$ with*

$$(1 - \delta)(L + 1) \;>\; (1 + \delta)(nd + 1) \tag{A.26}$$

there are P_0, P_1, Q_0, Q_1 in $\mathbb{Z}[X]$, not all zero, of degree at most L and with coefficients of absolute values at most $(2^{68nD+64L}B^{2n})^{1/\delta}$, such that

$$(P_0(x) + P_1(x)y)Q_*(x,y)^n = (Q_0(x) + Q_1(x)y)P_*(x,y)^n. \tag{A.27}$$

Proof We first claim that the difference of the two sides of (A.27), if not actually zero, can be made to have order at least $M = 2L + 4 + 2nd$ at some fixed point. This may recall the machinations of the Prologue and elsewhere. The choice of the point is not important, but we will use of course $\pi_0 = (0,2)$ for convenience. Thus we should substitute the power series $z, 2 + \psi(z)$ in (A.14) for x, y and equate to zero the corresponding coefficients, giving M linear equations. We have $N = 4(L+1)$ unknowns in P_0, P_1, Q_0, Q_1, and so by (A.26) the Siegel exponent

$$\varsigma = \frac{M}{N-M} < \frac{1}{\delta}. \tag{A.28}$$

However the equations might not be over $\mathbb{Z}$, due to possible denominators in (A.14). And indeed from (A.15) we see some. Inspection shows that they are all powers of 2. This is at first sight surprising, since for $k \geq 1$ the binomial coefficient

$$\binom{1/2}{k} = \frac{(-1)^{k-1}}{2^{2k-1}}\frac{1}{k}\binom{2k-2}{k-1}$$

appears to have k in its denominator. If this were so then it would actually pose no serious problem to our proof, but only because the least common multiple of all k between 1 and K grows at most exponentially in K, a well-known fact closely related to the weak form of the Prime Number Theorem used in Chapter 16. Mercifully it turns out that the

$$\frac{1}{k}\binom{2k-2}{k-1} = \binom{2k-2}{k-1} - \binom{2k-2}{k}$$

are well-known integers (the Catalan numbers).

It follows easily from (A.14) that 2^{4k} is a denominator for the Taylor coefficient of z^k in $2+\psi(z)$, in accordance with a famous Theorem of Eisenstein (see also Exercises A.30 and A.32). Thus $\chi = 2 + \psi(2^4 w)$ is in $\mathbb{Z}[[w]]$. Therefore so is χ^l ($l = 0, 1, 2, \ldots$); and we deduce that 2^{4M} is a common denominator for all the Taylor coefficients of all the $(2 + \psi(z))^l$ up to z^{M-1}. This is therefore a common denominator for all the linear equations. We seem to have gone all primeval once more; however the denominators in Eisenstein's Theorem are crucial to success here. We have made some fuss over divided derivatives $(1/k!)(\mathrm{d}/\mathrm{d}z)^k$, especially in the proof of Thue's Theorem working with mere polynomials; and more generally their values correspond just to the Taylor coefficients.

We need also an "archimedean" estimate for these coefficients. We could use a variety of methods; here is one suited for our special situation. As above it is enough to replace $2+\psi(z)$ by its truncation up to z^{M-1}. Replacing z by $-u$ in (A.14), we can easily see that the signs in (A.15) are alternating. Thus the length of the resulting polynomial Ψ is the limit for $0 < r \to 1$ of

$$4 - \Psi(-r) \leq 2 - \psi(-r) = 4 - \sqrt{4 - 3r - r^3} \leq 4.$$

Thus by Lemma A.3 the coefficients in the equations have absolute values at most

$$\max\{4(4\mathcal{L}(P_*))^n, 4(4\mathcal{L}(Q_*))^n\} \leq 4(8.2^{39D}\mathcal{B}^2)^n \leq (2^{44D}\mathcal{B}^2)^n$$

(see how convenient the length is?). So clearing denominators, we get from Proposition 8.3 the upper bound

$$(2^{4M}N(2^{44D}\mathcal{B}^2)^n)^{\varsigma} \leq (2^{68nD+64L}\mathcal{B}^{2n})^{1/\delta},$$

and the claim about high order vanishing follows.

We now show that in fact (A.27) holds. If not, then with $\rho = P(x,y)/Q(x,y)$ the function

$$\rho_n = (P_0(x) + P_1(x)y) - (Q_0(x) + Q_1(x)y)\rho^n \neq 0$$

would have "too many zeroes for its degree". Here we use the Sum Formula (A.22) directly to get a contradiction. We have in terms of valuations

$$|\rho_n|_\pi \leq \max\{1, |x|_\pi\}^L \max\{1, |y|_\pi\} \max\{1, |\rho|_\pi\}^n$$

for any π. Taking the product and using Proposition A.2, we get on the right $e^{2L}e^3e^{2nd}$ and so $\sum_\pi \mathrm{ord}_\pi\, \rho_n \geq -2L - 3 - 2nd$, far from a contradiction. However from the crucial (A.24), the order at $(0,2)$ is by construction at least $M = 2L + 4 + 2nd$, and so all we have to do is take this into account (which may remind the reader of the proofs of Theorem 14.8 and Lemma 16.2 with the special order at p). That establishes the present lemma. □

So far we have not mentioned the point (ξ, η). We remark that we can assume P is also non-zero there, else Theorem A.1 is trivial. Thus ρ is regular and non-zero at (ξ, η).

It is now handy to assume

$$0 < \delta \leq \frac{1}{2} \tag{A.29}$$

and to fix L as

$$L = \left[\frac{1+\delta}{1-\delta}(nd+1)\right] \leq (1+4\delta)(nd+1) \qquad \text{(A.30)}$$

so that we can forget about (A.26).

Here comes the zero estimate, which may remind the reader of Lemma 12.7. Actually after all these pages my brain hurts from calculating constants, so wotthehell archy I will not bother with the more awkward c below, which will be independent of δ, n, L. But then how on earth will we arrive at our completely explicit theorem?

Lemma A.5 *For $\mu = Q_0(x)+Q_1(x)y$ we have $\mu \neq 0$ and its order w at (ξ,η) satisfies*

$$w \leq 24\delta nD + c. \qquad \text{(A.31)}$$

Proof If $\mu = 0$ then $Q_0 = Q_1 = 0$ forcing also $P_0 = P_1 = 0$ from (A.27), a contradiction.

We put $\lambda = P_0(x) + P_1(x)y$ and compare degrees in $\mu\rho^n = \lambda$ using Proposition A.2.

For $\mu\rho^n$ we count the zeroes. At (ξ,η) we have w and these are not cancelled by poles of ρ. Any zero of ρ is not at (ξ,η) by the above remark, and also not at ∞ by (A.23), and the total number with multiplicities is $2d$. So $\mu\rho^n$ has degree at least $(w+2nd)/2$. On the other hand by counting poles, just at ∞, we see that the degree of λ is at most $L+3/2$. Thus $w \leq 2L+3-2nd$ and the result follows at once from (A.30) on recalling $d \leq 3D$. □

Now we can finish the proof of Theorem A.1. By the definition of w, we get a leading term $\mu_0 z^w$ ($\mu_0 \neq 0$) when we substitute suitable power series for x, y into μ. Also as ρ is non-zero at (ξ,η), we get a similar term $\lambda_0 z^w$ ($\lambda_0 \neq 0$) for $\lambda = \mu\rho^n$. We proceed to estimate the height of $\gamma^n = \lambda_0/\mu_0$ for our ultimate prey

$$\gamma = \frac{P(\xi,\eta)}{Q(\xi,\eta)}.$$

Supposing for the moment $\eta \neq 0$, we recall that the power series involved come from $\xi + z$ and

$$\eta + \psi(z) = \sqrt{(\xi+z)^3 + 3(\xi+z) + 4},$$

which expands as in (A.14) to

$$\eta \sum_{k=0}^{\infty} \binom{1/2}{k} \eta^{-2k} (\xi_1 z + \xi_2 z^2 + \xi_3 z^3)^k$$

with

$$\xi_1 = 3\xi^2 + 3,\ \xi_2 = 3\xi,\ \xi_3 = 1.$$

This also fits with Eisenstein, and it is easy to bound the heights of the coefficients, although only relatively recently are suitable general results available in the literature.

For the $x^l y$ occurring in $Q_1(x)y$ in μ, we find that the coefficient of z^w is

$$\eta \sum_{i=0}^{l} \binom{l}{i} \xi^{l-i} \sum_{k=0}^{\infty} \binom{1/2}{k} \eta^{-2k} \sum_{i_1+i_2+i_3=k} C(i_1,i_2,i_3)\xi_1^{i_1}\xi_2^{i_2}\xi_3^{i_3} \tag{A.32}$$

also with $i+i_1+2i_2+3i_3 = w$ (and trinomial coefficients). In particular $k \leq w$, and from the Catalan remark this has the form $2^{-2w}\eta\omega_l$, where

$$\omega_l = \Omega_l(\xi, \eta^{-2}, \xi_1, \xi_2, \xi_3)$$

for a polynomial Ω_l over $\mathbb{Z}$ of degree at most l in the first variable and at most w in the other four. Also

$$\mathcal{L}(\Omega_l) \ \leq \ 2^{2w} \sum_{i=0}^{l} \binom{l}{i} \sum_{k=0}^{w} \left| \binom{1/2}{k} \right| \sum_{i_1+i_2+i_3=k} C(i_1,i_2,i_3).$$

The sum over k is

$$1 + \sum_{k=1}^{w} (-1)^{k-1} \binom{1/2}{k} 3^k.$$

Using $3^k \leq 6^w 2^{-k}$ we can then extend the sum to infinity, and we end up with

$$\mathcal{L}(\Omega_l) \leq c24^w 2^l.$$

We can apply similar (but simpler) arguments to $x^l = (\xi + z)^l$ in $Q_0(x)$ in μ, and we get the same estimates (with w replaced by $w+1$ for safety).

It follows that

$$\mu_0 = 2^{-2w}\eta\Omega(\xi, \eta^{-2}, \xi_1, \xi_2, \xi_3)$$

with

$$\mathcal{L}(\Omega) \leq 2c\mathcal{H} \sum_{l=0}^{L} 24^w 2^l \leq c\mathcal{H}.24^w 3^L$$

for the bound

$$\mathcal{H} = (2^{68nD+64L}\mathcal{B}^{2n})^{1/\delta}$$

in Lemma A.4.

Analogous things hold for λ_0. So to estimate the height of $\lambda_0/\mu_0 = \gamma^n$, we can use Proposition 14.10 in five variables to get

$$H(\gamma)^n \le c\mathcal{H}.24^w 3^L H^L \left(H(\eta^{-2})H(\xi_1)H(\xi_2)H(\xi_3)\right)^{w+1},$$

where $H = H(\xi)$ for brevity. Clearly

$$H(\eta^{-2}) = H(\xi^3 + 3\xi + 4) \le 8H^3, \; H(\xi_1) \le 6H^2, \; H(\xi_2) \le 3H, \; H(\xi_3) = 1,$$

so we find

$$H(\gamma)^n \le c(3H)^L (3456H^6)^{w+1}\mathcal{H}.$$

Thus

$$nh(\gamma) \le c + L(h+2) + (w+1)(6h+9) + \log \mathcal{H}$$

for $h = \log H = h(\xi)$. For the visible L we use

$$L \le nd + 4\delta nd + c \le nd + 12\delta nD + c$$

from the right-hand side of (A.30), and we treat the nd here with kid gloves as it will end up the main term. But for the invisible L in $\log \mathcal{H}$ we take the gloves off with $L \le 9nD + c$. Then from Lemma A.5 we get after a bit of calculation $nh(\gamma) \le ndh + \mathcal{E}$, where the error has the form

$$\mathcal{E} = \mathcal{C} + \delta\mathcal{X} + \frac{\mathcal{Y}}{\delta}$$

with $\mathcal{C} = 6nD + ch + c$ and

$$\mathcal{X} = 156nhD + 240nD, \; \mathcal{Y} = 644nD + 2n\log\mathcal{B} + c.$$

We now optimize the choice of δ as $\min\{\sqrt{\mathcal{Y}/\mathcal{X}}, 1/2\}$ compatible with (A.29). This gives in $\mathcal{E}$

$$\min\left\{\sqrt{\frac{\mathcal{Y}}{\mathcal{X}}}, \frac{1}{2}\right\}\mathcal{X} + \max\left\{\sqrt{\frac{\mathcal{X}}{\mathcal{Y}}}, 2\right\}\mathcal{Y} \le \sqrt{\mathcal{X}\mathcal{Y}} + \max\{\sqrt{\mathcal{X}\mathcal{Y}}, 2\mathcal{Y}\}$$

or say just $3\max\{\sqrt{\mathcal{X}\mathcal{Y}}, \mathcal{Y}\}$. Dividing by n and making $n \to \infty$, we see that our awkward constants just happen to disappear (so our question a few pages back is answered). We pick up an extra $6D$ from $\mathcal{C}$, and an estimate

$$h(\gamma) \le dh + 3\max\{\mathcal{Z}, 22\sqrt{D\mathcal{Z}}\sqrt{h+1}\}$$

for $\mathcal{Z} = 646D + 2\log\mathcal{B}$ (here we used $\sqrt{a} + \sqrt{b} \le \sqrt{2}\sqrt{a+b}$). As $\sqrt{h+1} \le \sqrt{h} + 1$ we can tidy up the error term to $3\mathcal{Z}(\sqrt{h}+1)$ and so complete the proof of Theorem A.1 (when $\eta \neq 0$). But we have actually obtained a sharper upper bound

$$dh(\xi) + c_0 \max\{D + \log \mathcal{B}, \sqrt{D(D + \log \mathcal{B})}\sqrt{h(\xi)}\}$$

with absolute c_0.

We ruled out $\eta = 0$ above, but we could do this case simply by using the appropriate power series, this time with $\xi + z^2$. Even more simply, we could now after all use Proposition 14.10 for two variables to give

$$h(\gamma) \leq \log \mathcal{B} + Dh.$$

As $\xi^3 + 3\xi + 4 = 0$ we deduce $H \leq 512$ from Proposition 14.1(ii). So $h \leq 3\sqrt{h}$ and the result follows in this case too. This really does complete the proof of Theorem A.1.

We may mention that a similar result holds even if $Q(\xi, \eta) = 0$ is allowed, as long as $\mathrm{ord}_\pi \rho \geq 0$ for $\rho = P(x, y)/Q(x, y)$ and $\pi = (\xi, \eta)$ (see Exercises A.19 and A.22).

We remark that the auxiliary polynomial in Lemma A.4 goes back to Siegel (1929) (see p. 248 and also the translation in Zannier (2014b) p. 56).

And see Schmidt (1990) for much more on Eisenstein's Theorem.

One may consult Néron (1965) for the relevant quadratic properties of the Néron–Tate height on Jacobians.

The expert on seeing Theorem A.1 may mutter to herself "Ah yes, the Height Machine." It is indeed true that this method delivers quite quickly (see Exercise A.23) the correct asymptotics in the two-sided form

$$|h(\gamma) - dh(\xi)| \leq c_{PQ}(\sqrt{h(\xi)} + 1)$$

with a c_{PQ} independent of ξ, η. However it does not easily calculate c_{PQ} in terms of P, Q: "the Height Machine doesn't do Lengths". Such a calculation was done by Habegger, whose result in appendix A of his Thesis (Habegger, 2007) should give in our situation a value for c_{PQ} not much bigger than $1000d \max\{d, h_{ABC}\}^{1/2}$ (if $d \geq 1$), where h_{ABC} is the height of the linear form (in the sense of Chapter 14) with coefficients those of the polynomials A, B, C in (A.9). Our one-sided proof avoids the $d^{3/2}$ inherent in that expression.

In fact there can be no lower bound for $h(\gamma)$ of the form

$$dh(\xi) - c(D + \log \mathcal{B})(\sqrt{h(\xi)} + 1)$$

analogous to Theorem A.1 with c absolute. This remains true even if restricted to $Q = 1$ and $P = P(X)$ (so no denominator and no elliptic curve). For the example $P(X) = (X - 2^d)X^{d-1}$ with $D = d$ and $\xi = 2^d$, the lower bound would become at least $d^2 \log 2 - cd^{3/2}$; but now $P(\xi) = 0$ so that the height we are trying to estimate is zero!

Finally to see that the Néron square root cannot be avoided, we need after all some mild elliptic curve theory. For $\pi = (\xi, \eta)$ on say $y^2 = x^3 + 3x + 4$ say with rational coordinates, we define $h(\pi) = h(\xi)$ (note that this abscissa ξ determines the ordinate η). We define also $h(\infty) = 0$. It is well-known (see for example Exercise 20.100, at least over $\mathbb{C}$) that there is a natural abelian (chord-and-tangent) group law on these π together with ∞ (here making a group isomorphic to $(\mathbb{Z}/2\mathbb{Z}) \times \mathbb{Z}$). It is also known that the Néron–Tate height

$$\hat{h}(\pi) = \lim_{r\to\infty} \frac{h(2^r\pi)}{4^r} \geq 0 \tag{A.33}$$

exists, that

$$\hat{h}(n\pi) = n^2\hat{h}(\pi) \tag{A.34}$$

(this is the analogue of $h(\alpha^n) = |n|h(\alpha)$ for the standard height), that $\hat{h}(\pi) = 0$ exactly when π has finite order (here just $(-1, 0)$ and ∞ – this extended to number fields is the analogue of the fact that $h(\alpha) = 0$ for $\alpha \neq 0$ exactly when α is a root of unity), and that $|\hat{h}(\pi) - h(\pi)|$ is bounded above, here by an absolute constant c (for most of this see Exercise A.18 and see also the usual suspects Silverman (1992) or Hindry and Silverman (2000) or Bombieri and Gubler (2006) for much more).

The function $(3x - 4y + 8)/x^2$, as in (A.8) above, arises naturally; its value at π is the abscissa of $\pi + \pi_0$ for $\pi_0 = (0, 2)$ (this can be seen with the help of (20.10) the addition formula). Thus for $\pi = n\pi_0$ we have

$$h(\beta) = h((n+1)\pi_0) \geq \hat{h}((n+1)\pi_0) - c = c_0(n+1)^2 - c$$

for $c_0 = \hat{h}(\pi_0)$. On the other hand

$$h(\xi) = h(n\pi_0) \leq \hat{h}(n\pi_0) + c = c_0 n^2 + c.$$

So

$$h(\beta) - h(\xi) \geq 2c_0 n - 2c. \tag{A.35}$$

Now π_0 is not of finite order, so $c_0 > 0$, and in fact

$$c_0 = 0.947442471034742214550989218086053629429127953511191 82\ldots$$

(thanks, Joe!) to the usual squillions of places (whose calculation is quite another kettle of fish, and whose irrationality has never been proved). Thus the right-hand side of (A.35) is of order at least $\sqrt{h(\xi)}$ as n, and so also $h(\xi) \geq c_0 n^2 - c$, tends to infinity. Therefore (A.11) is essentially best possible.

We should also mention that the arguments of this appendix apply to elliptic curves $y^2 = x^3 + ax + b$ (see Exercises A.24 and A.25), and even to general algebraic curves $F(x, y) = 0$, not necessarily elliptic (see for example Exercise A.31). Siegel's asymptotics apply here too (yet one more Siegel (1929) breakthrough). Incidentally the curves must be irreducible; for example from

$$(\xi - \eta^2)(\xi^2 - \eta) = 0$$

we don't know whether $h(\eta) = 2h(\xi)$ or $h(\eta) = h(\xi)/2$.

For more on the Height Machine see Hindry and Silverman (2000) as well as Bombieri and Gubler (2006). It too can be considered as part of the Runge legacy.

Exercises

Here $4a^3 + 27b^2 \neq 0$.

A.1 Find constants c_1, c_2, c_3, c_4 such that

$$h(\eta) \leq \frac{3}{2}h(\xi) + c_1, \quad h\left(\frac{\eta}{\xi}\right) \leq \frac{3}{2}h(\xi) + c_2,$$

$$h\left(\frac{3\xi - 4\eta + 7}{\xi^2}\right) \leq 2h(\xi) + c_3, \quad h\left(\frac{3\xi - 4\eta + 8}{\xi^2}\right) \leq 2h(\xi) + c_4$$

for all algebraic $\xi \neq 0$ and η with $\eta^2 = \xi^3 + 3\xi + 4$.

A.2 Show that there is a constant c such that

$$h\left(\frac{\eta}{\xi}\right) \geq \frac{3}{2}h(\xi) - c$$

for all algebraic $\xi \neq 0$ and η with $\eta^2 = \xi^3 + 3\xi + 4$ [*Hint*: Exercise 14.77].

A.3 Show that there is a constant c such that

$$h\left(\frac{3\xi - 4\eta + 7}{\xi^2}\right) \geq 2h(\xi) - c$$

for all algebraic $\xi \neq 0$ and η with $\eta^2 = \xi^3 + 3\xi + 4$ [*Hint*: norm].

A.4 Let P, Q be in $\mathbb{Z}[X, Y]$ with $Q(x, y) \neq 0$, let $A \neq 0, B, C$ be in $\mathbb{Z}[X]$ without common factor, and let A_1, A_2 be in $\mathbb{Z}[X]$ with $A_2 \neq 0$ and $A(x)\rho^2 + B(x)\rho + C(x) = 0$ for

$$\rho = \frac{P(x, y)}{Q(x, y)} = \frac{A_1(x) + A_2(x)y}{2A(x)}$$

on $y^2 = x^3 + 3x + 4$ and

$$d = \frac{1}{2}\max\{\deg A, \deg B, \deg C\}.$$

(a) Show that there are integers r, s such that A and $D = rB + sC$ have no common factor and

$$d = \frac{1}{2}\max\{\deg A, \deg D\}.$$

(b) Show that there is a constant c, depending only on A, B, C, D, such that

$$h\left(\frac{D(\xi)}{A(\xi)}\right) \geq 2dh(\xi) - c$$

for any algebraic ξ with $A(\xi) \neq 0$.

(c) If $\xi = N/2$ for an odd integer N, show that $\eta = \sqrt{\xi^3 + 3\xi + 4}$ is irrational.

(d) If N is sufficiently large in (c) show that $A(\xi) \neq 0$, $A_2(\xi) \neq 0$, $Q(\xi, \eta) \neq 0$, $Q(\xi, -\eta) \neq 0$ [*Hint*: $\text{ord}_\infty Q(x, y)$] and

$$\gamma = \frac{P(\xi, \eta)}{Q(\xi, \eta)}, \quad \gamma' = \frac{P(\xi, -\eta)}{Q(\xi, -\eta)}$$

are conjugate quadratic numbers.

(e) Deduce that there is a constant c, depending only on P, Q, A, B, A_1, A_2, such that

$$h\left(\frac{P(\xi, \eta)}{Q(\xi, \eta)}\right) \geq dh(\xi) - c$$

for all $\xi = N/2, \eta$ as above.

A.5 Let $P(X, Y)$ be in $\mathbb{C}[X, Y]$ with $\rho = P(x, y) \neq 0$ on $y^2 = x^3 + ax + b$, and let $\pi = (\xi, \eta)$.

(a) Show that $\text{ord}_\pi\, \rho \geq 0$.

(b) Show that $\text{ord}_\pi\, \rho \geq 1$ if and only if $P(\xi, \eta) = 0$.

(c) When $\eta \neq 0$ show that $\text{ord}_\pi\, \rho \geq 2$ if and only if

$$P(\xi, \eta) = 0, \quad \frac{\partial P}{\partial X}(\xi, \eta) + \left(\frac{3\xi^2 + a}{2\eta}\right)\frac{\partial P}{\partial Y}(\xi, \eta) = 0.$$

(d) What happens when $\eta = 0$ in (c)?

A.6 On $y^2 = x^3 + 3x + 4$ calculate $\text{ord}_\pi(16176x - 19840y + 38951)$ for every π.

A.7 On $y^2 = x^3 + 3x + 4$ calculate the degree of

$$\frac{16176x - 19840y + 38951}{3x - 4y + 8}.$$

A.8 On $y^2 = x^3 + ax + b$ let $\rho \neq 0$ be in $\mathbb{C}(x)$.
(a) If $\pi = (\xi, \eta)$ with $\eta \neq 0$ show that $\mathrm{ord}_\pi\, \rho = \mathrm{ord}_\xi\, R$.
(b) If $\pi = (\xi, \eta)$ with $\eta = 0$ show that $\mathrm{ord}_\pi\, \rho = 2\,\mathrm{ord}_\xi\, R$.
(c) If $\pi = \infty$ show that $\mathrm{ord}_\pi\, \rho = 2\,\mathrm{ord}_\infty\, R = -\,\mathrm{ord}_{z=0}\, R(1/z^2)$.

A.9 On $y^2 = x^3+ax+b$ let $\rho \neq 0$ be in $\mathbb{C}(x, y)$. Show that $\mathrm{ord}_\pi\,(\overline{\rho}) = \mathrm{ord}_{\overline{\pi}}(\rho)$ for every π.

A.10 Show that there is a constant c such that

$$h\left(\frac{3\xi - 4\eta + 8}{\xi^2}\right) \geq h(\xi) - c(\sqrt{h(\xi)} + 1)$$

for all algebraic $\xi \neq 0$ and η with $\eta^2 = \xi^3 + 3\xi + 4$ [*Hint*: use Theorem A.1 for $(3x + 4y + 8)/x^2$].

****

A.11 On $y^2 = x^3+ax+b$ let $\rho \neq 0$ be in $\mathbb{C}(x, y)$ with $\mathrm{ord}_\pi\, \rho \geq 0$ for every π. Show that ρ is in $\mathbb{C}$.

A.12 On $y^2 = x^3 + ax + b$ let $\rho \neq 0$ be in $\mathbb{C}(x, y)$ with $\mathrm{ord}_\pi\, \rho \geq 0$ for every $\pi \neq \infty$ and $\mathrm{ord}_\infty\, \rho \geq -1$. Show that ρ is still in $\mathbb{C}$.

A.13 On $y^2 = x^3 + ax + b$ let $\rho \neq 0$ be in $\mathbb{C}(x, y)$ with $\mathrm{ord}_\pi\, \rho \geq 0$ for every $\pi \neq \infty$ and $\mathrm{ord}_\infty\, \rho \geq -2$. Show that ρ is in $\mathbb{C} + \mathbb{C}x$.

A.14 On $y^2 = x^3 + ax + b$ let $\rho \neq 0$ be in $\mathbb{C}(x, y)$ with $\mathrm{ord}_\pi\, \rho \geq 0$ for every $\pi \neq \infty$ and $\mathrm{ord}_\infty\, \rho \geq -3$. Show that ρ is in $\mathbb{C} + \mathbb{C}x + \mathbb{C}y$.

A.15 On $y^2 = x^3 + ax + b$ let $\rho \neq 0$ be in $\mathbb{C}(x, y)$ with $\mathrm{ord}_\pi\, \rho \geq 0$ for every $\pi \neq \infty$. Show that ρ is in $\mathbb{C}[x, y]$.

A.16 For $\rho \neq 0$ in $\mathbb{C}(x, y)$ on $y^2 = x^3 + ax + b$, define $\omega(\rho) = \sum_\pi (\mathrm{ord}_\pi\, \rho)\pi$ in the free abelian group $\sum_\pi \mathbb{Z}\pi$. Show that $\pi_1 + \pi_2 = \pi_3$ on the elliptic curve if and only if there is ρ with $\omega(\rho) = \pi_1 + \pi_2 - \pi_3 - \infty$.

A.17 On $y^2 = x^3 + ax + b$ let ρ, σ be in $\mathbb{C}(x, y)$. Show that there is non-zero P in $\mathbb{C}[R, S]$ with $P(\rho, \sigma) = 0$ [*Hint*: Chapter 1].

A.18 On $y^2 = x^3 + 3x + 4$ the double $2(\xi, \eta)$ is given by $(\tilde{\xi}, \tilde{\eta})$, where

$$\tilde{\xi} = \frac{\xi^4 - 6\xi^2 - 32\xi + 9}{4(\xi^3 + 3\xi + 4)}$$

(compare (20.4) with $g_2 = -12, g_3 = -16$).

(a) Show that there is an absolute constant c such that $|h(2\pi) - 4h(\pi)| \le c$ for all π with algebraic coordinates [*Hint*: Exercise 14.77].

(b) Deduce that $\hat{h}(\pi) = \lim_{r\to\infty} h(2^r\pi)/4^r$ exists and satisfies $|\hat{h}(\pi) - h(\pi)| \le c/3$.

(c) Show that $\hat{h}(n\pi) = n^2\hat{h}(\pi)$ $(n = 1, 2, 4, 8, \ldots)$ (this for all n in $\mathbb{Z}$ lies a little deeper), and that $\hat{h}(\pi) = 0$ if and only if π has finite order.

A.19 On $y^2 = x^3 + ax + b$ let $\rho \neq 0$ be in $\mathbb{C}(x, y)$ with $\mathrm{ord}_\pi\, \rho \ge 0$ for some $\pi = (\xi, \eta)$. Show that there are P_*, Q_* in $\mathbb{C}[X, Y]$ with $\rho = P_*(x, y)/Q_*(x, y)$ and $Q_*(\xi, \eta) \neq 0$ [*Hint*: if $\mathrm{ord}_{\overline{\pi}}\, \rho \ge 0$ use equation (A.20) and if not, replace ρ by some $1/(\rho - c)$].

A.20 On $y^2 = x^3 + 3x + 4$ let $\rho \neq 0$ be in $\mathbb{Q}(x, y)$ with $\mathrm{ord}_\pi\, \rho \ge 0$ for some $\pi = (\xi, \eta)$. Show that there are P_*, Q_* in $\mathbb{Z}[X, Y]$ with $\rho = P_*(x, y)/Q_*(x, y)$ and $Q_*(\xi, \eta) \neq 0$.

A.21 On $y^2 = x^3 + 3x + 4$ we know that $\mathrm{ord}_\pi\, \rho \ge 0$ for $\rho = (3x - 4y + 8)/x^2$ and $\pi = (0, 2)$. Show that there are no P_* in $\mathbb{C}[X, Y]$ and Q_* only in $\mathbb{C}[X]$ with $\rho = P_*(x, y)/Q_*(x, y)$ and $Q_*(0, 2) \neq 0$.

A.22 Prove a version of Theorem A.1 with $Q(\xi, \eta) \neq 0$ replaced by $\mathrm{ord}_\pi\, \rho \ge 0$ for $\pi = (\xi, \eta)$.

A.23 Let P, Q, A, B, C be as in Exercise A.4.

(a) Show that there is a constant c, depending only on P, Q, A, B, C, such that

$$h\left(\frac{P(\xi, \eta)}{Q(\xi, \eta)}\right) \ge dh(\xi) - c(\sqrt{h(\xi)} + 1)$$

for all algebraic ξ, η with $\eta^2 = \xi^3 + 3\xi + 4$ and

$$Q(\xi, \eta) \neq 0, \quad Q(\xi, -\eta) \neq 0. \qquad (*)$$

(b) Remove the conditions $(*)$.

A.24 Let a, b be algebraic and let P, Q be in $\mathbb{C}[X, Y]$ with algebraic coefficients and $Q(x, y) \neq 0$ in the function field.

(a) Show that there are algebraic ξ_0, η_0 with $\eta_0^2 = \xi_0^3 + a\xi_0 + b$ and $Q(\xi_0, \eta_0) \neq 0$, $Q(\xi_0, -\eta_0) \neq 0$.

(b) Let d be the degree of $\rho = P(x, y)/Q(x, y)$, assume $|\rho|_\infty = 1$, and let K be a number field containing a, b, ξ_0, η_0 and the coefficients of P, Q. Show that there is C_0 such that for any δ with $0 < \delta \le 1$ and $L \ge 1, n \ge 1$ in $\mathbb{Z}$ with $(1 - \delta)(L + 1) > (1 + \delta)(nd + 1)$, there are P_0, P_1, Q_0, Q_1 in $\mathbb{Z}_K[X]$, not all zero, of degrees at most L and with coefficients of heights at most $C_0^{(n+L)/\delta}$, such that

$$(P_0(x) + P_1(x)y)Q(x,y)^n = (Q_0(x) + Q_1(x)y)P(x,y)^n$$

[*Hint*: use ord_{π_0} for $\pi_0 = (\xi_0, \eta_0)$ and either Proposition 13.2 or Exercise 14.88].

(c) Fix $\delta \leq 1/2$ and $L = \left[\frac{1+\delta}{1-\delta}(nd+1)\right]$, and let ξ, η be algebraic with $\eta^2 = \xi^3 + a\xi + b$ and $P(\xi,\eta) \neq 0$, $Q(\xi,\eta) \neq 0$. Show that for $\mu = Q_0(x) + Q_1(x)y$ we have $\mu \neq 0$ and its order w at (ξ,η) satisfies $w \leq 8\delta nd + 9$.

(d) Conclude that there is C, depending only on a, b, P, Q, such that

$$h\left(\frac{P(\xi,\eta)}{Q(\xi,\eta)}\right) \leq dh(\xi) + C(\sqrt{h(\xi)} + 1)$$

for all ξ, η as above.

A.25 In the situation of Exercise A.24 prove that there is C, depending only on a, b, P, Q, such that

$$h\left(\frac{P(\xi,\eta)}{Q(\xi,\eta)}\right) \geq dh(\xi) - C(\sqrt{h(\xi)} + 1)$$

for all ξ, η.

A.26 Calculate a value of c in Exercise A.10.

A.27 (a) Let u, v be complex numbers with

$$v(v^2 - 2u^2) = u. \qquad (*)$$

Show that

$$u = -\frac{\xi\eta}{2(\xi^2 - 2)}, \quad v = \frac{\eta}{\xi^2 - 2}$$

for some ξ, η with $\xi^2 \neq 2$ and

$$\eta^2 = \xi^3 - 2\xi \qquad (**)$$

[*Hint*: $\xi = 4u^2 - 2v^2, \eta = -8u^3 + 4uv^2 - 2v$]. This defines a birational isomorphism between the cubic curve $(*)$ and the elliptic curve $(**)$.

(b) If u, v are algebraic in (a) deduce that

$$|2h(u) - 3h(v)| \leq c(\sqrt{h(u)} + \sqrt{h(v)} + 1)$$

for some absolute constant c.

(c) Conclude with the help of Proposition 4.1 that there are at most finitely many (x, y) in $\mathbb{Z}^2$ with $y(y^2 - 2x^2) = x$. See also (4.19) – talk about a sledge-hammer!

(d) Show that even $|2h(u) - 3h(v)| \leq c$ in (b).

A.28 Let

$$F(X,Y) = aX^3 + bX^2Y + cXY^2 + dY^3 + eX^2 + fXY + gY^2 + hX + kY + l$$

in $\overline{\mathbb{Q}}[X,Y]$ (so e cannot be the base of the natural logarithms) be irreducible there, with $ad \neq 0$. Show that there is C such that

$$|h(u) - h(v)| \leq C(\sqrt{h(u)} + \sqrt{h(v)} + 1)$$

for all u, v in $\overline{\mathbb{Q}}$ with $F(u,v) = 0$.

A.29 Show that the irreducibility in Exercise A.28 is equivalent to the matrix

$$\begin{pmatrix} 0 & 3a & 0 & 2b & 2e & 0 & c & f & h & 0 \\ 0 & 0 & 3a & 0 & 2b & 2e & 0 & c & f & h \\ b & 2c & f & 3d & 2g & k & 0 & 0 & 0 & 0 \\ 0 & 0 & b & 0 & 2c & f & 0 & 3d & 2g & k \\ e & f & 2h & g & 2k & 3l & 0 & 0 & 0 & 0 \\ 0 & e & 0 & f & 2h & 0 & g & 2k & 3l & 0 \\ a & 0 & e & -c & 0 & h & -2d & -g & 0 & l \\ a & b & 0 & c & 0 & -h & d & 0 & -k & -2l \end{pmatrix}$$

having rank at most seven (Ruppert, 1986).

A.30 Let a, b be algebraic, let $0 \neq \rho = R(x,y)$ have algebraic coefficients in the function field of $y^2 = x^3 + ax + b$, and let π be ∞ or a point (ξ, η) with algebraic ξ, η such that $\mathrm{ord}_\pi\, \rho \geq 0$, so that $\Psi(z) = \sum_{k=0}^{\infty} \alpha_k z^k$ for the function Ψ in (A.12),(A.16) or (A.17). Show that there is a positive integer d such that the $d^{k+1}\alpha_k$ $(k = 0, 1, 2, \ldots)$ are algebraic integers.

A.31 Define $\mathrm{ord}_\pi\, \rho$ for any $\rho \neq 0$ and π on $y^2 = x^5 - 1$.

A.32 Let $\Psi(z) = \sum_{k=0}^{\infty} \alpha_k z^k$ in $\mathbb{C}[[z]]$ be algebraic over $\mathbb{Q}(z)$. Show that there is a positive integer d such that the $d^{k+1}\alpha_k$ $(k = 0, 1, 2, \ldots)$ are algebraic integers (Eisenstein).

A.33 Google "Catalan Numbers".

A.34 That's enough auxiliary polynomials. Ed.

References

Adolphson, A.C., Conrey, J.B., Ghosh, A., Yager, R.I. (eds.) (1987). *Analytic Number Theory and Diophantine Problems (Stillwater 1984)*, Progress in Math. **70**, Birkhäuser.

Amoroso, F., David, S. (1999). Le probléme de Lehmer en dimension supérieure, *J. reine angew. Math.* **513**, 145–179.

Amoroso, F., Dvornicich, R. (2000). A lower bound for the height in abelian extensions, *J. Number Theory* **80**, 260–272.

Amoroso, F., Mignotte, M. (1996). On the distribution of the roots of polynomials, *Ann. Inst. Fourier, Grenoble* **46**, 1275–1291.

Amoroso, F., Zannier, U. (2000). A relative Dobrowolski lower bound over abelian extensions, *Ann. Scuola Norm. Sup. Pisa Cl. Sci.* **29**, 711–727.

(eds.) (2003). *Diophantine Approximation (Cetraro Italy 2000)*, Lecture Notes in Math. **1819**, Springer.

André, Y. (1989). *G-Functions and Geometry*, Aspects of Math. **E13**, Vieweg.

Baker, H.F. (1907). *An Introduction to the Theory of Multiply Periodic Functions*, Cambridge University Press.

Baker, A. (1964). Rational approximations to $\sqrt[3]{2}$ and other algebraic numbers, *Quarterly J. Math.* **15**, 375–383.

(1966). Linear forms in the logarithms of algebraic numbers I, *Mathematika* **13**, 204–216.

(1967a). Linear forms in the logarithms of algebraic numbers II, *Mathematika* **14**, 102–107.

(1967b). Linear forms in the logarithms of algebraic numbers III, *Mathematika* **14**, 220–228.

(1968a). Linear forms in the logarithms of algebraic numbers IV, *Mathematika* **15**, 204–216.

(1968b). Contributions to the theory of diophantine equations. II. The diophantine equation $y^2 = x^3 + k$, *Phil. Trans. Roy. Soc. London* **263A**, 193–208.

(ed.) (1988). *New Advances in Transcendence Theory*, Cambridge University Press.

(1990). *Transcendental Number Theory*, Cambridge University Press.

(2012). *A Comprehensive Course in Number Theory*, Cambridge University Press.

Baker, A., Masser, D. (eds.) (1977). *Transcendence Theory: Advances and Applications*, Academic Press.

Baker, A., Wüstholz, G. (1993). Logarithmic forms and group varieties, *J. reine angew. Math.* **442**, 19–62.

(2007). *Logarithmic Forms and Diophantine Geometry*, Cambridge University Press.

Barré, K., Diaz, G., Gramain, F., Philibert, G. (1996). Une preuve de la conjecture de Mahler–Manin, *Inventiones Math.* **124**, 1–9.

Bays, M., Habegger, P. (2015). A note on divisible points on curves, *Trans. Amer. Math. Soc.* **367**, 1313–1328.

Bertrand, D., Masser, D. (1980). Linear forms in elliptic logarithms, *Inventiones Math.* **58**, 283–288.

Bertrand, D., Waldschmidt, M. (eds.) (1983). *Approximations Diophantiennes et Nombres Transcendants (Luminy 1982)*, Progress in Math. **31**, Birkhäuser.

Beukers, F. (1979). A note on the irrationality of $\zeta(2)$ and $\zeta(3)$, *Bull. London Math. Soc.* **11**, 268–272.

(1997). On a sequence of polynomials, *J. Pure Appl. Algebra* **117/118**, 97–103.

Bilu, Yu.F. (1997) Limit distribution of small points on algebraic tori, *Duke Math. J.* **89**, 465–476.

Bilu, Yu.F., Masser, D. (2006). A quick proof of Sprindzhuk's decomposition theorem. In *More Sets, Graphs and Numbers – A Salute to Vera Sòs and András Hajnal* (eds. E. Györi, G. Katona, L. Lovász), Bolyai Society Mathematical Studies **15**, Springer (pp. 25–32).

Bilu, Yu.F., Bugeaud, Y., Mignotte, M. (2014). *The Problem of Catalan*, Springer.

Bilu, Yu.F., Masser, D., Zannier, U. (2013). An effective "Theorem of André" for CM-points on a plane curve, *Math. Proc. Cambridge Phil. Soc.* **154**, 145–152.

Boas, R.P. (1954). *Entire Functions*, Academic Press.

Bombieri, E. (1970). Algebraic values of meromorphic maps, *Inventiones Math.* **10**, 267–287; **11**, 163–166.

(1982). On the Thue–Siegel–Dyson theorem, *Acta Math.* **148**, 255–296.

(1983). On Weil's "Théorème de décomposition", *Amer. J. Math.* **105**, 295–308.

Bombieri, E., Gubler, W. (2006). *Heights in Diophantine Geometry*, Cambridge University Press.

Bombieri, E., Pila, J. (1989). The number of integral points on arcs and ovals, *Duke Math. J.* **59**, 337–357.

Bombieri, E., Vaaler, J.D. (1983). On Siegel's Lemma, *Inventiones Math.* **73**, 11–32.

Bombieri, E., Zannier, U. (1995). Algebraic points on subvarieties of $\mathbb{G}_{\mathrm{m}}^n$, *Int. Math. Res. Notices* **7**, 333–347.

Bombieri, E., Masser, D., Zannier, U. (1999). Intersecting a curve with algebraic subgroups of multiplicative groups, *Int. Math. Research Notices* **20**, 1119–1140.

Boyd, D.W. (1992). Two sharp inequalities for the norm of a factor of a polynomial, *Mathematika* **39**, 341–349.

Brenner, J.L., Foster, L.L. (1982). Exponential diophantine equations, *Pacific. J. Math.* **101**, 263–301.

Brownawell, W.D. (1974). The algebraic independence of certain numbers related by the exponential function, *J. Number Theory* **6**, 22–31.

(1983). Zero estimates for solutions of differential equations. In Bertrand and Waldschmidt (1983) (pp. 67–94).

Brownawell, W.D., Masser, D. (1980) Multiplicity estimates for analytic functions I, *J. reine angew. Math.* **314**, 200–216.

Buck, R.C. (1948). Integral valued entire functions, *Duke Math. J.* **15**, 879–891.

Burger, E.B., Tubbs, R. (2004). *Making Transcendence Transparent*, Springer.

Carlitz, L. (1935). On certain functions connected with polynomials in a Galois field, *Duke Math. J.* **1**, 137–168.

Cassels, J.W.S. (1965). *An Introduction to Diophantine Approximation*, Tracts in Mathematics and Mathematical Physics **45**, Cambridge University Press.

(1991). *Lectures on Elliptic Curves,* LMS Student Texts **24**, Cambridge University Press.

Chowla, S., Selberg, A. (1967). On Epstein's zeta function, *J. reine angew. Math.* **227**, 86–110.

Chudnovsky, G.V. (1984). *Contributions to the Theory of Transcendental Numbers*, Math. Surveys and Monographs **19**, Amer. Math. Soc.

Cohen, P.B. (1984). On the coefficients of the transformation polynomials for the elliptic modular function, *Math. Proc. Camb. Phil. Soc.* **95**, 389–402.

Cohen, P.B., Zannier, U. (2000) Multiplicative dependence and bounded height, an example. In *Proc. Algebraic Number Theory and Diophantine Analysis Conference (Graz 1998)*, Walter de Gruyter (pp. 93–101).

Conway, J.H., Guy, R.K. (1996). *The Book of Numbers*, Springer.

Corvaja, P., Zannier, U. (2002). Some new applications of the subspace theorem, *Compositio Math.* **131**, 319–340.

(2013). Greatest common divisors of $u-1, v-1$ in positive characteristic and rational points on curves over finite fields, *J. Eur. Math. Soc.* **15**, 1927–1942

David, S., Philippon, P. (1999). Minorations des hauteurs normalisées des sous-variétés des tores, *Ann. Scuola Norm. Sup. Pisa Cl. Sci.* **28**, 489–543.

Dixon, A.L. (1908). The eliminant of three quantics in two independent variables, *Proc. London Math. Soc.* **6**, 468–478.

Dobrolowski, E. (1979). On a question of Lehmer and the number of irreducible factors of a polynomial, *Acta Arith.* **34**, 391–401.

Duverney, D. (1997). Sur les propriétés arithmétiques de la fonction de Tschakaloff, *Periodica Math. Hungarica* **35**, 149–157.

Dyson, F.J. (1947). The approximation to algebraic numbers by rationals, *Acta Math.* **79**, 225–240.

Eagle, A. (1958). *The Elliptic Functions as They Should Be*, Galloway and Porter.

Eichler, M. (1966). *Introduction to the Theory of Algebraic Numbers and Functions*, Academic Press.

Erdős, P., Turán, P. (1950). On the distribution of roots of polynomials, *Annals of Math.* **51**, 105–119.

Esnault, H., Viehweg, E. (1984). Dyson's Lemma for polynomials in several variables (and the theorem of Roth), *Inventiones Math.* **78**, 445–490.

Evertse J.H., Silverman, J.H. (1986). Uniform bounds for the number of solutions to $Y^n = f(X)$, *Math. Proc. Cambridge Phil. Soc.* **100**, 237–248.

Favre, C., Rivera-Letelier, J. (2006). Équidistribution quantitative des points de petite hauteur sur la droite projective, *Math. Annalen* **335**, 311–361.

Feldman, N.I. (1951). Approximation of certain transcendental numbers. II: The approximation of certain numbers connected with the Weierstrass $\wp$-function, *Izv. Akad. Nauk. SSSR* **15**, 153–176 (Russian); English transl. *Amer. Math. Soc. Trans.* **59** (1966), 246–270.

(1971). An effective refinement of the exponent in Liouville's theorem, *Izv. Akad. Nauk. SSSR* **35**, 973–990 (Russian); English transl. *Math. USSR Izv.* **5**, 985–1002.

Feldman, N.I., Nesterenko, Yu.V. (1997). *Number Theory IV, Transcendental Numbers*, Encyclopaedia of Math. Sci. **44** (eds. A.N. Parshin, I.R. Shafarevich), Springer.

Galochkin, A.I. (1974). Lower bounds of polynomials in the values of a certain class of analytic functions, *Mat. Sbornik* **95**, 396–417 (Russian); English transl. *Math. USSR Sb.* **24**, 385–407.

(1980). Transcendence measure of values of functions satisfying certain functional equations, *Mat. Zametki* **27**, 175–183 (Russian); English transl. *Math. notes Acad. Sci. USSR* **27**, 83–88.

Garcia, A., Voloch, J.F. (1988). Fermat curves over finite fields, *J. Number Theory* **30**, 345–356.

Gelfand, I.M., Kapranov, M.M., Zelevinsky, A.V. (1994). *Discriminants, Resultants, and Multidimensional Determinants*, Birkhäuser.

Gelfond, A.O. (1934). Sur quelques résultats nouveaux dans la théorie des nombres transcendants, *Comptes Rendus Acad. Sci. Paris*, **199**, 259.

(1960). *Transcendental and Algebraic Numbers*, Dover.

Ghosh, A., Ward, K. (2015). The number of roots of polynomials in a prime field, *Int. Math. Res. Notices* **4**, 898–926.

Goss, D. (1996). *Basic Structures of Function Field Arithmetic*, Ergebnisse Math. **35**, Springer.

Gramain, F. (1978). Fonctions entiéres arithmétiques. In *Séminaire P. Lelong–H. Skoda (Analyse) 1976/77*, Lecture Notes in Math. **694**, Springer (pp. 96–125).

(1981). Sur le théorème de Fukasawa–Gelfond, *Inventiones Math.* **63**, 495–506.

Gramain, F., Mignotte, M., Waldschmidt, M. (1986). Valeurs algébriques de fonctions analytiques, *Acta Arith.* **47**, 97–121.

Grant, D. (1990). Formal groups in genus two, *J. reine angew. Math.* **411**, 96–121.

Greenhill, A.G. (1892). *The Applications of Elliptic Functions*, Macmillan.

Grytczuk, A., Schinzel, A. (1991). On Runge's theorem about diophantine equations. In *Coll. Math. Soc. János Bolyai* **60** (Sets, Graphs and Numbers), North-Holland (pp. 329–356).

Habegger, P. (2005). The equation $x + y = \alpha$ in multiplicatively independent unknowns, *Acta Arith.* **119**, 349–372.

(2007). *Heights and Multiplicative Relations on Algebraic Varieties*, Doctoral Thesis, Basle.

Hall, R.R. (1971). On pseudo-polynomials, *Mathematika* **18**, 71–77.

Halphen, G.H. (1886). *Traité des Fonctions Elliptiques et de leurs Applications, I.* Gauthier-Villars.

Hančl, J., Tijdeman R. (2004). On the irrationality of Cantor series, *J. reine angew. Math.* **571**, 145–158.

Heath-Brown, D.R. (1996). An estimate for Heilbronn's exponential sum. In *Analytic Number Theory (Proceedings of a Conference in Honor of Heini Halberstam Volume 2)*, Progr. Math. **139**, Birkhäuser (eds. B.C. Berndt, H.G. Diamond, A.J. Hildebrand) (pp. 451–463).

Heath-Brown, D.R., Konyagin, S. (2000). New bounds for Gauss sums derived from *k*th powers, and for Heilbronn's exponential sum, *Quart. J. Math.* **51**, 221–235.

Hensel, K. (1905). Über die arithmetischen Eigenschaften der algebraischen und transzendenten Zahlen, *Jahresbericht der Deutschen Mathematiker-Vereinigung* **14**, 545–558.

Hilliker, D.L., Straus, E.G. (1983). Determination of bounds for the solutions to those binary diophantine equations that satisfy the hypotheses of Runge's theorem, *Trans. Amer. Math. Soc.* **280**, 637–657.

Hindry, M., Silverman, J.H. (2000). *Diophantine Geometry*, Graduate Texts in Math. **201**, Springer.

Jones, G.O., Wilkie, A.J. (eds.) (2015). *O-minimality and Diophantine Geometry*, LMS Lecture Note Series **421**, Cambridge University Press.

Klein, F. (1956). *Lectures on the Icosahedron and the Solution of Equations of the Fifth Degree*, Dover.

Kmošek, M. (1979). *Continued Fraction Expansion of some Irrational Numbers*, Master's thesis, Uniwersytet Warszawski (in Polish).

Kowalski, E. (2010). Lecture notes and related documents, available online at www.math.ethz.ch/kowalski/exp-sums.html

Kronecker, L. (1857). Zwei Sätze über Gleichungen mit ganzzahligen Coefficienten, *J. reine angew. Math.* **53**, 173–175.

Kühne, L. (2013). An effective result of André–Oort type II, *Acta Arith.* **161**, 1–19.

Lang, S. (1996a). *Introduction to Transcendental Numbers*, Addison-Wesley.

(1996b). *Introduction to Diophantine Approximations*, Addison-Wesley.

(1973). *Elliptic Functions*, Addison-Wesley.

(1978). *Elliptic Curves, Diophantine Analysis*, Grundlehren der Math. Wiss. **231**, Springer.

(1983). *Fundamentals of Diophantine Geometry*, Springer.

(1993). *Algebra*, Addison-Wesley.

Laurent, M. (1983). Minoration de la hauteur de Néron–Tate. In *Séminaire de théorie des nombres, Paris 1981/1982*, Progr. Math **38**, Birkhäuser (ed. M.-J. Bertin) (pp. 137–151).

Lehmer, D.H. (1933). Factorization of certain cyclotomic functions, *Annals of Math.* **34**, 461–479.

Leitner, D.J. (2011). Two exponential diophantine equations, *J. Théorie des Nombres Bordeaux* **23**, 479–487.

Loher, T., Masser, D. (2004). Uniformly counting points of bounded height, *Acta Arithmetica* **111**, 277–297.

Loxton, J.H., van der Poorten, A.J. (1978). Algebraic independence properties of the Fredholm series, *J. Australian Math. Soc.* **A26**, 31–45.

Macaulay, F.S. (1916). *The Algebraic Theory of Modular Systems*, Tracts in Mathematics and Mathematical Physics **19**, Cambridge University Press.

Mahler, K. (1953). On the approximation of π, *Koninkl. Nederl. Akad. Wet.* **A56**, 30–42 or *Indagationes Math.* **15**, 30–42.

(1964). Inequalities for ideal bases in algebraic number fields, *J. Australian Math. Soc.* **4**, 425–448.

(1968). Applications of a theorem by A.B. Shidlovsky, *Phil. Trans. Roy. Soc. London* **A305**, 149–173.

(1975). On the transcendency of the solutions of a special class of functional equations, *Bull. Australian Math. Soc.* **13**, 389–410.

(1976). *Lectures on Transcendental Numbers*, Lecture Notes in Math. **546**, Springer.

Masser, D. (1975). *Elliptic Functions and Transcendence*, Lecture Notes in Math. **437**, Springer.

(1977). Division fields of elliptic functions, *Bull. London Math. Soc.* **9**, 49–53.

(1980a). Sur les fonctions entières à valeurs entières, *Comptes Rendus Acad. Sci. Paris* **291**, 1–4.

(1980b). Problem 6479, *Amer. Math. Monthly* **93**, 486–488.

(1981). A note on Baker's Theorem. In *Recent Progress in Analytic Number Theory* (eds. H. Halberstam and C. Hooley), Academic Press (pp. 153–158).

(1983). On certain functional equations in several variables. In Bertrand and Waldschmidt (1983) (pp. 173–190).

(2003). Sharp estimates for Weierstrass elliptic functions, *J. d'Analyse Math.* **90**, 257–302.

(2011). Rational values of the Riemann zeta function, *J. Number Th.* **131**, 2037–2046.

Masser, D., Vaaler J.D. (2008). Counting algebraic numbers with large height I. In *Diophantine Approximation – Festschrift for Wolfgang Schmidt* (eds. H.-P. Schlickewei, K. Schmidt, R.F. Tichy), Developments in Mathematics **16**, Springer (pp. 237–243).

Masser, D., Wolbert, J. (1993). On products of polynomials, *Proc. Amer. Math. Soc.* **117**, 593–599.

Masser, D., Wüstholz, G. (1986). Algebraic independence of values of elliptic functions, *Math. Annalen* **276**, 1–17.

(1990). Estimating isogenies on elliptic curves, *Inventiones Math.* **100**, 1–24.

Masser, D., Zannier, U. (2010). Torsion anomalous points and families of elliptic curves, *Amer. J. Math.* **132**, 1677–1691.

(2015). Torsion points on families of abelian surfaces and Pell's equation over polynomial rings (with Appendix by V. Flynn), *J. European Math. Soc.* **17**, 2379–2416.

Mignotte, M. (1979). Approximations des nombres algébriques par des nombres algébriques de grand degré, *Ann. Fac. Sci Toulouse* **1**, 165–170.

(1989). Sur un théorème de M. Langevin, *Acta Arith.* **54**, 81–86.

Mihăilescu, P. (2004). Primary cyclotomic units and a proof of Catalan's conjecture, *J. reine angew Math.* **572**, 167–195.

Miller, W. (1982). Transcendence measures by a method of Mahler, *J. Australian Math. Soc.* **32**, 68–78.

Mitkin, D.A. (1992). An estimate for the number of roots of some comparisons by the Stepanov method, *Mat. Zametki* **51**, 52–58, 157 (Russian); English transl. *Math. Notes* **51**, 565–570.

Mordell, L.J. (1923). On the integer solutions of the equation $ey^2 = ax^3 + bx^2 + cx + d$, *Proc. London Math. Soc.* **21**, 415–419.

Myerson, G. (1986). How small can a sum of roots of unity be?, *Amer. Math. Monthly* **93**, 457–459.

Néron, A. (1965). Quasi-fonctions et hauteurs sur les variétés abéliennes, *Annals of Math.* **82**, 249–331.

Nesterenko, Yu.V. (1983). Estimates for the number of zeros of certain functions. In Bertrand and Waldschmidt (1983) (pp. 263–269).

(1996). Modular functions and transcendence questions, *Mat. Sb.* **187/9**, 65–96 (Russian); English transl. *Sb. Math.* **187/9**, 1319–1348.

Nesterenko, Yu.V., Philippon, P. (eds.) (2001). *Introduction to Algebraic Independence Theory*, Lecture Notes in Math. **1752**, Springer.

Nishioka, K. (1986). Proof of Masser's conjecture on the algebraic independence of values of Liouville series, *Proc. Japan Acad.* **A62**, 219–222.

(1996). *Mahler Functions and Transcendence*, Lecture Notes in Math. **1631**, Springer.

Niven, I. (1956). *Irrational Numbers*, The Carus Math. Monog. **11**, Math. Assoc. Amer.

Osgood, C.F. (ed.) (1973). *Diophantine Approximation and its Applications*, Academic Press.

Perelli, A., Zannier, U. (1981). Su un teorema di Pólya, *Boll. U. Mat. Italiana* **18A**, 305–307.

(1984). On recurrent mod p sequences, *J. reine angew. Math.* **348**, 135–146.

Petsche, C. (2005). A quantitative version of Bilu's equidistribution theorem, *Int. J. Number Theory* **1**, 281–291.

Philibert, G. (1997). Un lemme de zéros modulaire, *J. Number Theory* **66**, 306–313.

Philippon, P. (1983). Variétés abéliennes et indépendance algébrique II: Un analogue abélien du théorème de Lindemann–Weierstrass, *Inventiones Math.* **72**, 389–405.

(ed.) (1992). *Approximations Diophantiennes et Nombres Transcendants (Luminy 1990)*, Walter de Gruyter.

Pila, J. (2004). Integer points on the dilation of a subanalytic surface, *Quarterly J. Math.* **55**, 207–223.

(2011). O-minimality and the André–Oort conjecture for $\mathbb{C}^n$, *Annals of Math.* **173**, 1779–1840.

Pila, J., Wilkie A. (2006). The rational points of a definable set, *Duke Math. J.* **133**, 591–616.

Pólya, G. (1915). Über ganzwertige ganze Funktionen, *Rend. Cont. Math. Palermo* **40**, 1–16.

Pontreau, C. (2005). Minoration effective de la hauteur des points d'une courbe de $\mathbb{G}_m^2$ définie sur $\mathbb{Q}$, *Acta Arith.* **120**, 1–26.

van der Poorten, A.J. (1979). A proof that Euler missed... Apéry's proof of the irrationality of $\zeta(3)$, *Math. Intelligencer* **1**, 195–203.

Rankin, R. (1977). *Modular Forms and Functions*, Cambridge University Press.

Rausch, U. (1985). On a theorem of Dobrowolski about the product of conjugate numbers, *Colloq. Math.* **50**, 137–142.

Rémond, G. (2002). Sur les sous-variétés des tores, *Compositio Math.* **134**, 337–366.

Roth, K.F. (1955). Rational approximations to algebraic numbers, *Mathematika*, **2**, 1–20.

Runge, C. (1887). Über ganzzahlige Lösungen von Gleichungen zwischen zwei Veränderlichen, *J. reine angew. Math.* **100**, 425–435.

Ruppert, W. (1986). Reduzibilität ebener Kurven, *J. reine angew. Math.* **369**, 167–191.

Rusza, I.Z. (1971). On congruence preserving functions (Hungarian), *Mat. Lapok*, **22**, 125–134.

Schanuel, S.H. (1979). Heights in number fields, *Bull. Soc. Math. France* **107**, 433–449.

Schinzel, A. (1969). An improvement of Runge's theorem on diophantine equations, *Commentarii Pontif. Acad. Sci.* **2**, 1–9.

(1982). *Selected Topics on Polynomials*, University of Michigan Press.

(2000). *Polynomials with Special Regard to Reducibility*, Encyclopaedia of Math. and its Applications **77**, Cambridge University Press.

Schmidt, W.M. (1976). *Equations over Finite Fields, an Elementary Approach*, Lecture Notes in Math. **536**, Springer.

(1980). *Diophantine Approximation*, Lecture Notes in Math. **785**, Springer.

(1990). Eisenstein's theorem on power series expansions of algebraic functions, *Acta Arith.* **56**, 161–179.

(1991). *Diophantine Approximations and Diophantine Equations*, Lecture Notes in Math. **1467**, Springer.
Schneider, T. (1957). *Einführung in die Transzendenten Zahlen*, Grundlehren der Math. Wiss. **81**, Springer.
Serre, J-P. (1988). *Algebraic Groups and Class Fields*, Graduate Texts in Math. **117**, Springer.
Shallit, J. (1979). Simple continued fractions for some irrational numbers, *J. Number Theory* **11**, 209–217.
Shanks, D. (1974). Incredible identities, *Fibonacci Quart.* **12**, 271–280.
Shidlovsky, A.B. (1989). *Transcendental Numbers,* Walter de Gruyter.
Shkredov, I.D. (2014). On exponential sums over multiplicative subgroups of medium size, *Finite Fields and Their Applications* **30**, 72–87.
Shparlinsky, I. (1991). Estimates for Gauss sums, *Mat. Zametki* **50**, 122–130 (Russian); English transl. *Math. Notes* **50**, 740–746.
Siegel, C.L. (1921). Approximation algebraischer Zahlen, *Math. Zeit.*, **10**, 173–213; *Ges. Abh. I*, 6–46.
(1929). Über einige Anwendungen Diophantischer Approximationen, *Abh. Preuss. Akad. Wiss. Phys.-math. Kl.*, No. 1; *Ges. Abh. I*, 209–266.
(1935). Über die Classenzahl quadratischer Zahlkörper, *Acta Arith.* **1**, 83–86; *Ges. Abh. I*, 406–409.
(1949). *Transcendental Numbers*, Princeton.
(1955). Meromorphe Funktionen auf kompakten analytischen Mannigfaltigkeiten, *Nachr. Akad. Wiss. Göttingen, Math.-phys. Kl.* **4**, 71–77; *Ges. Abh. III*, 216–222.
Silverman, J.H. (1992). *The Arithmetic of Elliptic Curves*, Graduate Texts in Math. **106**, Springer.
Sprindzuk, V.G. (1993). *Classical Diophantine Equations in Two Unknowns*, Lecture Notes in Math. **1559**, Springer.
Stark, H. (1973). Effective estimates of solutions of some diophantine equations, *Acta Arith.* **24**, 251–259.
Stewart, C.L. (1978). Algebraic integers whose conjugates lie near the unit circle, *Bull. Soc. Math. France* **106**, 169–176.
Stöhr, K.-O., Voloch, J.F. (1986). Weierstrass points and curves over finite fields, *Proc. London Math. Soc.* **52**, 1–19.
Stolarsky, K.B. (1974). *Algebraic Numbers and Diophantine Approximation*, Pure and Applied Math. **26**, Marcel Dekker.
Strassen, V. (1974). Polynomials with rational coefficients which are hard to compute, *SIAM J. Comp.* **3**, 128–149.
Surroca, A. (2006). Valeurs algébriques de fonctions transcendantes, *Int. Math. Research Notices* **ID 16834**, 1–31.
Szpiro, L., Ullmo, E., Zhang, S. (1997). Equirépartition des petits points, Inventiones Math. **127**, 337–347.
Thue, A. (1909). Über Annäherungswerte algebraischer Zahlen, *J. reine angew. Math.* **135**, 284–305.
Thunder, J.L. (1995). Siegel's lemma for function fields, *Michigan Math. J.* **42**, 147–162.
Tijdeman, R. (1971a). On the number of zeros of general exponential polynomials, *Indagationes Math.* **33**, 1–7.
(1971b). On the algebraic independence of certain numbers, *Indagationes Math.* **33**, 146–162.
Viola, C. (1985). On Dyson's lemma, *Ann. Scuola Norm. Sup. Pisa Cl. Sci.* **12**, 105–135.
Waldschmidt, M. (1973). Solution du huitième problème de Schneider, *J. Number Theory* **5**, 191–202.
(1974). *Nombres Transcendants*, Lecture Notes in Math. **402**, Springer.
(1978). Pólya's theorem by Schneider's method, *Acta Math. Acad. Sci. Hungaricae* **31**, 21–25.
(1987). *Nombres Transcendants et Groupes Algébriques*, Astérisque **69–70**.

(2000). *Diophantine Approximation on Linear Algebraic Groups*, Grundlehren der Math. Wiss. **326**, Springer.

Whittaker, E.T., Watson, G.N. (1965). *A Course of Modern Analysis*, Cambridge University Press.

Wilkie, A.J. (2004). Diophantine properties of sets definable in o-minimal structures, *J. Symbolic Logic* **69**, 851–861.

Wüstholz, G. (1983). Über das Abelsche Analogon des Lindemannschen Satzes I, *Inventiones Math.* **72**, 363–388.

(ed.) (1987). *Diophantine Approximation and Transcendence Theory*, Lecture Notes in Math. **1290**, Springer.

(1989a). Multiplicity estimates on group varieties, *Annals of Math.* **129**, 471–500.

(1989b). Algebraische Punkte auf analytischen untergruppen algebraischer Gruppen, *Annals of Math.* **129**, 501–517.

(ed.) (2002). *A Panorama of Number Theory or the View from Baker's Garden*, Cambridge University Press.

Yu, H.B. (1999). Note on Heath-Brown's estimate for Heilbronn's exponential sum, *Proc. Amer. Math. Soc.* **127**, 1995–1998.

Zagier, D. (1993). Algebraic numbers close to both 0 and 1, *Math. Comp.* **61**, 485–491.

Zannier, U. (1996). On periodic mod p sequences and G-functions, *Manuscripta Math.* **90**, 391–402.

(1998). Polynomials modulo p whose values are squares (elementary improvements on some consequences of Weil's bounds), *L'Enseignement Math.* **44**, 95–102.

(2003). *Some Applications of Diophantine Approximation to Diophantine Equations*, Forum, Udine.

(ed.) (2007). *Diophantine Geometry*, Edizione della Normale **4** Scuola Normale Pisa.

(2012). *Some Problems of Unlikely Intersections in Arithmetic and Geometry (with Appendices by D. Masser)*, Annals of Math. Studies **181**, Princeton University Press.

(2014a). *Lecture Notes on Diophantine Analysis*, Edizione della Normale **8** Scuola Normale Pisa.

(ed.) (2014b). *On Some Applications of Diophantine Approximations (with Commentary by C. Fuchs and U. Zannier)*, Edizione della Normale **2** Scuola Normale Pisa.

Zhang, S. (1992). Positive line bundles on arithmetic surfaces, *Annals of Math.* **136**, 569–587.

Index

www.ingramcontent.com/pod-product-compliance
Lightning Source LLC
LaVergne TN
LVHW101457200926
848483LV00001B/150
9781107061576